AF356534

Power Converters, Electric Drives and Energy Storage Systems for Electrified Transportation and Smart Grid

Power Converters, Electric Drives and Energy Storage Systems for Electrified Transportation and Smart Grid

Editors

Lucian Mihet-Popa
Sergio Saponara

MDPI • Basel • Beijing • Wuhan • Barcelona • Belgrade • Manchester • Tokyo • Cluj • Tianjin

Editors

Lucian Mihet-Popa
Engineering
Østfold University College
Fredrikstad
Norway

Sergio Saponara
Electronic Engineering
University of Pisa
Pisa
Italy

Editorial Office
MDPI
St. Alban-Anlage 66
4052 Basel, Switzerland

This is a reprint of articles from the Special Issue published online in the open access journal *Energies* (ISSN 1996-1073) (available at: www.mdpi.com/journal/energies/special_issues/e_transportation_smartgrid).

For citation purposes, cite each article independently as indicated on the article page online and as indicated below:

LastName, A.A.; LastName, B.B.; LastName, C.C. Article Title. *Journal Name* **Year**, *Volume Number*, Page Range.

ISBN 978-3-0365-1937-1 (Hbk)
ISBN 978-3-0365-1936-4 (PDF)

Contents

About the Editors

Lucian Mihet-Popa

Lucian Mihet-Popa received his Habilitation (2015) and Ph.D. degree (2002) in Electrical Engineering, a Master's degree in Electric Drives and Power Electronics and a Bachelor's degree in Electrical Engineering, from the Politehnica University of Timisoara-Romania. Since 2016, he has worked as Full Professor in Energy Technology at Oestfold University College in Norway. Dr. Lucian Mihet-Popa has published more than 150 papers in national and international journals and conference proceedings, and 12 books. Since 2017, he has been a Guest Editor for five Special Issues for the MDPI *Energies* and *Applied Sciences* Journals. Professor Mihet-Popa has participated in more than 15 international grants/projects, such as FP7, EEA and Horizon 2020, and has been awarded more than 10 national research grants. He is also the head of the Research Lab "Intelligent Control of Energy Conversion and Storage Systems and one of the Coordinators of the Master Program in "Green Energy Technology.

Sergio Saponara

Sergio Saponara is an Italian scientist, engineer and entrepreneur, active in the fields of electronics and automotive engineering. He is Full Professor of Electronics at the University of Pisa, Italy, where he is also President of the B.Sc. and M.Sc. degrees in Electronic Engineering. He got his Master and Ph.D. degrees cum laude in Electronic Engineering in 1999 and 2003, respectively. As an entrepreneur, he co-founded the company IngeniArs in 2014, and is the winner of several innovation prizes, such as the H2020 SME Instrument. He founded the Summer School in Enabling Technologies for Industrial IoT in 2016, and is also its director. The organisation was awarded by the IEEE CAS society in 2017 and 2018. In 2020, he was appointed director of the Automotive Electronics and Powertrain Electrifications specialization course for retraining in the field of vehicle electrification for the Vitesco Technologies (Continental) group.

Preface to "Power Converters, Electric Drives and Energy Storage Systems for Electrified Transportation and Smart Grid"

It is expected that by 2050, around 65%–70% of the world's population to live in urban areas. As our cities become bigger and smarter, this trend leads to new opportunities for specialized and flexible electric vehicles (EVs) designs, new vehicle architectures in order to ensure urban-readiness, such as compatibility with charging infrastructure, appropriate range, different implementation levels of infrastructure, and smart technology. In the next-generation of EVs, power and efficiency are critical. These factors drive the need for new power electronic converters, optimization of powertrain design and power management solutions to optimize system efficiency. Based on these assumptions, the huge interest coming from industry and research centers, and following our previous successful Special Issue with more than 20 papers published, which acquired in the last 3 years more than 400 citations and 50,000 views, we decided to open a new Special Issue. The Special Issue, entitled "Power Converters, Electric Drives and Energy Storage Systems for Electrified Transportation and Smart Grid" on MDPI *Energies* presents 10 accepted papers from 18 submitted, as well as an Editorial, with authors from Europe, US, Asia, and Africa. The published papers are related to the emerging trends in solar power, energy storage, electric drives and power electronic converters based on specific optimization and control methods and algorithms, with focus on electric vehicles and charging stations-based RES towards smart/micro grids. An extensive exploitation of renewable energy sources is foreseen for smart grid, as well as a close integration with the energy storage and recharging issues of the electrified transportation systems. Innovation at both algorithmic and hardware (i.e., power converters, electric drives, electronic control units, energy storage modules, and charging stations) levels are also proposed.

Lucian Mihet-Popa, Sergio Saponara
Editors

Editorial

Power Converters, Electric Drives and Energy Storage Systems for Electrified Transportation and Smart Grid Applications

Lucian Mihet-Popa [1,*] and **Sergio Saponara** [2]

[1] Faculty of Engineering, Østfold University College, Kobberslagerstredet 5, 1671 Fredrikstad, Norway
[2] Department of Information Engineering, University of Pisa, Via G. Caruso 16, 56122 Pisa, Italy; sergio.saponara@unipi.it
* Correspondence: lucian.mihet@hiof.no

Citation: Mihet-Popa, L.; Saponara, S. Power Converters, Electric Drives and Energy Storage Systems for Electrified Transportation and Smart Grid Applications. *Energies* **2021**, *14*, 4142. https://doi.org/10.3390/en14144142

Academic Editor: José Matas

Received: 18 June 2021
Accepted: 30 June 2021
Published: 9 July 2021

Publisher's Note: MDPI stays neutral with regard to jurisdictional claims in published maps and institutional affiliations.

The proposed special issue (SI) has invited submissions related to renewable energy, energy storage, power converters and electric drive systems for electrified transportation and smart grid applications [1–10]. The particular topics of interest have included:

- New emerging technologies for power converters, electric drives and energy storage;
- Aging mechanisms of power converters, electric drives and energy storage devices;
- Electronic control units for energy storage system (ESS) monitoring and management;
- Online estimation of state-of-charge (SoC) and state-of-health (SoH);
- Power electronic converters for renewable energy sources (RES);
- Fast chargers and smart chargers for electric-vehicles, including wireless power transfer;
- Integration of charging infrastructures in the smart grid for e-transportation;
- Predictive diagnostic for renewables and ESS;
- Methods for design and verification of hardware (HW) and software (SW) for energy storage and renewables;
- Embedded systems, machine learning (ML), artificial intelligence (AI) and deep neural network (DNN) for energy storage, conversion and management;
- Integration of Internet of Things (IoT) and digitalization into e-transportation.

Research and technology transfer activities in ESS, such as batteries and super/ultra-capacitors, are essential for the success of electric transportation and to foster the use of RES. ESS are the key to increase the adoption of RES in the smart grid. However, major challenges have yet to be solved, such as the design of high-performance and cost-effective ESS, the on-line estimation of SoC/SoH)of batteries and super/ultra-capacitors, the estimation of aging effects, the design and optimization of fast chargers and the integration within the smart grid of the charging infrastructure for electrified transportation [11–13]. The strategic interest for this R&D activity is proved by the rise of initiatives such as the Battery 2030+ initiative or the European Battery Alliance [14], where a mixed effort of the European commission, industries and research organizations aims at developing an innovative, sustainable and competitive battery "ecosystem" in Europe.

Power converters and electric drives also need optimization in terms of increased efficiency and implementation of predictive diagnostic features. Beside the HW parts, the role of the SW is also increasing, and new design and verification methods have to be investigated to achieve high functional safety levels. Due to the increasing role of Information and Communication Technology (ICT) in smart grid and electrified transportation, toward an Internet of Energy scenario, cybersecurity is also becoming a key issue.

The main objective of the 10 manuscripts published in this SI is, hence, to provide timely solutions for the design and management of ESS, of RES and of the relevant power electronics converter topologies and their control and modulation techniques. The accepted articles addressed these issues from the low component level, up to the integration of all these sub-systems within the smart grid for e-transportation and smart/green cities.

The SI includes, after a strict review process, 10 papers, of which nine are original research papers [1–9] and one is a comprehensive review paper [10].

The first work [1], entitled "Minimization of Cross-Regulation in PV and Battery Connected Multi-Input Multi-Output DC to DC Converter", written by a group of authors from India and South Africa, Vibha Kamaraj et al., deals with a digital model predictive controller (DMPC) for a multi-input multi-output (MIMO) DC-DC converter interfaced with renewable energy resources in a hybrid system. This paper proposes a controller, which increases the speed of response maintaining the output stable by regulating the load voltage independently. To prove the efficacy of the proposed DMPC controller, simulations followed by the experimental results are executed on a hybrid system consisting of a dual-input dual-output (DIDO) positive Super-Lift Luo converter (PSLLC) interfaced with a photovoltaic (PV) renewable energy resource.

The second paper [2], entitled "Voltage-Balancing Strategy for Three-Level Neutral-Point-Clamped Cascade Converter under Sequence Smooth Modulation", by Le Yu et al., from China, mainly discusses the open-circuit fault in the DC-side of the three-level neutral-point clamped cascaded converters (3LNPC-CC). A sequence smooth modulation (SSM) optimized by the sequence pulse modulation to keep the DC-side voltage balance while the 3LNPC-CC suffers an open-circuit fault from the DC-side is proposed. The SSM found an efficient switch-state path through a 3D cube model and simplified the path from thousands of switch states. The SSM avoids the complex calculation in the voltage-balancing modulation, while its dynamic characteristics were less influenced. At the same time, the modulation changes the voltage level smoothly and balances the fault DC-side voltage effectively.

The third paper [3], entitled "Technical and Economic Analysis of One-Stop Charging Stations for Battery and Fuel Cell EV with Renewable Energy Sources", written by Saumya Bansal et al., a group of authors from the Netherlands, Denmark, Norway and China, investigates a technical and economic analysis of a one-stop charging station for battery electric vehicles (BEV) and fuel cell electric vehicles (FCEV). The hybrid optimization model for electric renewables (HOMER) SW and the heavy-duty refueling station analysis model (HDRSAM) are used to conduct the case study for a one-stop charging station. A total of 42 charging station scenarios by considering two systems (a grid-connected system and an off-grid connected system) are analyzed. For each system three different charging station designs (design A—hydrogen load; design B—an electrical load, and design C—an integrated system consisting of both hydrogen and electrical load) are set up for analysis. Furthermore, seven potential wind turbines with different capacity are selected from the HOMER database for each system. A total of 18 scenarios are analyzed with variations in the hydrogen delivery option, production volume and hydrogen dispensing option. The optimal solution from HOMER for a lifespan of 25 years is integrated into design C with the grid-connected system. The optimal solution design consists of tube trailer as hydrogen delivery with a cascade dispensing option at 350 bars together with a high production volume.

The fourth paper [4], entitled "Design of Adaptive Controller Exploiting Learning Concepts Applied to a BLDC-Based Drive System", by P. Dini and S. Saponara, from the University of Pisa-Italy, proposes a novel and innovative control architecture, which takes its ideas from the theory of adaptive control techniques and the theory of statistical learning at the same time. The main idea was to divide the architecture of the adaptive controller into three different levels, considering the architecture of a classical neural network with several hidden levels. The design of the control system is reported from both a rigorous and an operational point of view. As an application example, the proposed control technique is applied on a second-order non-linear system. The authors consider a servo-drive based on a brushless DC (BLDC) motor, whose dynamic model considers all the non-linear effects related to the electromechanical nature of the electric machine itself, and also an accurate model of the switching power converter. The reported example shows the capability of the control algorithm to ensure trajectory tracking while allowing for disturbance rejection with different disturbance signal amplitudes.

The fifth paper [5], entitled "Trusted Simulation Using Proteus Model for a PV System: Test Case of an Improved HC MPPT Algorithm", by Abdelilah Chalh et al., from Morocco and Saudi Arabia, presents a trusted simulation of a PV system designed with the Proteus SW. The proposed PV simulator can be used to verify and evaluate the performance of MPPT algorithms with a closer approximation to the real implementation. The main advantage of this model is related to the fact that the same code for the MPPT algorithm can be used in the simulation and the real implementation. In contrast, when using a SW program/simulation tool, like PSIM or MATLAB/Simulink, the code of the algorithm must be rewritten once the real experiment begins, because these tools do not provide a microcontroller or an electronic board in which our algorithm can be implemented and tested in the same way as the real experiment. A modified Hill-Climbing (HC) algorithm is also introduced. The proposed algorithm can avoid the drift problem posed by conventional HC under a fast variation in insolation. The simulation results show that this method presents good performance in terms of efficiency (99.21%) and response time (10 ms), which improved by 1.2% and 70 ms, respectively, compared to the conventional HC algorithm.

The sixth paper [6], written by A. Plesca and Lucian Mihet, from the Technical University of Iasi-Romania and Østfold University College-Norway, entitled "Thermal Analysis of Power Rectifiers in Steady-State Conditions", proposes a new mathematical model to calculate the junction and the case temperature in power diodes as part of a three-phase bridge rectifier used in electric traction applications, which supplies an inductive-resistive load. The new thermal model may be used to investigate the thermal behavior of the power diodes in a steady-state regime for various values of the tightening torque, direct current through the diode, airflow speed and load parameters (resistance and inductance). The obtained computed values were compared with 3D thermal simulation results and experimental tests.

The seventh paper [7], entitled "Microcontroller-Based Strategies for the Incorporation of Solar to Domestic Electricity", by Mabunda and Joseph, from the University of Johannesburg-South Africa, deals with innovative methods to reduce the reliance on national grid energy and to supplement this source of energy with alternative methods. A microcontroller is used to monitor the energy consumed by household equipment and then decide, based on the power demand and available solar energy. In this research, a special circuit was also designed to control geyser power and align it to the capacity of the RES. This geyser control circuit includes a Dallas temperature sensor and a triode for an alternating current (TRIAC) circuit that is included to control the output current drawn from a low-power, RES. Alternatively, two heating elements may be used instead of the TRIAC circuit. The first heating element is powered by solar to maintain the water temperature and to save energy. The second heating element is powered by national grid power and is used for the initial heating, and therefore saves water heating time. The strategy used was to add a programmed microcontroller-based control circuit and a low power element. The current is the controlled element of the geyser circuit thereby photovoltaic (PV) energy was used to save the energy geysers consume from the domestic electricity source when they are not in use.

The eight work [8], authored by Dini and Saponara, entitled "Cogging Torque Reduction in Brushless Motors by a Nonlinear Control Technique", addresses the problem of mitigating the effects of the cogging torque in PM synchronous motors, particularly brushless motors, which is a main issue in precision electric drive applications. A method for mitigating the effects of the cogging torque is proposed, based on the use of a nonlinear automatic control technique known as feedback linearization, that is ideal for underactuated dynamic systems. The aim of this work was to present an alternative to classic solutions based on the physical modification of the electrical machine to try to suppress the natural interaction between the PMs and the teeth of the stator slots. Such modifications of electric machines are often expensive because they require customized procedures, while the proposed method does not require any modification of the electric drive. With respect to other algorithmic-based solutions for cogging torque reduction, the proposed control technique is scalable to different motor parameters, deterministic and robust, and hence easy to use and verify for safety-critical applications. As an application case example, the work reports the reduction of the oscillations for the angular position control of a PM synchronous motor vs. classic proportional-integrative (PI) cascaded control.

The ninth work [9], entitled "Spatio-Temporal Model for Evaluating Demand Response Potential of Electric Vehicles in Power-Traffic Network", by Chen et al., a group of authors from China and UK, introduces a composite methodology that takes into account the dynamic road network (DRN) information and fuzzy user participation (FUP) for obtaining spatio-temporal projections of demand response potential from electric vehicles (EV) and the EV aggregator. A dynamic traffic network model taking over the traffic time-varying information is developed by graph theory, and a trip chain based on a housing travel survey is set up, where the Dijkstra algorithm is employed to plan the optimal route of EVs in order to find the travel distance and travel time of each trip of EVs. To demonstrate the uncertainties of the EVs' travel pattern, a simulation analysis is conducted using the Monte Carlo method. Subsequently, the authors suggest a fuzzy logic-based approach to uncertainty analysis that starts with investigating EV users' subjective ability to participate in a DR event, and develop the FUP response mechanism, which is constructed by three factors including the remaining dwell time, remaining SOC and incentive electricity pricing. The FUP is used to calculate the real-time participation level of a single EV. Finally, the authors use a simulation example with a coupled 25-node road network and 54-node power distribution system to demonstrate the effectiveness of the proposed method.

Last but not least, the tenth contribution [10] is a comprehensive review paper by S. Vadi et al., a group of authors from Turkey (Gazi University), Denmark (Aalborg University) and Norway (Østfold University College), entitled "A Review on Optimization and Control Methods Used to Provide Transient Stability in Microgrids". The authors propose a comprehensive review on optimization and control methods in Microgrids, with DERs such as PV systems and wind turbines, and storage systems. The paper also points out the fact that the microgrids are an efficient source in terms of inexpensive, clean and renewable energy for distributed RES that are connected to the existing grid, but these RES can also cause many difficulties to the microgrid due to their characteristics. These difficulties mainly include voltage collapses, voltage and frequency fluctuations and phase difference faults in both islanded and grid-connected operation modes. That is why the stability of the microgrid structure is necessary for providing transient stability using intelligent optimization methods to eliminate the abovementioned difficulties that affect power quality. This paper also highlights some optimization and control techniques that can be used to provide transient stability in the islanded or grid-connected operation modes of a microgrid-based RES.

In conclusion, the SI entitled "Power Converters, Electric Drives and Energy Storage Systems for Electrified Transportation and Smart Grid" has accepted for publication 10 original research papers, among which one comprehensive review paper, related to the emerging trends in solar power, energy storage, power converters and electric drives. These open-access publications already received by now more than 50 citations (with approx. 9000 views).

This SI follows the previous successful SI [13] in *Energies*, entitled "Energy Storage Systems and Power Conversion Electronics for E-Transportation and Smart Grid", which published 21 papers from 40 submitted.

The published articles provide an overview of the most recent research advances in Microgrids, exploiting the opportunities offered by the use of RES, BESS, power converters, innovative control and energy management strategies. These publications have also been addressing both algorithmic-level and HW-level works, where the focus was on applications such as transportation electrification (full EV and hybrid EV (HEV), mainly for automotive and railway scenarios) and the evolution of the microgrid to a smart grid, with a massive use of RES. Moreover, a close integration is foreseen between the smart electric grid and the electrified vehicles due to the need of an efficient and fast recharging infrastructure. Adaptive control methods for brushless DC motor (BDCM) Drives as well as voltage balancing strategies for cascade power converters with smooth modulation techniques have also been highlighted.

Based on our previous records, obtained from this SI's publications and the previous one, we have already decided, together with MDPI *Energies* Editors, to launch a new SI on a similar topic.

Funding: This research received no external funding.

Acknowledgments: The authors are grateful to the MDPI publisher for the possibility to act as guest editors of this special issue and want to thank the editorial staff of Energies for their kind co-operation, patience and committed engagement.

Conflicts of Interest: The authors declare no conflict of interest.

References

1. Kamaraj, V.; Chellammal, N.; Chokkalingam, B.; Munda, J.L. Minimization of Cross-Regulation in PV and Battery Connected Multi-Input Multi-Output DC to DC Converter. *Energies* **2020**, *13*, 6534. [CrossRef]
2. Yu, L.; Peng, X.; Zhou, C.; Gao, S. Voltage-Balancing Strategy for Three-Level Neutral-Point-Clamped Cascade Converter under Sequence Smooth Modulation. *Energies* **2020**, *13*, 4969. [CrossRef]
3. Bansal, S.; Zong, Y.; You, S.; Mihet-Popa, L.; Xiao, J. Technical and Economic Analysis of One-Stop Charging Stations for Battery and Fuel Cell EV with Renewable Energy Sources. *Energies* **2020**, *13*, 2855. [CrossRef]
4. Dini, P.; Saponara, S. Design of Adaptive Controller Exploiting Learning Concepts Applied to a BLDC-Based Drive System. *Energies* **2020**, *13*, 2512. [CrossRef]
5. Chalh, A.; El Hammoumi, A.; Motahhir, S.; El Ghzizal, A.; Subramaniam, U.; Derouich, A. Trusted Simulation Using Proteus Model for a PV System: Test Case of an Improved HC MPPT Algorithm. *Energies* **2020**, *13*, 1943. [CrossRef]
6. Plesca, A.; Mihet-Popa, L. Thermal Analysis of Power Rectifiers in Steady-State Conditions. *Energies* **2020**, *13*, 1942. [CrossRef]
7. Mabunda, N.E.; Joseph, M.K. Microcontroller-Based Strategies for the Incorporation of Solar to Domestic Electricity. *Energies* **2019**, *12*, 2811. [CrossRef]
8. Dini, P.; Saponara, S. Cogging Torque Reduction in Brushless Motors by a Nonlinear Control Technique. *Energies* **2019**, *12*, 2224. [CrossRef]
9. Chen, L.; Zhang, Y.; Figueiredo, A. Spatio-Temporal Model for Evaluating Demand Response Potential of Electric Vehicles in Power-Traffic Network. *Energies* **2019**, *12*, 1981. [CrossRef]
10. Vadi, S.; Padmanaban, S.; Bayindir, R.; Blaabjerg, F.; Mihet-Popa, L. A Review on Optimization and Control Methods Used to Provide Transient Stability in Microgrids. *Energies* **2019**, *12*, 3582. [CrossRef]
11. Mihet-Popa, L.; Saponara, S. Toward Green Vehicles Digitalization for the Next Generation of Connected and Electrified Transport Systems. *Energies* **2017**, *11*, 3124. [CrossRef]
12. Un-Noor, F.; Padmanaban, S.; Mihet-Popa, L.; Mollah, M.N.; Hossain, E. A Comprehensive Study of Key Electric Vehicle (EV) Components, Technologies, Challenges, Impacts, and Future Direction of Development. *Energies* **2017**, *10*, 1217. [CrossRef]
13. Special Issue "Energy Storage Systems and Power Conversion Electronics for E-Transportation and Smart Grid". Available online: https://www.mdpi.com/journal/energies/special_issues/e_transportation_smart_microgrid (accessed on 8 December 2018).
14. The European Battery Alliance. Available online: https://ec.europa.eu/growth/industry/policy/european-battery-alliance_it (accessed on 9 December 2018).

 *energies*

Article

Technical and Economic Analysis of One-Stop Charging Stations for Battery and Fuel Cell EV with Renewable Energy Sources

Saumya Bansal [1], Yi Zong [2,*], Shi You [2], Lucian Mihet-Popa [3] and Jinsheng Xiao [4]

[1] Institute of Engineering, Hanze University of Applied Sciences, 9747 AS Groningen, The Netherlands; sbanzal93@gmail.com
[2] Center for Electric Power and Energy, Technical University of Denmark, 4000 Roskilde, Denmark; sy@elektro.dtu.dk
[3] Faculty of Engineering, Oestfold University College, 1671 Fredrikstad, Norway; lucian.mihet@hiof.no
[4] School of Automotive Engineering, Wuhan University of Technology, Wuhan 430070, China; jinsheng.xiao@whut.edu.cn
*   Correspondence: yizo@elektro.dtu.dk

Received: 4 May 2020; Accepted: 1 June 2020; Published: 3 June 2020

Abstract: Currently, most of the vehicles make use of fossil fuels for operations, resulting in one of the largest sources of carbon dioxide emissions. The need to cut our dependency on these fossil fuels has led to an increased use of renewable energy sources (RESs) for mobility purposes. A technical and economic analysis of a one-stop charging station for battery electric vehicles (BEV) and fuel cell electric vehicles (FCEV) is investigated in this paper. The hybrid optimization model for electric renewables (HOMER) software and the heavy-duty refueling station analysis model (HDRSAM) are used to conduct the case study for a one-stop charging station at Technical University of Denmark (DTU)-Risø campus. Using HOMER, a total of 42 charging station scenarios are analyzed by considering two systems (a grid-connected system and an off-grid connected system). For each system three different charging station designs (design A-hydrogen load; design B-an electrical load, and design C-an integrated system consisting of both hydrogen and electrical load) are set up for analysis. Furthermore, seven potential wind turbines with different capacity are selected from HOMER database for each system. Using HDRSAM, a total 18 scenarios are analyzed with variation in hydrogen delivery option, production volume, hydrogen dispensing option and hydrogen dispensing option. The optimal solution from HOMER for a lifespan of twenty-five years is integrated into design C with the grid-connected system whose cost was $986,065. For HDRSAM, the optimal solution design consists of tube trailer as hydrogen delivery with cascade dispensing option at 350 bar together with high production volume and the cost of the system was $452,148. The results from the two simulation tools are integrated and the overall cost of the one-stop charging station is achieved which was $2,833,465. The analysis demonstrated that the one-stop charging station with a grid connection is able to fulfil the charging demand cost-effectively and environmentally friendly for an integrated energy system with RESs in the investigated locations.

Keywords: battery operated electric vehicles; fuel cell electric vehicles; one-stop charging station; renewable energy sources

---

1. Introduction

We are living in a very exciting period because global energy systems are going through a major transformation. One of the main reasons for it is the climate shift which has made it necessary to shift from non-renewable resources to renewable resources in order to build a low-carbon, climate-safe future. In recent years, the demand of renewable energy systems has increased. After the 2015, United Nations Climate Change Conference (COP21) agreement in Paris, many countries have issued a statement regarding the ban on the use of the internal combustion engines (ICE) [1]. For example, in 2018, Denmark announced that it will ban the sale of new cars with ICE by 2030 and hopes to have one million electric and hybrid cars on the roads by then [2]. Electric vehicles (EV) will play a major part in smart grids with a high penetration of renewables [3–5]. The number of EVs and fuel cell vehicles has been on the rise in the past few years. It is also predicted that by 2020 the total cost of a personal use EV could be the same as an ICE [6,7]. However, even if the cost of the EVs goes down in the future, shortage of available charging stations and limited distance travelled on a single charge are the major reasons impeding the purchase of these EVs, as seen in Figure 1 [1]. Hence, it is quite important to develop an optimal design for an integrated charging station which is able to meet the needs.

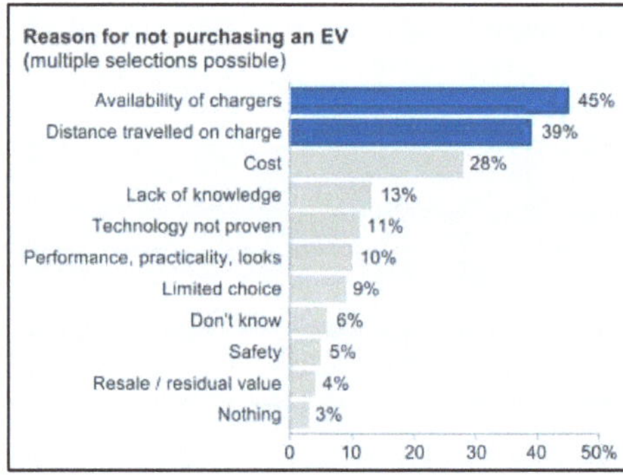

Figure 1. Top reasons for not purchasing an electric vehicle (EV) [1].

Denmark has a well-established natural gas grid, and here hydrogen and biogas can play a complementary role in terms of providing stored, renewable energy in large quantities to the Danish energy system. Furthermore, the European Union has provided funding of EUR 40 million for the production of 600 hydrogen buses, of which one-third will be provided to Denmark [8]. With this green transition development of hydrogen refueling stations is quite imperative. Simultaneously, fuel cell electric vehicles (FCEV) are receiving more focus and stronger support. The charging solutions for both battery electric vehicles (BEVs) and FCEVs are developed and operated independently as they are often seen as two competitive technologies. These two types of charging stations can be integrated into a one-stop charging station, acting as an interface to an integrated energy system, which includes electricity, transportation, and gas. This one-stop charging station will cater to both the BEVs and FCEVs. These circumstances make Denmark an ideal place to demonstrate the power-to-gas solutions.

To transform DTU-Risø campus into 100% renewable campus in the coming years, a case study was conducted by using a fleet of battery-operated electric cars. Adding to this, an integrated charging station for both electric and fuel cell vehicles will bring it one step closer to achieving the aim. To achieve this goal, an example will be set up for other campuses/community to convert into a climate-friendly future integrated energy system with high penetration of renewable energy sources (RESs). The main contribution of this paper is to find an optimal solution for the charging of fuel cell electric vehicles and battery-operated vehicles by answering the following questions:

- What is the minimum capacity of the renewable energy required to meet the load demand?
- What is the minimum cost of the system to achieve the charging demand of a particular region?
- Will the increase in the capacity of the renewables help in decreasing the cost of the system?
- Which renewable configuration (grid connectivity or off-grid) provides the most economical solution?

- Will developing an integrated energy system for both fuel cell vehicles and battery-operated vehicles be more cost-effective as compared to the stand-alone systems?

So far, many researchers people have used the hybrid optimization model for electric renewables (HOMER) software to find an optimal solution for an electrical load with combinations of different energy sources for producing energy. For example, Edwin Moses in his research used HOMER to find an optimal solution for providing energy to a small village. In this paper the author used wind turbines, Photovoltaics (PV) and bio-diesel generator for producing energy for an electrical load for a small remote rural village. The optimal solution included the use of a hybrid system consisting of solar PV and a bio-diesel generator for producing energy for the village [9]. David Restrepo used the HOMER simulation for a scenario-based optimization. His work consisted of analysis of two different scenarios with the same load requirements for a location in Medellin, Columbia. The first case considered renewable energy sources and a diesel generator but in a stand-alone configuration. The second case also considered renewable sources but in grid-tied configuration. Both cases showed that PVs are significant contributions for power generation [10].

The work done by previous researchers showed that a multi scenario-based analysis with different source of renewables to obtain an optimal solution can be done. However, it does not include a charging station as they just use HOMER for power generation and a maximum of two scenario-based optimization analysis. The analysis in this paper takes advantage of two different models-HOMER and heavy-duty refueling station analysis model (HDRSAM), which were integrated to find an optimal solution for a one-stop charging station. The major differences which are included in this paper are the use of different types of loads—hydrogen load and electrical load; use of hundred percent renewables for energy production. Different design scenarios with different system connection setup and use of different types of wind turbines sets the research in this paper aside from the previous work done.

The rest of the paper is structured as follows: The system setup description is briefly introduced in Section 2. Section 3 presents the simulation tools used to analyze and design the system. The economic and technical results are described and discussed in Section 4. Finally, a conclusion is drawn in Section 5, followed by the discussion on future research.

2. System Setup Description

The investigated system included three types of charging stations: hydrogen refueling station, electrical charging station, and an integrated station with a charging load demand representing the load profile. The charging demand was for one fuel cell bus and fifteen battery-operated electric cars. The hybrid optimization model for electric renewables (HOMER) developed by the National Renewable Energy Laboratory was the tool used for simulation and optimization for energy production in the study case [11] HOMER has a large inbuilt database of different components, such as photovoltaics, wind turbines, hydro, reformers, batteries, electrolyzer, hydrogen tank, grid, boilers, thermal load controllers, generators, etc. Analysis with HOMER requires information on resources, economic constraints, and control methods. It also requires inputs on component types, their numbers, costs, efficiency, longevity, etc. To find the optimal system set up different configurations of components that the system should have, were optimized. All these configurations were simulated, the infeasible ones were discarded and only the feasible ones are presented, the system with the lowest total net present cost being the optimal system configuration. HOMER only takes into account the energy production and not the charging station part. To integrate the charging station part in the analysis, a heavy-duty refueling station analysis model (HDRSAM) was used to analyses the charging station for the fuel cell bus by optimization to find out the least cost refueling configuration from various refueling station combinations and demand profiles [12]. It included the data on station configuration, component technologies and cost of interest to government agencies and industry stakeholders. It was assumed that a specific charging station for the battery-operated cars was not required and was treated only as an electric load. The use of these two models was done to get a multi integrated system that included the whole well to wheel process. The final cost of the system was found by adding the optimal solutions from the two case analyses. The system design flowchart is shown in the Figure 2 The design criteria for finding the optimal solution for charging stations are defined below.

- All the electricity came from local renewables, and no electricity was bought from the grid.
- Excess electricity could be sold back to the grid in grid-connected systems.

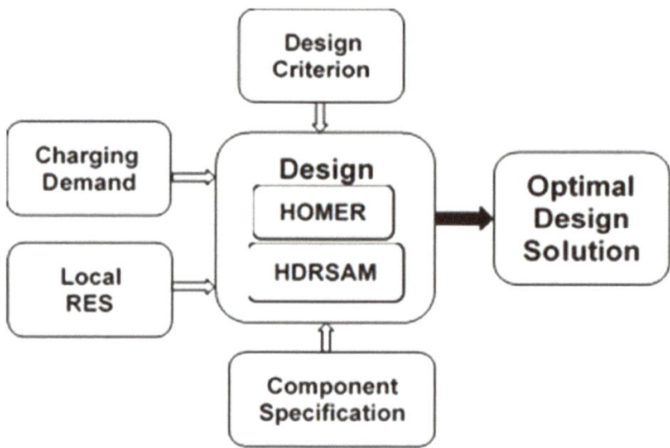

Figure 2. The flowchart of the system design.

Further, the inputs required for the analysis were the charging demand (hydrogen load and the electrical load profiles), components specifications, and the available local energy sources. These inputs were then used in the design of models HOMER and HRSADM to obtain an optimal solution.

2.1. Locations and Local RESs

The Danish Technical University (DTU)-Risø campus is located near Roskilde in Denmark (55°41′32.03″ N 12°05′52.21″ E) and the DTU-Lyngby campus is about 14 km north of Copenhagen (55°46′20.96″ N 12°30′06.32″ E). The driving distance between these two campuses is around 42 km, as shown in Figure 3.

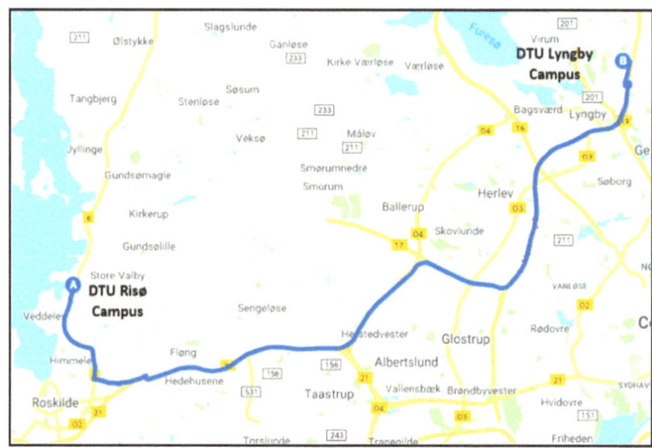

Figure 3. Location of the two Danish Technical University (DTU) campuses.

The shuttle bus/cars service ran between the two campuses during the day time. The electric cars were used by small groups of students and staff travelling within the Risø campus and also between the two campuses. The monthly average RESs generation data (left axis) and clearness index (right axis) are shown in the Figure 4. The average wind speed for the location is 5.49 m/s and the solar global horizontal irradiance is 2.9 kWh/m^2. The average clearness index (0-cloudy, 1-Sunny) is 0.455. In HOMER the clearness index is defined as a measure of the clearness of the atmosphere. It is the fraction of the solar radiation that is transmitted through the atmosphere to strike the surface of the Earth. It is a dimensionless number between 0 and 1. The clearness index has a high value under clear, sunny conditions, and a low value under cloudy conditions [11].

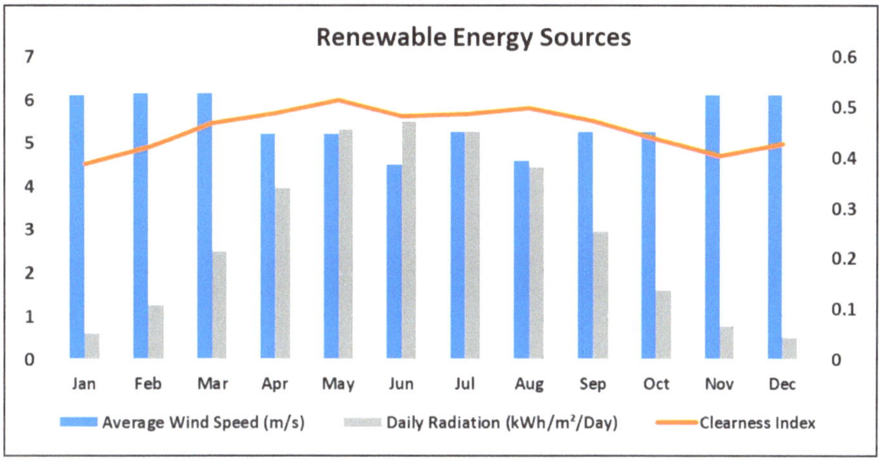

Figure 4. The average monthly renewable energy source (RES) generation data at DTU-Risø.

2.2. Load Profile

A typical fuel cell bus takes 40 kgs of hydrogen and can travel 440 km with a full tank [13]. The fuel cell bus was used between the two campuses and made five round trips through the day and the charging of the fuel cell bus took place over the night from 20:00 to 08:00 so that it could be used during the day. The load profile of 40 kg/day was spread equally between the twelve hours giving an average load of 3.33 kg/h.

While each battery-operated car had battery energy of 30 kWh/day, the total electrical load was 450 kWh/day for 15 EVs. It was also assumed that the charging was DC system type charging. Five cars were charged during the night from 20:00 to 08:00, and ten cars were charged during the day time from 08:00 to 20:00. During night time, the load of five cars was spread equally over twelve hours giving us the average load of 12.5 kW. While during the day time the load of ten cars was spread equally over twelve hours giving the average of 25 kW. The average load for the whole day was 18.75 kW. The hourly load profiles of one day are shown in the Figure 5.

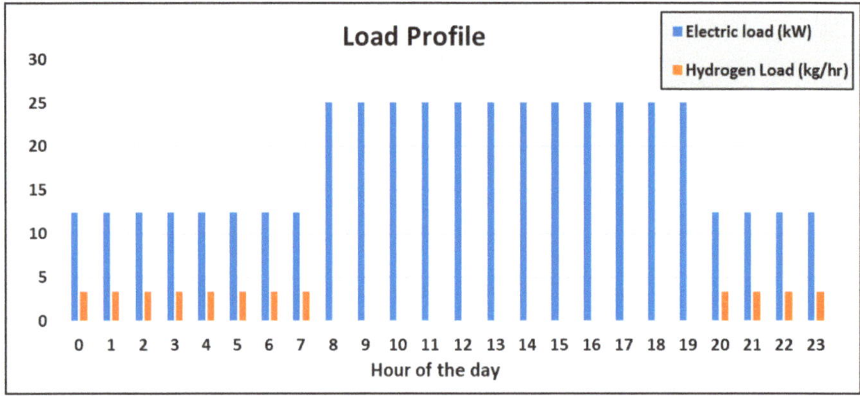

Figure 5. Hydrogen and electrical load.

3. System Analysis Using HOMER Software and HDRSAM Model

Using HOMER software, the analysis at DTU-Risø campus was conducted for three design scenarios–design A (hydrogen load for a fuel cell bus), design B (electrical load for battery-operated electric cars), and design C (both hydrogen and electrical load for electrical cars and fuel cell bus). Design C is basically an integrated one-stop charging station which caters to both the BEVs and FCEVs. Each design further had two system configurations: grid-connected and stand-alone/off-grid.

The basic configuration for the three designs is shown Figure 6. In all the designs, wind energy was considered to be the primary source of energy. An electrolyzer driven by RES (wind in this case) was used for hydrogen production to fulfil the hydrogen load for a fuel cell bus. A hydrogen tank was used for storing the excess hydrogen. When the energy produced by the wind was not enough to fulfil the required hydrogen load, hydrogen from the tank was used. In case of grid connection configuration, a grid was present to which the excess electricity could be sold. In the off-grid system, the grid component was not used. Other components of the system were a power electronic converter to convert the current from AC to DC and a generic Li-ion battery for energy storage.

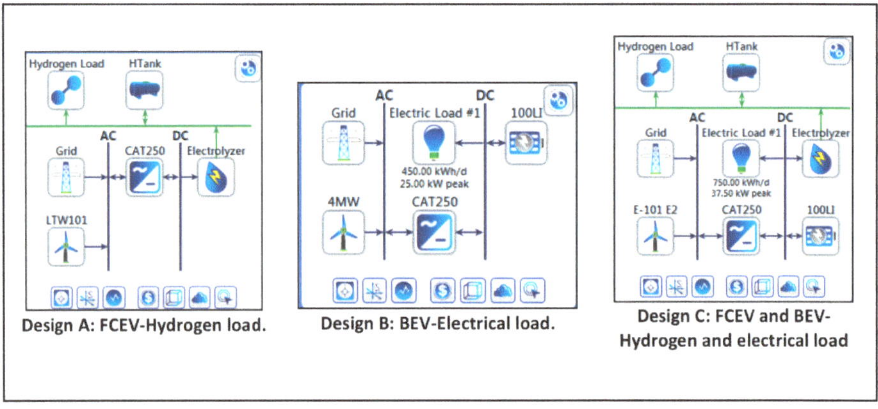

Figure 6. System configuration for different designs in the hybrid optimization model for electric renewables (HOMER).

3.1. Component Specifications

3.1.1. Wind Turbine

Seven wind turbines selected for system analysis are listed in the Table 1. Wind turbines ranging from 500 kW to 1500 kW were investigated. The lifetime for each wind turbine was considered to be 25 years for each scenario. The wind turbines in our analysis had an alternating electrical bus connection. The onshore wind installation cost in Denmark is \$1.6 million per MW, while the operation and maintenance cost are around \$35 per kW [14].

Table 1. List of the investigated wind turbines.

Wind Turbine Model	Turbine Size (kW)	Hub Height (m)
Windflow 45	500	38
Vestas V47	660	50
Enercon E-53	800	73
Leitwind 80	850	80
EWT DW 61	900	75
Leitwind 90	1000	97.5
Leitwind 86	1500	100

3.1.2. Electrolyzer and Hydrogen Storage

The size of the electrolyzer used in the analysis varied from 100 kW to 600 kW. The efficiency of the electrolyzer was 85% and the lifetime was considered to be 15 years. The capital cost of the electrolyzer used in our study was \$368 per kW, and the replacement cost was taken as 50% of capital cost, \$184 per kW, and the operational and maintenance cost was 10% of capital cost, \$35 per kW [15]. The size of the hydrogen tank varied from 100 kg to 400 kg. The capital cost of the hydrogen tank was taken to be \$623 per kg, the replacement cost was the same as the capital cost, and the operational and maintenance cost was taken to be 5% of the capital cost \$35 per kg [16].

3.1.3. Battery Storage System and Power Converter

A Li-ion battery of capacity 100 kWh, with a nominal capacity of 167 Ampere-hour (Ah), round trip efficiency of 90% was used for electricity storage in design B and C. The capital cost of the battery per quantity was considered to be \$15,000, the replacement cost was also the same as the capital cost, and the operation and maintenance cost was estimated at \$1000 per battery per year [6].

In addition, a caterpillar bidirectional power converter CATBDP250 was used. The size of the convertor varied from 250 kW to 750 kW. The lifetime and efficiency of this converter were estimated at twenty-five years and 96%, respectively. The capital cost of the convertor was \$94 per kW, the replacement cost was the same as the capital cost and the operational and maintenance cost was \$10 per kW [17].

3.1.4. Grid

A simple model of the grid was used; this operation mode allows specifying a constant power price, sellback price and sale capacity. For an off-grid system, this system is not added. The price for the electricity for the industry in Denmark for the year 2017 was 0.095 \$/kWh, while the grid sellback price was taken 0.045 \$/kWh [18]. The cost of all the components is listed in the Table 2.

Table 2. Cost of the components.

	Capital Cost ($)	Replacement Cost ($)	O & M Cost ($)
Hydrogen Tank (per kg)	623	623	35
Electrolyzer (per kW)	368	184	35
Converter (per kW)	94	94	10
Battery (per quantity)	15,000	15,000	1000
Wind Turbine ($/MW)	1,666,670	1,666,670	35,000
PV Module ($/kW)	1200	1200	18

The total net present cost of the system is the present value of all the costs the system incurs over its lifetime. The project lifetime was taken to be twenty-five years, while the nominal discount rate and the inflation rate were kept at 8% and 2%, respectively. The cost of the system included all the capital cost, replacement cost, Operation & Maintenance(O&M) cost, fuel cost, emission cost, cost of buying power from the grid. It also included the earnings from selling of the excess energy if a grid connection was present. This was the main economic output, the value by which it ranked all the system configurations in the optimized results. The net present cost or the life cycle cost of individual components was calculated and then of the system as a whole. The nominal cost of each component was calculated by adding the capital cost, replacement cost, salvage cost, O&M cost, and the fuel cost. Salvage cost is the residual value of the power system components at the end of the project lifetime. The salvage cost is different for each component. This amount is subtracted from the sum of the capital cost, replacement cost, and O&M cost to get the total cost of each component. HOMER multiples the nominal cost with the discount factor for each year and then sums all the cost for each year, to get the discounted total for each component. The discount factor and the salvage cost were calculated by HOMER [11].

3.2. Optimization Variables

By using of HOMER, all possible combinations of system types can be investigated in a single run, and then sorts the systems according to the optimization variable of choice. The search space in HOMER helped us to define the capacity or the quantities of the various components used. HOMER uses these values to simulate all of the system configurations which are feasible. HOMER simulates every configuration in the search space and only shows the feasible solutions. The feasible solutions are then presented from the lowest to highest net cost of the system. Optimization variables used in the analysis included the number of wind turbines and number of batteries, electrolyzer capacity (kW), converter capacity (kW), and hydrogen tank (kg), which were the key optimized values to answer the research questions in the Section 1.

3.3. Sensitivity Variables

A sensitivity analysis can be performed by inputting multiple values for a particular input variable. HOMER repeats the optimization process for each of value assigned to the variables. Three sensitivity variables (capacity shortage, maximum unmet hydrogen load factor, and renewable fraction) were used in the analysis. Capacity shortage is the shortfall that occurs between the required operating capacity and the actual operating amount that the system can provide. HOMER considers any system that experiences unmet load as infeasible, only if a system is able to meet the demand, the system is termed as feasible. The default value of the capacity shortage was zero but was varied from zero percent to 100% in the analysis. This parameter was only for the battery-operated vehicles which were represented by the electrical load. Similarly, for fuel cell bus another factor called unmet hydrogen load was taken into account, the same principle as capacity shortage implied for this, but for fuel cell buses represented as hydrogen load. The default value for this parameter was zero, but was varied from zero to 100% to check how much of the hydrogen load could be fulfilled by a particular type of renewable. The renewable energy fraction was varied from 98% to 100%; however, as discussed in the research question and the design criterion that all the energy came from renewable energy sources making the renewable energy fraction fixed to 100%. This also took into account that no energy was being bought from the grid. The optimal solution will be the one where the capacity shortage is zero percent, unmet hydrogen load is zero percent, and the renewable fraction is hundred percent. By doing this, it ensures that all the energy is from the renewables. The sensitivity analysis will help in determining whether a scenario is able to fulfill a load profile or not. If not, then by what percentage was the load profile not fulfilled for both capacity shortage and hydrogen unmet load.

3.4. HDRSAM Analysis

The HDRSAM model was used for the fuel cell bus refueling analysis. The number of years for analysis, amount of hydrogen per vehicle and the discount rate were kept the same as in HOMER. The schedule for refueling the busses was also kept same as that of the load profile used in HOMER. In this way, HDRSAM could be integrated into the HOMER model simulation by adding a new module function of refueling analysis. The charging of the bus took place during the night. Two types of station options were present—a gaseous hydrogen station and a liquid hydrogen station. For our analysis only the gaseous hydrogen station was considered. While the dispensing option varied from 350 or 700 bar cascade dispensing, 700 bar booster compressor, 350 or 700 bar vaporization or compression and 350 bar cryo-pump dispensing, the production volume of the components varied from high, medium to low. Overall, there were more than hundreds of input parameters in the HDRSAM model. They were kept to their default value for our analysis.

A total of 18 scenarios were analyzed with variation in hydrogen delivery option (tube trailer or pipeline delivery), production volume (low, mid or high), hydrogen dispensing option (cascade dispensing or booster compressor dispensing), and hydrogen dispensing option (350 bar or 700 bar). The general economic assumptions are listed in the Table 3. It was assumed that the station started in 2019, with a one-year construction period. The fueling rate was taken as the default value of 1.8 Kg/min. It was also assumed that the station was utilized as 100% of its capacity through the year.

Table 3. General assumption for the heavy-duty refueling station analysis (HDRSAM) model.

Assumed start-up year	2019
Construction Period (year)	1
Desired year dollars for cost estimates	2016
Real after-tax discount rate (%)	0.08
Analysis period (years)	25
Max. dispensed amount per vehicle(kg)	10
Refueling rate (kg/min)	1.8
Vehicle fill time (min)	5.6
Vehicle lingering time (min)	5

The model outputs were three key economic figures, the cost of the hydrogen, station capital cost, and time to positive return on investment. The hydrogen cost that is presented here was only for owning and operating the station. The components whose cost was included in the system were dispenser, storage, compressor/pump, refrigeration, electrical, controls, and others.

4. Simulation Results and Discussion

4.1. HOMER Analysis Results

A total of 42 scenarios were analyzed using HOMER. Seven different wind turbines (WT) with capacity ranging from 0.5 MW to 1.5 MW were used in each design, A, B, and C, each presenting different load profiles with two systems grid and off-grid, as shown in Figure 7.

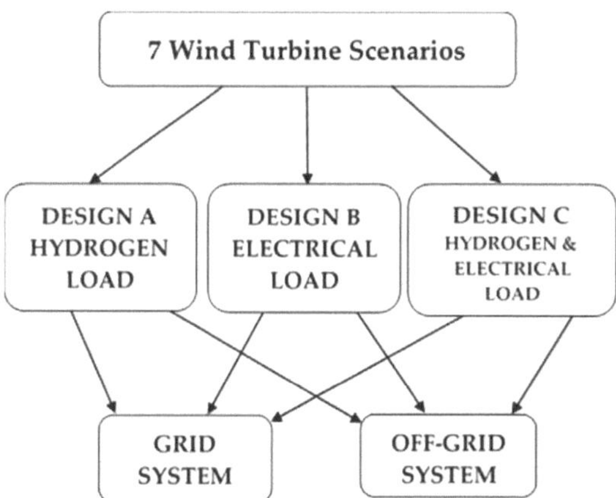

Figure 7. Scenarios' flowchart for the HOMER analysis.

A comparison of design A, B, and C for the grid-connected systems is shown in the Figure 8. The general trend of increasing system cost with increasing wind turbine capacity can be observed. However, the remaining variation of the system cost can be explained by other factors, such as the grid connectivity and the annual energy production, which was dependent on the wind turbine type. The red dots on the curves in Figure 8 indicate that the load profile was not being fulfilled by the renewable sources and required additional energy from the grid to fulfill the load profile. For the design A, 0.5 MW and 0.66 MW wind turbine, the hydrogen unmet load was 20% and 5%, respectively. Similarly, for design C, 0.5 MW and 0.66 MW wind turbines, the electrical load was met but the hydrogen unmet load was 25% and 10%, respectively. As these scenarios did not fulfill the load profile, they were considered infeasible.

For the scenarios where the demand was being fulfilled, the irregularity of the cost peak for the 0.9 MW wind turbine can be explained by the fact that the 0.9 MW wind turbine produced less energy and hence, less payback from the grid, raising the overall cost of the system. Since all scenarios were grid connected, the excess electricity could be completely sold back to the grid, which helped reduce the overall system cost. This trend can be clearly seen at the decreasing cost between the 0.8 MW and 0.85 MW wind turbine and between the 0.9 MW and 1 MW wind turbine.

The answers to the research questions listed in Section 1 can be obtained from Figure 8. It can be observed that for design A and C, 0.8 MW wind turbine system was sufficient to fulfil the load profile. However, the optimal solution was found by using a 1 MW wind turbine system. For design B, a 0.5 MW turbine system was sufficient to fulfil the load profile. However, the optimal solution was found by using a 1 MW turbine system. The optimal solution for all the designs was provided by a 1 MW turbine system. The cost for design A, B, and C was $719,422, $289,161, and $986,065, respectively. It can be seen that the cost of the system decreased as well as increased when going from 0.5 MW to 1.5 MW, depending on the energy produced by the wind turbine. It also showed that the increase in the capacity of the renewables could help in decreasing the cost of the systems for the grid-connected setup. For the optimal design, the net system cost of design C was 2.23% less than the combined cost of the design A and design B.

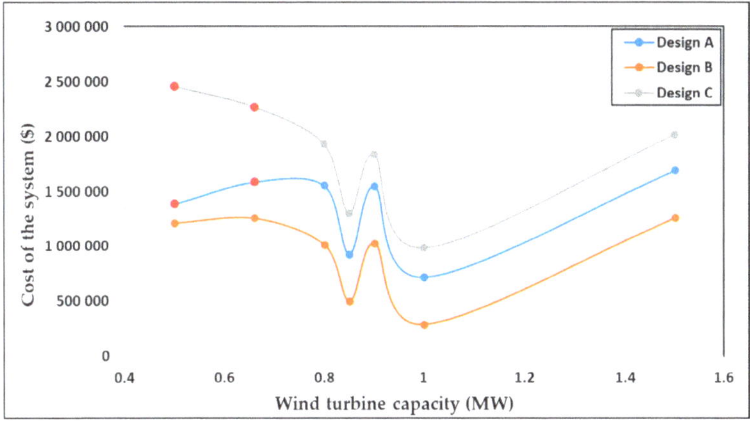

Figure 8. Comparison of grid connected system for design A, B, and C.

The Figure 9 shows the comparison of the three designs when the systems did not have a grid connection. It can also be seen that the system cost for all three designs almost increased linearly with increasing size of the wind turbine. Since there was no grid connection, no excess electricity could be sold.

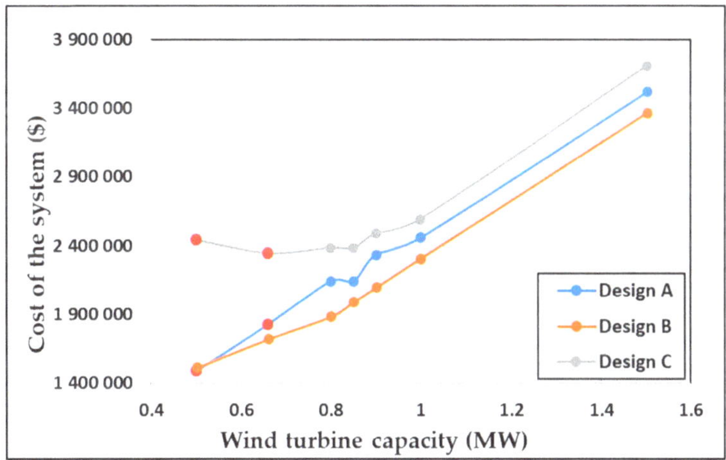

Figure 9. Comparison of the off-grid system for design A, B, and C.

The red dot in the Figure 9 indicates the scenarios where the load profile (charging demand) was not fulfilled by the renewable sources and hence, was not the optimal solution. The unmet load percentage for these scenarios was the same as in grid connection. The irregularity between the cost of the system for 0.8 MW WT and 0.85 MW WT for design A and C is because of the size of the hydrogen tank which was required. For design A, a 200 kg hydrogen tank was required for the 0.8 MW WT system, and a 100 kg hydrogen tank was required for the 0.85 MW WT system. For design C, a 300 kg hydrogen tank was required for the 0.8 MW WT system, and a 200 Kg hydrogen tank was required for the 0.85 MW WT system. It can be seen from the results of the designs A and C that the load profiles (charging demand) could be fulfilled by using 0.8 MW WT. However, the optimal solution was given by the use of a 0.85 MW WT, while the optimal solution for design B was given by use of the system with capacity of 0.5 MW WT. The optimal cost system of design A, B, and C was $2,141,395, $1,512,511, and $2,381,317, respectively. For the optimal design, the net system cost of design C was 34.8% less than the combined cost of the design A and design B.

A comparison between the grid connection and off-grid connection system for the optimal scenario is illustrated in Figure 10. It can be seen that in all three designs having a grid-connected system was significantly more feasible than an off-grid connection system, which can answer the 4th research question listed in Section 1

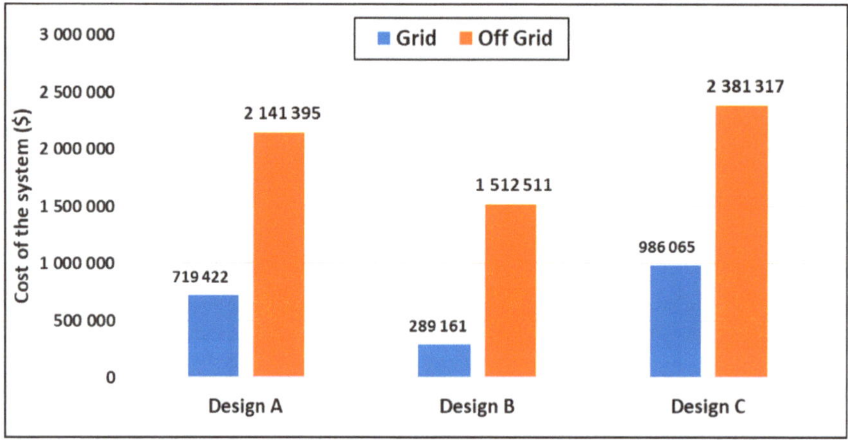

Figure 10. Comparison of grid/off-grid connection for optimal designs.

The off-grid system for design A, B and C was 2.97, 5.23, and 2.41 times higher than grid-connected systems. It can also be observed that the cost of the grid-connected system for design C was less than the combined sum of the designs A and design B, hence making it an optimal design solution.

4.2. Results of HDRSAM

The HDRSAM analysis was performed for getting an estimate cost of the charging station for fuel cell busses. Eighteen scenarios were analyzed and the total investment cost is shown in Figure 11. The results were divided into groups of three, where the cost of the system decreased when the production volume of the components changed from low, mid to high. This trend was followed across all the scenarios. It can be seen that the optimal solution design consisted of tube trailer as hydrogen delivery option with cascade dispensing option at 350 bar with high production volume. The cost of the system was $452,148.

The whole well to wheel process cost was calculated by adding the cost of the optimal solution from HOMER and HDRSAM. The optimal solution form HOMER was of design C grid-connected system ($2,381,317) and for HDRSAM, the optimal solution cost was $452,148. The overall cost of the system was $2,833,465. The charging station comprised only 15.95% of the total cost of the system, while the remaining cost was for the production of energy. Hence, developing an integrated energy system for both fuel cell vehicles and battery-operated vehicles is more cost effective than an individual stand-alone system.

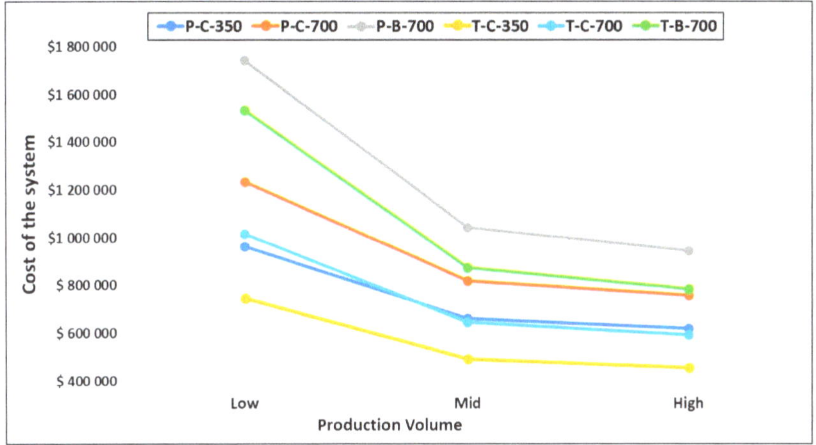

Figure 11. HDRSAM Anlysis: Comparison of the investment cost of the systems. In the legend, the letters and numbers (from left to right) refer to the type of hydrogen delivery option (P = pipeline, T = Tube trailer), type of dispensing (C = cascade dispensing, B = booster dispensing) and dispensing option (350 or 700 bar)).

4.3. Carbon Dioxide Emissions' Reduction

Transportation is one of the biggest contributors of greenhouse gas emissions, generating almost 27% of the emissions [19]. For a diesel-powered bus during combustion, the carbon dioxide emission is 1290 gms/km [17]. While for a fuel cell bus during the combustion, no emissions are made. While for cars the emissions are 120 gms/km [20]. An estimate carbon dioxide emissions reduction was calculated. By using the battery-operated vehicles and fuel cell electric vehicles a total of 373,307.4 kg of CO_2 emissions could be prevented annually at the DTU-Risø campus.

5. Conclusions

To make the whole Danish energy system independent of fossil fuels in 2050, the one-stop charging station for EV and fuel cell cars was studied, and it showed the goal for 100% renewable campus for DTU-Risø is quite achievable. Location plays quite an important role in determining where a charging station is developed. It influences the average wind speed and the global horizontal irradiance which are used in determining the annual wind and solar energy production. Wind resource in Denmark bear a better potential as compared to solar resources. According to HOMER analysis, a grid connected system is always better option than an off-grid system for achieving an economical optimal design solution. It ensures that the excess electricity produced is not wasted and can be used for other purposes, such as for providing electricity to local end-users. It also helps in reducing the cost of the system because of the payback from the grid. The minimum renewable capacity of 0.8 MW for design A and design C and 0.5 MW for design B is required to fulfill the charging load profiles for a grid connected system, and the costs of these systems were $1,552,331, $1,215,106, and $1,929,283 for design A, B, and C, respectively. However, it can also be seen with the increase in the capacity of the renewables a lower cost of these system can be achieved. The optimal solution was a grid connected 1 MW WT system for all three designs, A, B, and C. The cost of these systems being $719,422, $289,161, and $986,065 for design A, B, and C, respectively. Even though the difference in the cost of design C and the combined cost of design A and design B was quite small, an integrated system for both BEV and FCEV is an optimal design solution. From the HDRSAM analysis, it can be seen that the optimal solution design consisted of tube trailer as the hydrogen delivery option with cascade dispensing option at 350 bar with high production volume. The cost of the system was $452,148. The whole well-to-wheel process cost was calculated by adding the cost of the optimal solution from HOMER and HDRSAM. The cost of the whole system from producing the energy and to delivering the energy was $2,833,465. A charging station comprised only about 15% of the whole system. It can be concluded that a renewable energy system with a grid connection for an integrated multi energy system can be a feasible solution from both technical and economical point of view for one-stop charging stations.

When the target of one stop charging station is met, it will offer the global EV industry and its associated energy sectors, a more renewable-based, cost-effective, user-friendly, and grid-friendly energy charging infrastructure solution. With the integration of these two types of charging stations, a reduction in the infrastructure and operation expenses, with grid support for distribution and transmission, will offer the users a one-stop charging station for both battery charging and hydrogen refueling. Further studies should focus on developing a robust, optimal multi-objective and multi-stage decision strategies for controlling and managing multi-energy processes associated with the one-stop charging stations.

Author Contributions: Conceptualization, S.B., S.Y. and J.X.; Methodology, S.B. and S.Y.; Software, Simulation; Simulation, S.B.; Data curation, S.B.; Writing—original draft, S.B.; Resources and Supervision, S.Y. and Y.Z.; Writing—review and editing, Y.Z. and L.M.-P.; Project administration and Funding acquisition, Y.Z. Review, J.X. All authors have read and agreed to the published version of the manuscript.

Funding: This work is supported by the Proactive Energy Management Systems for Power-to-Heat and Power-to-Gas Solutions (PRESS) project granted by the Danish Agency for Science and Higher Education (No. 8073-00026B).) and by the Enhancing wind power integration through optimal use of cross-sectoral flexibility in an integrated multi-energy system (EPIMES) project granted by the Danish Innovation Funding (No. 5185-00005A).

Acknowledgments: I would like to thank the administration and the technical group at the Technical University of Denmark for their support.

Conflicts of Interest: The authors declare no conflict of interest.

References

1. World Economic Forum. *Electric Vehicles for Smarter Cities: The Future of Energy and Mobility*; World Economic Forum: Koroni, Switzerland, 2018.
2. Morgan, S. Euractiv. 03 October 2018. Available online: https://www.euractiv.com/section/electric-cars/news/denmark-to-ban-petrol-and-diesel-car-sales-by-2030/ (accessed on 10 January 2019).
3. Un-Noor, F.; Padmanaban, S.; Mihet-Popa, L.; Mollah, M.N.; Hossain, E. A comprehensive study of key electric vehicle (EV) components, technologies, challenges, impacts, and future direction of development. *Energies* **2017**, *10*, 1217. [CrossRef]
4. Tan, K.M.; Ramachandaramurthy, V.K.; Yong, J.Y.; Padmanaban, S.; Mihet-Popa, L.; Blaabjerg, F. Minimization of load variance in power grids—Investigation on optimal vehicle-to-grid scheduling. *Energies* **2017**, *10*, 1880. [CrossRef]
5. Camacho, O.M.F.; Mihet-Popa, L.; Mihet-Popal, L. Fast charging and smart charging tests for electric vehicles batteries using renewable energy. *Oil Gas Sci. Technol. Rev. IFP Energ. Nouv.* **2014**, *71*, 13. [CrossRef]
6. Berckmans, G.; Messagie, M.; Smekens, J.; Omar, N.; Vanhaverbeke, L.; Van Mierlo, J. Cost projection of state of the art lithium-ion batteries for electric vehicles up to 2030. *Energies* **2017**, *10*, 1314. [CrossRef]
7. Popa, L.M.; Saponara, S. Toward green vehicles digitalization for the nextgeneration of connected and electrifiedtransport systems. *Energies* **2018**, *11*, 3124. [CrossRef]
8. 200 New Hydrogen Buses to Denmark, State of Green. 04 October 2018. Available online: https://stateofgreen.com/en/partners/state-of-green/news/200-new-hydrogen-buses-to-denmark-fosters-sustainable-transportation/ (accessed on 5 December 2018).
9. Kumari, J.; Subathra, P.; Moses, J.E.; Shruthi, D. Economic analysis of hybrid energy system for rural electrification using HOMER. In Proceedings of the 2017 International Conference on Innovations in Electrical, Electronics, Instrumentation and Media Technology (ICEEIMT), Coimbatore, India, 3–4 February 2017; pp. 151–156.
10. Restrepo, D.; Restrepo, B.; Grisales, L.T. Microgrid analysis using homer: A case study. *DYNA* **2018**, *85*, 129–134. [CrossRef]

11. Energy, H. Homer Pro, Homer Energy. Available online: https://www.homerenergy.com/products/pro/docs/latest/index.html (accessed on 5 August 2018).

12. Laboratory, A.N. Heavy-Duty Refueling Station Analysis Model (HDRSAM), Argonne National Laboratory. Available online: https://hdsam.es.anl.gov/index.php?content=hdrsam (accessed on 8 August 2018).

13. Lozanovski, A.; Whitehouse, N.; Ko, N.; Whitehouse, S. sustainability assessment of fuel cell buses in public transport. *Sustainability* **2018**, *10*, 1480. [CrossRef]

14. International Renewable Energy Agency. *Renewable Power Generation Costs in 2017*; International Renewable Energy Agency: Abu Dhabi, UAE, 2018.

15. DOE Technical Targets for Hydrogen Production from Electrolysis. Available online: https://www.energy.gov/eere/fuelcells/doe-technical-targets-hydrogen-production-electrolysis (accessed on 10 May 2018).

16. Steward, D.; Saur, G.; Penev, M.; Ramsden, T. *Lifecycle Cost Analysis of Hydrogen Versus Other Technologies for Electrical Energy Storage*; NREL: Golden, CO, USA, 2009.

17. Li, W.; He, K.; Wang, Y. Cost comparison of AC and DC collector grid for integration of large-scale PV power plants. *J. Eng.* **2017**, *2017*, 795–800. [CrossRef]

18. Denmark: Industrial Prices of Electricity 2008–2018, Satista. Available online: https://www.statista.com/statistics/595800/electricity-industry-price-denmark/ (accessed on 14 May 2018).

19. Todts, W. *CO2 Emissions from Cars: Facts*; European Federation for Transport and Environment AISBL: Brussels, Belgium, 2018.

20. Cockroft, C.J.; Owen, A.D. *Hydrogen Fuel Cell Buses: An Economic Assessment*; Murdoch University and University of New South Wales: Perth, Sydney, Australia, 2008.

Article

Minimization of Cross-Regulation in PV and Battery Connected Multi-Input Multi-Output DC to DC Converter

Vibha Kamaraj [1,*], N. Chellammal [1], Bharatiraja Chokkalingam [1,*] and Josiah Lange Munda [2]

[1] Department of Electrical and Electronics Engineering, SRM Institute of Science and Technology, Chennai 603 203, India; chellamn@srmist.edu.in

[2] Department of Electrical Engineering, Tshwane University of Technology, Pretoria 0001, South Africa; MundaJL@tut.ac.za

* Correspondence: vibhak@srmist.edu.in (V.K.); bharatiraja@gmail.com (B.C.); Tel.: +91-9042701695 (B.C.)

Received: 14 October 2020; Accepted: 24 November 2020; Published: 10 December 2020

Abstract: This paper proposes a digital model predictive controller (DMPC) for a multi-input multi-output (MIMO) DC-DC converter interfaced with renewable energy resources in a hybrid system. Such MIMO systems generally suffer from cross-regulation, which seriously impacts the stability and speed of response of the system. To solve the contemporary issues in a MIMO system, a controller is required to attenuate the cross-regulation. Therefore, this paper proposes a controller, which increases speed of response and maintains stable output by regulating the load voltage independently. The inductor current and the capacitor voltage of the proposed converter are considered as the controlling parameters. With the aid of Forward Euler's procedure, the future values are computed for the instantaneous values of controlling parameters. Cost function defines the control action by the predicted values that describe the system performance and establish optimal condition at which the output of the system is required. This allows proper switching of the system, thereby helping to regulate the output voltages. Thus, for any variation in load, the DMPC ensures steady switching operation and minimization of cross-regulation. To prove the efficacy of proposed DMPC controller, simulations followed by the experimental results are executed on a hybrid system consisting of dual-input dual-output (DIDO) positive Super-Lift Luo converter (PSLLC) interfaced with photovoltaic renewable energy resource. The results thus obtained are compared with the conventional PID (proportional integrative derivative) controller for validation and prove that the DMPC controller is able to control the cross-regulation effectively.

Keywords: renewable energy; MIMO systems; cross-regulation; positive Super-Lift Luo converter; PID controller; digital model predictive controller (DMPC)

---

1. Introduction

With the depletion of fossil fuels and its subsequent ecological impacts, renewable energy resources are in greater demand in power industry [1]. Renewable energy sources (RES) in a hybrid system lack reliability and have lower efficiency and greater fluctuation. Thus, energy storage devices like fuel cells, batteries, and super-capacitors need to be integrated with RES for stable operation. It is relevant to interface the multi-input multi-output (MIMO) converter with RES along with energy storage devices [2–4]. Thus, MIMO DC-DC converters are utilized for multiple supply voltage applications. The MIMO dc-dc converter produces a favorable result [5] with regard to cost, size, and power efficiency. Furthermore, development of MIMO DC-DC converters finds wide applications in smart devices, electric vehicles, fuel cell power generation, etc., [6]. Despite its merits, the MIMO system has a cross-regulation problem. As multiple loads share a single inductor, the load change in one output will affect other outputs. Thus, cross-regulation arises, which in turn affects the stability and versatility of the system.

Numerous studies are conducted in DC-to-DC converters to deal with cross-regulation issues. Time-multiplexing control method in discontinuous conduction mode (DCM) is presented in [7,8]. However, the proposed converter requires a large peak current to meet the heavy load demand. To reduce such a large peak current in DCM, the freewheeling switching method is proposed in pseudo-continuous conduction mode (PCCM) [9]. However, the converter produces a large peak current under heavy load conditions, by adding a supplementary switch, leading to lower efficiency, and control complexity. A voltage comparator circuit [10] is used to meet the voltage demand by adjusting the duty cycle of switches for higher load side, but at the same time, the stability of lower load side voltage is reduced. The dual-mode and pulse width modulation (PWM) base methods are proposed for the particular circuit model in [11,12], which are not compatible with the goal of increasing the number of inputs and the outputs. The digital control method is proposed in [13], in which output voltage utilizes separate regulation for two modes such as common mode and differential mode. However, cross-regulation is suppressed only by adding adaptive gain compensation, moreover, the voltage of m branch depends linearly on (m-1) branch. This method fails to regulate the supply voltage of m branch. In [14], the author proposes a control technique that uses reference current to regulate the duty cycle to drive each output. Due to the difference in bandwidth of the control loop, cross-regulation affects the final output. Cross derivative state feedback method is proposed in [15], where the small-signal model is designed based on inductor current and capacitor voltage. However, the converters are sensitive to change in input and in circuit parameters. In [16], the author investigates multivariable controller to reduce the cross-regulation. The decoupling method is designed to break DIDO (dual-input dual-output) system to SISO (single-input single-output) systems to satisfy the load demand. However, due to its structure, the output load voltage in one branch is always lower than the other branch. Output current feed-forward control is proposed in [17]. This method decouples the transfer function of the system, thus eliminating cross-regulation. However, this method has disadvantages in satisfying the design requirement and involves tedious calculations. The hysteresis method is used in [18], which requires additional circuits; nevertheless, the efficiency is reduced due to current flow in the freewheeling diode.

In [19], a linear small-signal AC analysis is done for dual series connected outputs, which makes the system complex in designing the controllers. In paper [20], SIDO (single-input dual-output) boost converter with digital PWM (pulse width modulation) produces high efficiency but the system operates only for low power applications. In paper [21], the author proposes a deadbeat control method for a single-inductor MIMO system. The above paper utilizes output voltage regulation (OVR) and input current regulation (ICR) method instantaneously to reduce cross-regulation. However, the circuit used to develop ICR and OVR increases the system complexity. In paper [22], the pulse delay control (PDC) method is employed to examine the capability of cross-regulation in a MIMO system, which exhibits a large range of control, but the ripple current in the inductor is not same for all cases and the state averaging method is not applied to all different cases. The linear quadratic controller (LQR) method for SIDO and DIDO system is developed to achieve good steady-state response, transient response, stable line, load regulation, and reduced cross-regulation [23]. Genetic algorithm (GA) is designed effectively to determine the gain values for the conventional controller and weighting matrix for LQR but it makes the system inaccurate if the values of the parameters are high. Based on the previous discussion, it is observed that the cross-regulation problem is an added constraint in MIMO systems. Thus, to reduce the cross-regulation, different methods and their features are discussed.

Model predictive control (MPC) is one of the main approaches in present system control. It has had a major impact on power electronics applications. MPC efficiently controls different types of converters such as rectifier [24], inverter [25], and chopper [26,27]. MPC can handle multiple states and switches with a single controller, whereas in a conventional proportional integrative derivative

(PID) controller, it is difficult to design the multi-structure with a single controller. The experiment proves that MPC has a faster dynamic response even in a nonlinear system [28,29]. These features exploit the advantage of using MPC in a MIMO system to reduce cross-regulation.

Considering these aspects, a digital model predictive controller (DMPC) is suggested in this paper to suppress the cross-regulation problem. The positive Super-lift Luo converter (PSLLC) [30–32] is a power electronic interface between inputs and outputs of a hybrid RES-based MIMO system, and cross-regulation is observed in an open-loop case and minimized with the help of DMPC. The key contribution of the proposal is as follows:

- The paper proposes the photovoltaic (PV) and battery connected MIMO positive Super-Lift Luo converter with high-voltage transfer gain, high power density, high efficiency, reduced ripple voltage, and current.
- Development of a PID controller for a DIDO hybrid energy system.
- Development of a PID controller DMPC for a DIDO hybrid energy system.
- The controller performance is analyzed and compared with a conventional PID controller to check the extent of reducing the cross-regulation and the time delay. The remaining structure of the paper comprises the following: Section 2 describes the working, and provides a state-space model of proposed converter. Section 3 discusses the comparisons of conventional PID controller with DMPC. Section 4 elaborates the design procedure of component selection. Section 5 discusses the simulation results and hardware results of the designed converter. Section 6 provides the conclusion of the analysis carried out in this paper.

2. Proposed Converter Topology

This section presents the general structure of the MIMO positive Super-Lift Luo converter (PSLLC) that is utilized in hybrid energy systems (HES) under consideration.

2.1. Converter Description

The architecture of a MIMO-PSLLC converter is shown in Figure 1a. With respect to the input side, there are "n" input supply units which are represented by voltage sources of $V_{in,1}$, $V_{in,2}$, $V_{in,3}$, ..., $V_{in,n}$ and current sources of $I_{in,1}$, $I_{in,2}$, $I_{in,3}$, ...,$I_{in,n}$. These supply units can be of any type of RES like PV, rectifying wind energy or energy storage unit like battery, super capacitor or fuel cell etc. The switches corresponding to input side are $S_{in,1}$, $S_{in,2}$, $S_{in,3}$, ..., $S_{in,n}$ and the output side are $S_{out,1}$, $S_{out,2}$, $S_{out,3}$, . . . , $S_{out,n}$. For nth load Rn, the respective voltage, current, and filter capacitor are $V_{out,n}$, $I_{out,n}$ and C_n. From Figure 1a, it can be seen that the switches are connected with the diode to form the forward conducting bidirectional blocking (FCBB) mode. FCBB ensures unidirectional current flow and avoids the interface between the sources and loads. When the switch $S_{Control}$ (Figure 1a) is set to ON, inductor and capacitor are magnetized and charge simultaneously. When the switch is set to OFF period, inductor and the capacitor become demagnetized and discharge to the load. Based on the time multiplexing method, the input source and the load are connected through the PSLLC which can be significantly used for multiple loads.

However, to utilize renewable energy and battery as input sources, and to examine the cross-regulation effect, the MIMO system is restructured as a dual-input dual-output (DIDO) system, as shown in Figure 1b. A photovoltaic (PV) module and battery are the two inputs of the converter. To charge the battery by PV source through the converter, switch S_2 is used, and to discharge the battery, switch S_1 is employed and switch S_3 acts as a control switch for the converter. Inputs are in turn connected to the converter, which generates two output voltages V_{01} and V_{02} respectively. Capacitors C_{01} and C_{02} are in parallel to output voltage V_{01} and V_{02} for load resistances R_{01} and R_{02}, respectively.

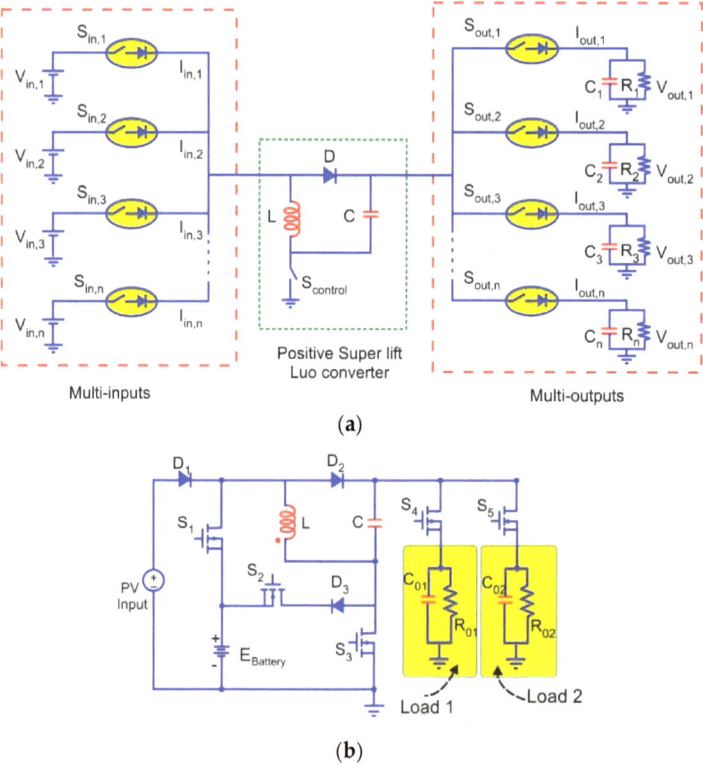

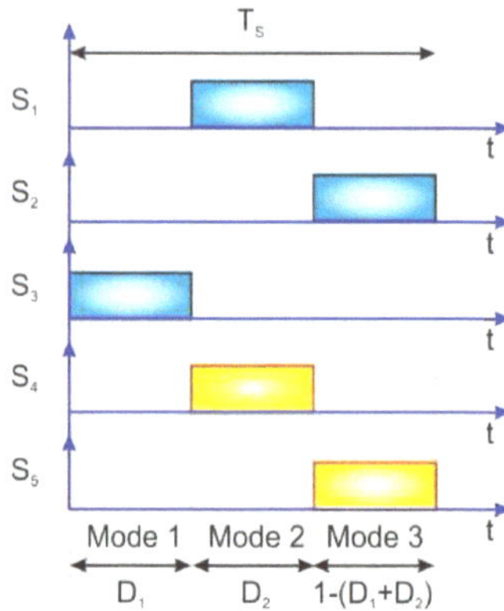

Figure 1. (**a**) Architecture of multi-input multi-output positive Super-Lift Luo converter (MIMO-PSLLC). (**b**) Dual-input dual-output (DIDO) structure of PSLLC converter.

A Timing diagram of various modes are shown in Figure 2, where T_s is represented as the switching period. Referring to the diagram of PSLLC converter circuit, S_3 behaves as complementary switch for S_1, S_2 and S_4, S_5 act as load switches.

Figure 2. Timing diagram for various modes of DIDO-PSLLC converter.

Three modes of operation are considered for the proposed converter. In the first mode, S_3 is ON and inductor L magnetizes for the period of $0 < t < D_1 T_S$. The time period $D_1 T_S < t < D_2 T_S$ is considered as the second mode, where switch S_1 and S_4 are ON and and inductor L demagnetizes during this period. In the third mode, switch S_2 and S_5 are ON for the period of $D_2 T_S < t < D_3 T_S$, where D_3 can be expressed as $1 - (D_1 + D_2)$. Inductor L demagnetizes in this period and works in continuous conduction mode.

2.2. State Space Analysis of PSLLC Converter

To obtain a mathematical model of the proposed DIDO system, a state-space representation is utilized. Using Kirchoffs' voltage law and state-space averaging approach, performance equations that describe the converter are obtained.

Mode 1: During this mode switch, S_3 is ON and S_1, S_2, S_4 and S_5 are OFF. From Figure 3 it is understood that the inductor L and capacitor C become parallel connected and are energized respectively for the period of $D_1 T_s$. Capacitor C_{01} and C_{02} are discharged through the load, assuming capacitor C_{01} and C_{02} are charged initially. The state-space equation for inductor current and capacitor voltage are represented in Equation (1).

$$\left.\begin{array}{c} L\frac{di_L}{dt} = V_{in} \\ C\frac{dV_c}{dt} = \frac{V_{in}}{R} - i_L - \frac{V_c}{R} \\ C_{01}\frac{dV_{c01}}{dt} = -\frac{V_{c01}}{R_{01}} \\ C_{02}\frac{dV_{c02}}{dt} = -\frac{V_{c02}}{R_{02}} \end{array}\right\} \tag{1}$$

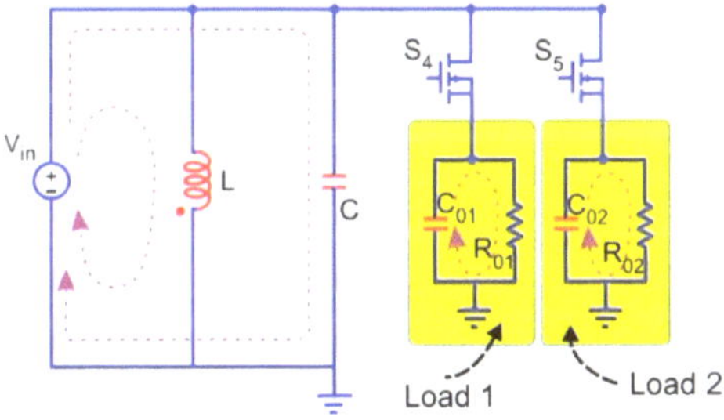

Figure 3. Equivalent circuit diagram of PSLLC converter during D_1 period.

Mode 2: During this mode, the switches S_2, S_3, S_5 are OFF and S_1, S_4 are ON. Referring to Figure 4, inductor and capacitor de-energize to cater for the load power. Capacitor C_{O1} charges for the period of $D_2 T_s$ and C_{O2} discharges through the load. The state-space equation for the inductor current and capacitor voltage are represented in Equation (2).

$$\left.\begin{array}{c} L\frac{di_L}{dt} = V_{in} + V_C - V_{C01} \\ C\frac{dV_C}{dt} = -i_L \\ C_{01}\frac{dV_{C01}}{dt} = i_L - \frac{V_{C01}}{R_{01}} - I_1 \\ C_{02}\frac{dV_{C02}}{dt} = -\frac{V_{C02}}{R_{02}} \end{array}\right\} \tag{2}$$

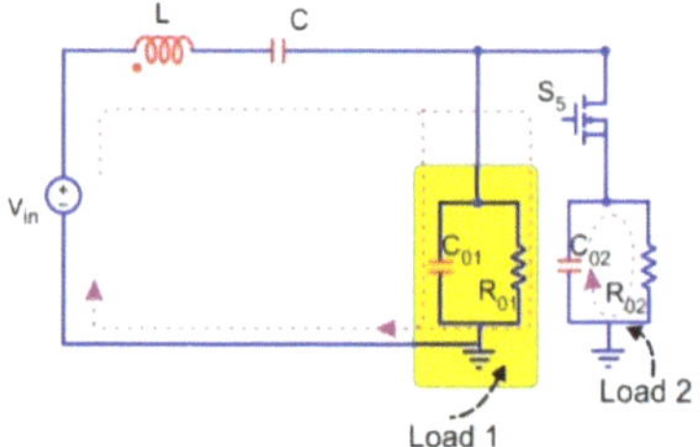

Figure 4. Equivalent circuit diagram of PSLLC converter during D_2 period.

Mode 3: While switching OFF the switch S_3, inductor L and capacitor C will be connected in series and therefore de-energized. Referring to Figure 5, as S_2 is ON, the battery is charged through inductor. As S_5 is ON, capacitor C_{O2} charges for the period of $(1 - (D_1 + D_2)) T_s$. The state-space equation for inductor current and capacitor voltage are represented in Equation (3).

$$\left.\begin{aligned}
L\frac{di_{L1}}{dt} &= V_{in} - V_B \\
C\frac{dV_C}{dt} &= -i_L - \frac{V_B}{R} \\
C_{01}\frac{dV_{C01}}{dt} &= -\frac{V_{C01}}{R_{01}} \\
C_{02}\frac{dV_{C02}}{dt} &= -i_L + \frac{V_B}{R} - \frac{V_{C02}}{R_{02}} - i_2
\end{aligned}\right\} \tag{3}$$

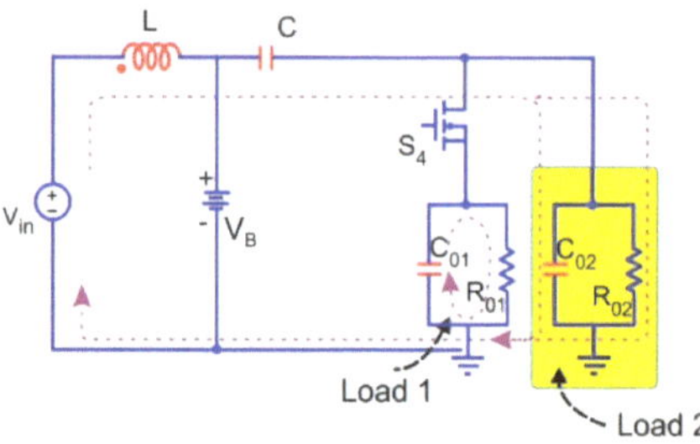

Figure 5. Equivalent circuit diagram of proposed converter during $(1 - (D_1 + D_2))$ period.

Thus, the three modes of operations can be summarized as mentioned in Table 1.

Table 1. Summary of switching states and its modes of operation.

Modes	Charging and Discharging for Battery		Control Switch	Load Switches		Converter Parameters		Remarks
	S_1	S_2	S_3	S_4	S_5	L	C	
Mode 1	OFF	OFF	ON	OFF	OFF	Charging	Charging	L, C → forms a parallel connection and gets charged. C_{01} and C_{02} → discharges to the load.
Mode 2	ON	OFF	OFF	ON	OFF	Discharging	Discharging	L, C → forms a series connection and discharges to the load. C_{01} → charge. C_{02} → discharge to the load.
Mode 3	OFF	ON	OFF	OFF	ON	Discharge to the load and charges the battery	Discharging	L, C → discharges. C_{01} → discharges to the load. C_{02} → charges. Battery → gets charged through inductor.

Representation of State-Space Analysis

Thus, from the modes of Equations (1)–(3), state-space equations are derived by determining the state and output of the system. The system is modeled using capacitor voltage and inductor currents over a switching period T_s. The state model is represented generally by the Equation (4).

$$\left.\begin{array}{l}\dot{X} = AX + BU \\ Y = CX + DU\end{array}\right\} \tag{4}$$

Thus, the representation of state variable (X), input variable (U) and output matrix are given in Equation (5).

$$A = \begin{bmatrix} 0 & \frac{D_2}{L_1} & \frac{-D_2}{L_1} & 0 \\ \frac{1-2(D_1+D_2)}{C} & \frac{-D_1}{RC} & 0 & 0 \\ \frac{D_2}{C_{01}} & 0 & \frac{-1}{R_{01}C_{01}} & 0 \\ \frac{-(1-(D_1+D_2))}{C_{02}} & 0 & 0 & \frac{-1}{R_{02}C_{02}} \end{bmatrix} \quad B = \begin{bmatrix} \frac{1}{L_1} & \frac{-1}{L_1} & 0 & 0 \\ \frac{D_1}{RC} & \frac{-1}{RC} & 0 & 0 \\ 0 & 0 & \frac{-1}{C_{01}} & 0 \\ 0 & 0 & 0 & \frac{-1}{C_{02}} \end{bmatrix} \tag{5}$$

$$C = \begin{bmatrix} 0 & 0 & 1 & 0 \\ 0 & 0 & 0 & 1 \end{bmatrix} \quad D = \begin{bmatrix} 0 & 0 \\ 0 & 0 \end{bmatrix}$$

The obtained output voltage equations with respect to duty cycle and input voltages are stated in Equations (6) and (7).

$$V_{01} = \left[\frac{(1+D_2)}{D_2}\right]V_{in} - \left[\frac{1}{D_2} + \frac{1}{D_1}\right]V_B \tag{6}$$

$$V_{02} = \left[\frac{(D_1+D_2-1)}{D_1}\left(\frac{1}{D_2} + \frac{1}{D_1}\right)\right]V_{in} - \left[\frac{D_1+D_2-1}{D_1^2} + \frac{D_1+D_2-1}{D_1 D_2} + 1\right]V_B \tag{7}$$

2.3. Small Signal Modeling of PSLLC Converters

As the practical system is nonlinear, it is necessary to consider the nonlinearities present in the system. Therefore, combining the perturbations which have small variations over a switching period along with the steady state parameters, the state space variables can be rewritten as: $d_1 = D_1 + \hat{d}_1$, $d_2 = D_2 + \hat{d}_2$, $v_{in} = V_{in} + \hat{v}_{in}$, $v_B = V_B + \hat{v}_B$, $i_1 = I_1 + \hat{i}_1$, $i_2 = I_2 + \hat{i}_2$, $v_C = V_C + \hat{v}_C$, $v_{01} = V_{01} + \hat{v}_{01}$, $v_{02} = V_{02} + \hat{v}_{02}$, $i_L = I_L + \hat{i}_L$. Neglecting the higher order components, the steady state equation can be transformed as $\dot{\hat{x}} = A\hat{x} + B\hat{u}$ and equating steady state element equal to zero, the matrices thus obtained are represented in Equation (8).

$$\frac{d}{dt}\begin{bmatrix} \hat{i}_L \\ \hat{v}_C \\ \hat{v}_{01} \\ \hat{v}_{02} \end{bmatrix}$$

$$= \begin{bmatrix} 0 & \frac{D_2}{L} & \frac{-D_2}{L} & 0 \\ \frac{-(D_1+D_2)}{C} & \frac{-D_1}{RC} & 0 & \frac{1-(D_1+D_2)}{R_{02}C} \\ \frac{D_2}{C_{01}} & 0 & \frac{-1}{R_{01}C_{01}} & 0 \\ \frac{1-(D_1+D_2)}{C_{02}} & 0 & 0 & \frac{-1}{R_{02}C_{02}} \end{bmatrix}\begin{bmatrix} \hat{i}_L \\ \hat{v}_C \\ \hat{v}_{01} \\ \hat{v}_{02} \end{bmatrix}$$

$$+ \begin{bmatrix} \frac{1}{L} & \frac{-1}{L} & 0 & 0 & 0 & \frac{V_C-V_{01}}{L} \\ \frac{D_1}{RC} & \frac{-1}{RC} & 0 & 0 & -\left[\frac{2I_L}{C} + \frac{V_C}{RC}\right] & \frac{-2I_L}{C} \\ 0 & 0 & \frac{-1}{C_{01}} & 0 & 0 & \frac{I_L}{C_{01}} \\ 0 & \frac{1}{RC_{02}} & 0 & \frac{-1}{C_{02}} & \frac{-I_L}{C_{02}} & \frac{-I_L}{C_{02}} \end{bmatrix}\begin{bmatrix} \hat{v}_{in} \\ \hat{v}_B \\ \hat{i}_1 \\ \hat{i}_2 \\ \hat{d}_1 \\ \hat{d}_2 \end{bmatrix} \tag{8}$$

With the designed values of inductor L, Capacitors C, C_{01}, C_{02}, the transfer functions of the output voltage with respect to control variables can be derived as shown in the Equations (9)–(12).

$$G_{11}(s) = \frac{V_{01}}{\hat{d}_1} = \frac{-1.128 \times 10^{10}s - 2.096 \times 10^{13}}{s^4 + 2090s^3 + 3.627s^2 - 2.614 \times 10^8 s - 1.012 \times 10^{10}} \tag{9}$$

$$G_{12}(s) = \frac{V_{01}}{\hat{d}_2} = \frac{-8.5 \times 10^4 s^3 - 1.59308s^2 - 1.394 \times 10^{10}s - 5.208 \times 10^{13}}{s^4 + 2090s^3 + 3.62 \times 10^5 s^2 - 2.614 \times 10^8 s - 1.012 \times 10^{10}} \tag{10}$$

$$G_{21}(s) = \frac{V_{02}}{\hat{d}_1} = \frac{-8.5 \times 10^4 s^3 - 1.734 \times 10^8 s^2 - 3.902 \times 10^{10}s - 7.953 \times 10^{12}}{s^4 + 2090s^3 + 3.62 \times 10^5 s^2 - 2.614 \times 10^8 s - 1.012 \times 10^{10}} \tag{11}$$

$$G_{22}(s) = \frac{V_{02}}{\hat{d}_2} = \frac{-8.5 \times 10^4 s^3 - 1.959 \times 10^8 s^2 - 7.848 \times 10^{10}s - 1.94 \times 10^{13}}{s^4 + 2090s^3 + 3.62 \times 10^5 s^2 - 2.614 \times 10^8 s - 1.012 \times 10^{10}} \tag{12}$$

Using an adjunct polynomial scheme, the higher order transfer functions $G_{11}(s)$, $G_{12}(s)$, $G_{21}(s)$, and $G_{22}(s)$ can be reduced to second order transfer functions as shown in Equations (13)–(16).

$$G_{11}(s) = \frac{V_{01}}{\hat{d}_1} = \frac{1.128 \times 10^{10}s - 3.848 \times 10^6}{s^2 - 722.09s - 27955.8} \tag{13}$$

$$G_{12}(s) = \frac{V_{01}}{\hat{d}_2} = \frac{s + 3736}{s^2 - 722.09s - 27955.8} \tag{14}$$

$$G_{21}(s) = \frac{V_{02}}{\hat{d}_1} = \frac{8.5 \times 10^4 s - 0.16 \times 10^6}{s^2 - 722.09s - 27955.8} \tag{15}$$

$$G_{22}(s) = \frac{V_{02}}{\hat{d}_2} = \frac{8.5 \times 10^4 s - 0.478 \times 10^8}{s^2 - 722.09s - 27955.8} \tag{16}$$

The Bode plot is drawn for the transfer functions [13–16], which shows the existence of cross-regulation in the proposed converter.

2.4. Effect of Cross-Regulation in an Open Loop MIMO Structure of PSLLC Converter

To obtain a stable system, it is necessary to examine the Bode plot for the transfer functions depicted by Equations (13)–(16). The bode plots of transfer functions relating output voltage to duty cycle are obtained and represented in Figure 6 for second order system of $G_{11}(s)$, $G_{12}(s)$, $G_{21}(s)$, and $G_{22}(s)$, where blue graph is the direct transfer function for output voltages and the red graph indicates the cross-coupling transfer function. The Bode plots of coupled transfer function show the phase margin P_m as −90 and indicates that the system is unstable. The existence of coupling between the output voltages makes the closed-loop system destabilized, and tuning becomes difficult. Thus, elimination of interaction using decoupling methodology for voltage control is discussed in Section 3.

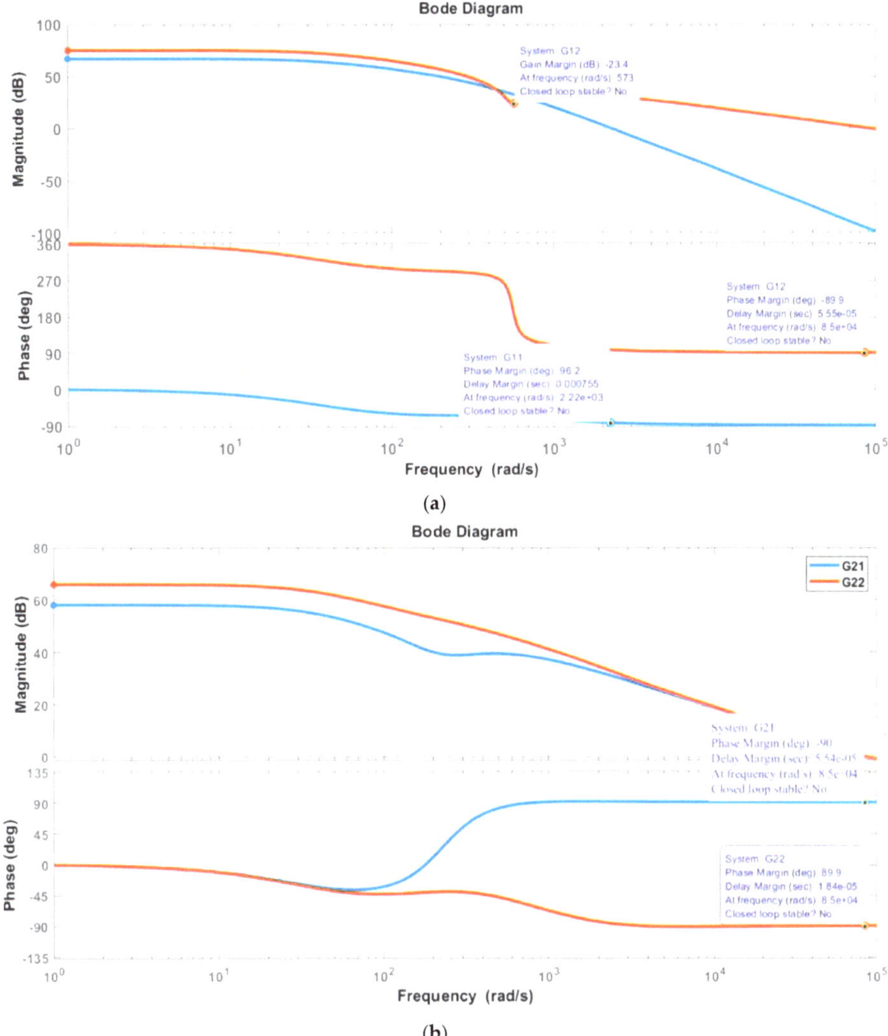

Figure 6. Bode plots of (**a**) controlled transfer function G_{11}, G_{22} of proposed converter (**b**) cross-coupled transfer function G_{21}, G_{12}.

3. Close Loop Controller

To suppress the cross-regulation and to regulate the line and load voltages within the operating limit, it is necessary to design a controller. Therefore, in this paper, two different controllers, a conventional PID controller and a DMPC controller, are discussed.

3.1. Conventional Controller

3.1.1. Decoupling Method

As the proposed converter has a strong coupling effect between input and output, there is nonlinearity, which makes tuning difficult for the individual loop. In the decoupling technique, cross-coupled loops are segregated and considered as individual loops (a MIMO system is converted into SISO system) to reduce the complexity in the coupled network. By separating the loops, the PID controller can be independently controlled for better performance. Figure 7 shows the DIDO structure, representing d_1 and d_2 as input variables and V_{01} and V_{02} as the output variables. The input and output variables are related to the transfer function, as shown in Equations (17) and (18). The change in d_1 affects V_{01} and V_{02}. Similarly, the change in d_2 affects V_{02} and V_{01}. This coupling effect makes the system unstable. Therefore, the decoupling method is preferred, to overcome the interaction between the control variables.

$$V_{01} = G_{11}(S)\hat{d}_1 + G_{12}(s)\hat{d}_2 \tag{17}$$

$$V_{02} = G_{21}(S)\hat{d}_1 + G_{22}(s)\hat{d}_2 \tag{18}$$

$$G(S) = \begin{bmatrix} G_{11}(s) & G_{12}(s) \\ G_{21}(s) & G_{22}(s) \end{bmatrix} \tag{19}$$

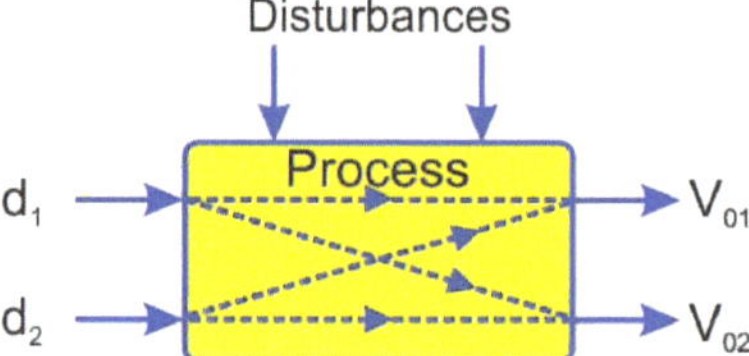

Figure 7. DIDO structure with input and output variable.

To determine the best pairing for the MIMO system from Equation (19) and to measure process interactions, relative gain array (RGA) and Niederlinski index (NI) tools are employed. RGA requires that negative pairing not be selected, and that the pair which is approximately equal to one be chosen. If the NI < 0 system is integral unstable and if NI = 0, a favorable interaction is possible. The formulae to calculate RGA are mentioned in Equations (20)–(22).

$$\Lambda = \begin{bmatrix} \lambda_{11} & \lambda_{12} \\ \lambda_{21} & \lambda_{22} \end{bmatrix} \tag{20}$$

$$\lambda_{ij} = \frac{\text{open loop gain}}{\text{closed loop gain}} \tag{21}$$

$$NI = \frac{|K|}{\Pi_i K_{ii}} \tag{22}$$

Figure 8 shows the decoupled network of the DIDO system. It indicates that by adding $-G_{21}/G_{22}$ in the network, G_{21} term is eliminated, which in turn affects the second output. Similarly, by adding $-G_{12}/G_{11}$ in the network, G_{12} term is eliminated, which affects the first output. Thus, multi-loop interactions are nullified and two individual loops are obtained which are independent of each other by transforming the transfer function matrix into a diagonal one as shown in Equation (23).

$$G' = XU^{-1}G^{-1}; U = \begin{bmatrix} d_1 & d_2 \end{bmatrix}^T \tag{23}$$

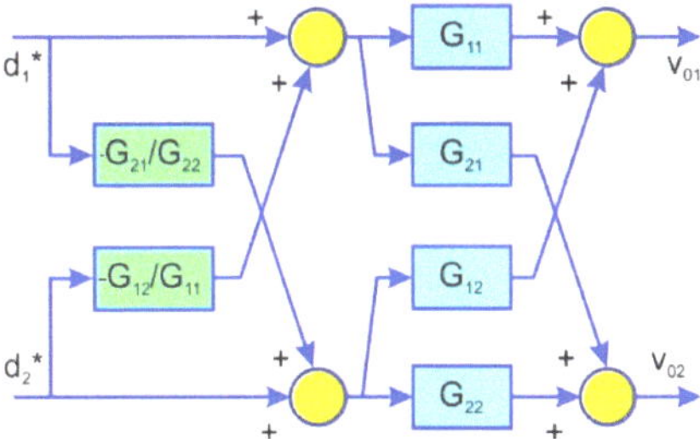

Figure 8. Decoupled network.

From Figure 8, G′(s) can be calculated using a diagonal matrix, and are expressed as Equations (24)–(26).

$$G'(s) = \begin{bmatrix} 1 & -\dfrac{G_{12}(s)}{G_{11}(s)} \\ -\dfrac{G_{21}(s)}{G_{22}(s)} & 1 \end{bmatrix} \tag{24}$$

$$G_1(s) = \frac{V_{01}}{\hat{d}_1} = G_{11}(s) - \frac{G_{21}(s)}{G_{22}(s)} G_{12}(s) \tag{25}$$

$$G_2(s) = \frac{V_{02}}{\hat{d}_2} = G_{22}(s) - \frac{G_{12}(s)}{G_{11}(s)} G_{21}(s) \tag{26}$$

By using adjunct polynomial scheme for reduction of higher order system to second order system, the second order transfer function is expressed as shown in Equations (27) and (28).

$$G_1(s) = \frac{V_{01}}{\hat{d}_1} = \frac{5.179 \times 10^6 s + 1.0297 \times 10^{-3}}{s^2 + 10.558s - 22.695} \tag{27}$$

$$G_2(s) = \frac{V_{02}}{\hat{d}_2} = \frac{6.466 \times 10^3 s - 0.4258}{s^2 + 11.714s + 6.4287e - 5} \tag{28}$$

Thus, the DIDO controller is converted to two individual PID controllers with SISO model.

3.1.2. Design of PID Controller

To obtain the desired regulated voltage, and to enhance the performance, the PID controller is tuned using the Ziegler and Nichols method. By tuning, the gain of proportional and integral value of the conventional controller is designed to vindicate the cross-regulation effect based on the requirements of load disturbance. Controller output C(s), with respect to the error E(s), is expressed in transfer function in the Equation (29).

$$\frac{C(s)}{E(s)} = K_p\left(1 + \frac{K_i}{s} + K_d s\right) \tag{29}$$

For the individual loop of $G_1(s)$ and $G_2(s)$, the controller transfer function is obtained as stated in Equations (30) and (31).

$$\frac{C_1(s)}{E_1(s)} = 2.801\left(1 + \frac{1.748e15}{s} + 0s\right) \tag{30}$$

$$\frac{C_2(s)}{E_2(s)} = 3.052\left(1 + \frac{2.099e15}{s} + 0s\right) \tag{31}$$

The PID controller reduces the cross-regulation. However, it has a variety of drawbacks. For the DIDO system, the system has to be divided into two SISO systems and thus two PID Loop are applied, which increase the complexity of the system. The future trajectory of outputs cannot be determined, as PID emphasizes its effect on error signal rather than on the controlling variables. Moreover, PID takes time to reach a stable operation and suffers from overshoot conditions without proper damping, which affects the steady state error. The PID controller needs to be tuned for every different case or change in system dynamics. This is not ideal to regulate cross-regulation. To overcome all these drawbacks, DMPC is designed.

3.2. Digital Model Predictive Controller (DMPC)

Digital model predictive controller (DMPC) depends on dynamic models that can be developed by system identification. The cost function defines the system behavior of the variables, such as inductor current and the capacitor voltage, that can be used to anticipate the function of desired output voltage. The instantaneous values I_L, V_C are represented by $I(k)_L$ and $V(k)_C$. These values are fed from the system and are used to find out the predicted values of I_L, V_C, which are represented as $I(k+1)_L$ and $V(k+1)_C$. As mentioned in the literature, the predictive values can be obtained by a variety of state estimation methods like Kalmann filter, Lagrangian extrapolation, Newton-Raphson method, and Euler's method. While constructing the predictive model, the controlled variables, that is, V_C and I_L must be measured with the aim of attaining the discrete-time models. The discrete-time model is used to predict the future value of controlled variables at the kth sampling instant. Several discretization methods are used in order to obtain a discrete-time model appropriate for the predictive calculation. Considering that the load can be modelled as a lower order system, the discrete-time model can be obtained by forward Euler's method, which has a simple approximation of the derivative. However, for more complex systems, this approximation may introduce errors into the model and a more accurate discretization method is required. The other methods are used for higher order derivatives. By constructing the sequence of successive approximations, estimated values are obtained accurately. According to the forward Euler approach, the future values of controlled variables are estimated by considering the current values of the system inputs as shown in Equation (32):

$$(k+1) = x(k) + T_s(f(x(k), u(k))) \tag{32}$$

where T_s is sampling time and $f(x(k), u(k))$ is change in state variables obtained from state-space analysis. Applying this formula, the values of $I(k+1)_L$ and $V(k+1)_C$ are obtained. Using the predicted values of control variables, cost function is represented as shown in Equation (33):

$$J = \left\| I(k+1) - x(k+1) \right\|^2 \tag{33}$$

where $(k+1)$ is the reference value and $x(k+1)$ is the anticipated value of the controlled variable considered from the discretized system model. If the system has more than one main control variable, then the cost function can include control parameters X_1 and X_2 with the help of weighting factor λ as shown in Equation (34):

$$J = \left\| X_1(k+1)_{ref} - X_1(k+1)_p + \lambda(X_2(k+1)_{ref} - X_2(k+1)_p) \right\|^2 \tag{34}$$

Thus, for the proposed converter, the controlling parameters are I_L and V_C, therefore, the cost function is replaced as mentioned in Equation (35).

$$J = \left\| I_L(k+1)_{ref} - I_L(k+1)_p + \lambda(V_C(k+1)_{ref} - V_c(k+1)_p) \right\|^2 \tag{35}$$

The cost function J is reduced for the entire control horizon i.e., the duration at which the plant is to be controlled. This produces the optimal condition at which the output of the system is required. For switching combinations, the proposed converter has two switches (MOSFETs) which control the load voltage; as a result the following set S is derived with respect to various possible switching combinations: $S = \{(0,0) \ (0,1) \ (1,0) \ (1,1)\}$.

The value of S (1) to S(n) is allocated on the basis of our optimization of cost function, so when the value of J is min, use S (1) combination and then gradually increase to S(n) for the max value of J. This entire process is repeated for the complete duration of prediction horizon and control horizon. This thus allows proper switching of the converter, which helps in regulating the output voltages. Therefore, for any variation in load, the DMPC ensures a steady switching operation such that the load variations are handled with ease, thus minimizing cross-regulation. This algorithm is implemented in MATLAB Simulink. The DMPC controller is employed by replacing the PID controller. The inputs of DMPC controller are I_L (inductor current), V_{01} (output voltage across load1), V_{02} (output voltage across load2), V_{in1} (PV Input voltage), V_{in2} (battery Input voltage), V_C (capacitor voltage), and t (simulation time). The outputs of the DMPC controller are the switching pulses for two switches across the load output. Figure 9 shows the flowchart of the DMPC algorithm.

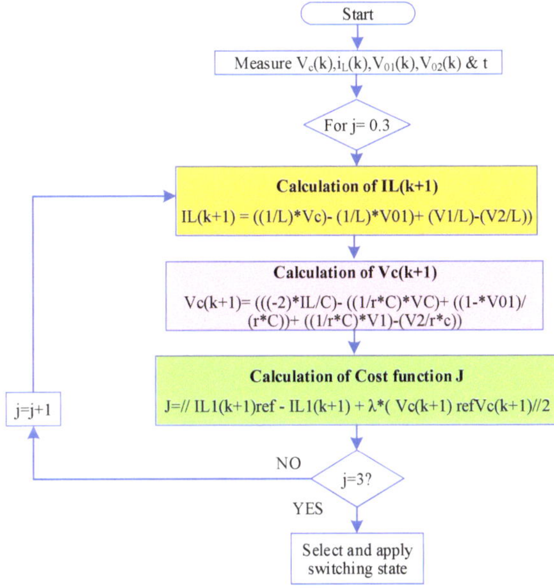

Figure 9. Flow Chart of digital model predictive controller (DMPC) Algorithm.

A control scheme for DMPC applied to the power electronic interface circuit (PSLLC) converter is shown in Figure 10. In this scheme, the output voltage and the current are used as measured variable which is used in the model to estimate the predicted output for n possible actuations. These predicted values and the reference values are evaluated and the error is minimized by the cost function J. The optimal switching state S is selected, which is applied to load switch in the PSLLC converter. Thus, for any variation in one load, the proposed controller allows the proper switching state to the converter, thereby reducing the cross-regulation. Thus, the following points are summarized for the DMPC controller:

- The instantaneous value of inductor current and capacitor voltage are considered as the reference values which are represented as $I^*(k)_L$ and $V^*(k)_C$.
- From the load side, the instantaneous values of inductor current and the capacitor voltage are measured and signified as $I(k)_L$ and $V(k)_C$.
- These values are fed to the predictive model to find the predicted values of $I(k + 1)_L$ and $V(k + 1)_C$.
- The error obtained by the reference value and the predicted values are measured and minimized by the cost function J for the entire control horizon.

- The optimal actuation is attained by minimizing the cost function J, and the corresponding switching state is produced, which controls the load switch effectively, thus minimizing the cross-regulation.

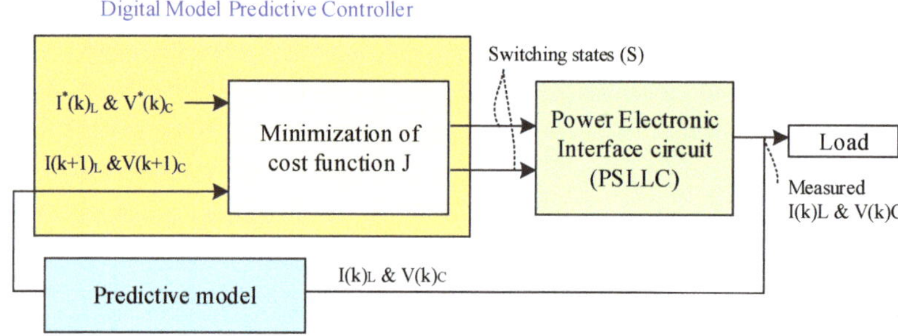

Figure 10. Control scheme of DMPC.

4. Design Procedure

The components used in the proposed converter are designed for the maximum power range of 250 W. The input sources are considered as fixed DC input and battery input. The following points are considered for the converter design as shown in Figure 11.

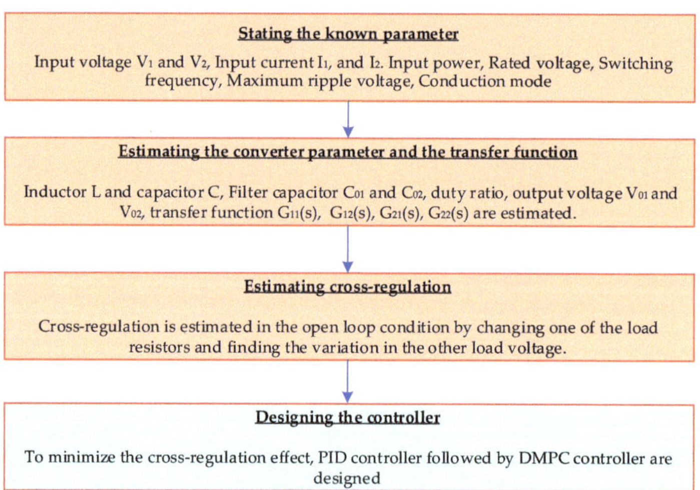

Figure 11. Flow Chart of design procedure.

To determine the parameters of the components used in proposed converter, the following specifications are considered.

(a) The input voltage V_1 and V_2—40 V, 20V
(b) The input current I_{in} —4 A
(c) Ripple voltage—±0.1%
(d) Ripple current—±5%
(e) Switching frequency—100 KHz
(f) Conduction mode—continuous

(g)   Duty cycle are taken in the range of (0.2–0.7)

(h)   The output voltage of the converter is in the range of (60–65) V

(i)   The output current is in the range of (2–3) A

(j)   The minimum value of the inductor is calculated using the formula: $L \geq \frac{V_{in}D_1}{F_s\Delta i\_L} = 1$ mH

(k)   The minimum value of the capacitance is calculated using the formula: $C \geq \frac{\frac{V_{01}}{R}D_1}{F_s\Delta V_{01}}$ where $C = C_{01}$ $= C_{02} = 680$ μF

(l)   The presence of cross-regulation in the system in MATLAB Simulink in open loop condition is verified and estimated. By means of a circuit breaker, a step change across the load R_{01} is introduced during the simulation runtime, using parallel resistance connection. This load variation causes a subsequent variation in the voltage and current across R_{02} which is calculated as follows:

Cross-regulation (%) = $(V_{02peak} - V_{o2drop}) \times 100/V_{02peak} = (93.5 - 90) \times 100/93.5\% = 3.74\%$

Based on this design of the converter, the proposed controller is simulated and tested for different load conditions.

5. Performance Analysis Based on Simulation and Experimental Results

The proposed converter shown in Figure 1b is simulated in MATLAB-Simulink platform. The DIDO PSLLC converter is designed to a set of specifications mentioned in Table 2. Based on the availability of industrial products, the input source and the load specifications are considered. The inductor and capacitor values are selected based on the voltage ripple calculations. Different case studies for line regulation and load regulation are carried out.

Table 2. Specification of Converter for simulation and prototype.

V_{in} (V)	V_B (V)	L (mH)	C (μF)	C_{01} (μF)	C_{02} (μF)	Freq. (KHz)	R_1 (Ω)	R_2 (Ω)	D_1	D_2
40	20	1	680	680	680	100	25	20	0.5	0.2

5.1. Simulation of PID Controller

For steady state operation, the proposed converter has $V_{01} = 103$V, $V_{02} = 94$ V, and current $I_{01} = 4.08$ A, $I_{02} = 4.63$ A at $R_1 = 25$ Ω, and $R_2 = 20$ Ω. Different cases are discussed for the evaluation of cross-regulation in the proposed converter using the PID controller. To analyze the cross-regulation effect, a breaker is introduced for the load change in one output and the corresponding changes are observed in the other output.

Case 1: Higher step change at load output 1: Under steady state operation, load current I_{o1}, I_{o2} are found to be 4.087 A, 4.63 A respectively. At t = 5 s, the load resistance R_1 is stepped up from 25 Ω to 50 Ω and the results are displayed in Figure 12. It can be seen that the output current I_{01} decreases from 4.08 A to 2.117 A as the load increases. Due to decremented change in I_{01} current, the other load current I_{02} increases from 4.63 A to 4.728 A, and the output voltage V_{01} and V_{02} are modulated with respect to change in load. The load decrement in I_{01} does not affect the second load current I_{02}, and it is regulated with the difference of 0.098 A, while the cross-regulation is calculated as 0.02 (or 2%). Thus the controller effectively suppresses the cross-regulation for the higher load change but the step change in the first load modulates the output voltage and produces a ripple in $V_{02.}$

Case 2: Moderate step increase at load output 1: At t = 5 s, the load resistance R_1 is stepped up from 25 Ω to 30 Ω and the results are displayed in Figure 13. It is clearly seen that the output current I_{01} decreases from 4.08 A to 3.445 A as the load increases. Due to decremented change in I_{01} current, the other load current I_{02} increases from 4.63 A to 4.66 A, and the output voltage V_{01} and V_{02} are modulated with respect to change in load. The full load decrement in I_{01} does not affect the second load current I_{02}, and it is regulated with the difference of 0.03 A, while the cross-regulation is calculated as 0.008 (or 0.8%). Thus, the proposed controller is able to suppress the cross-regulation for the moderate load change but it produces ripples in the output voltage V_{02}.

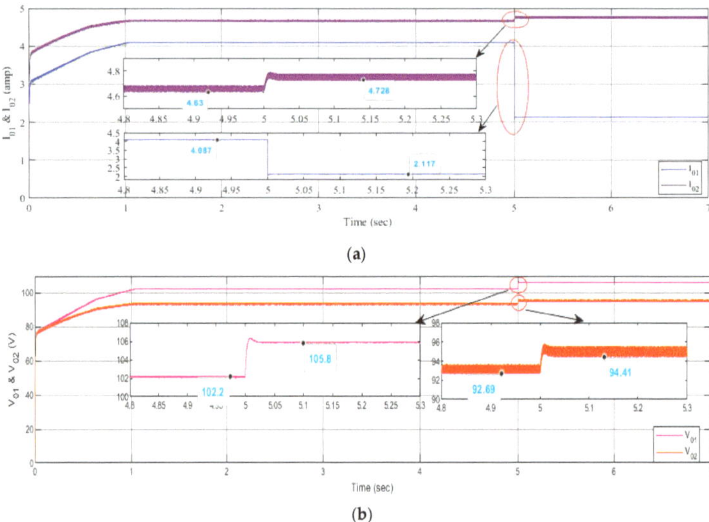

Figure 12. Simulated results of higher step change at load output 1 using proportional integrative derivative (PID) controller (**a**) Output current I_{01} and I_{02} (**b**) Output voltage V_{01} and V_{02}.

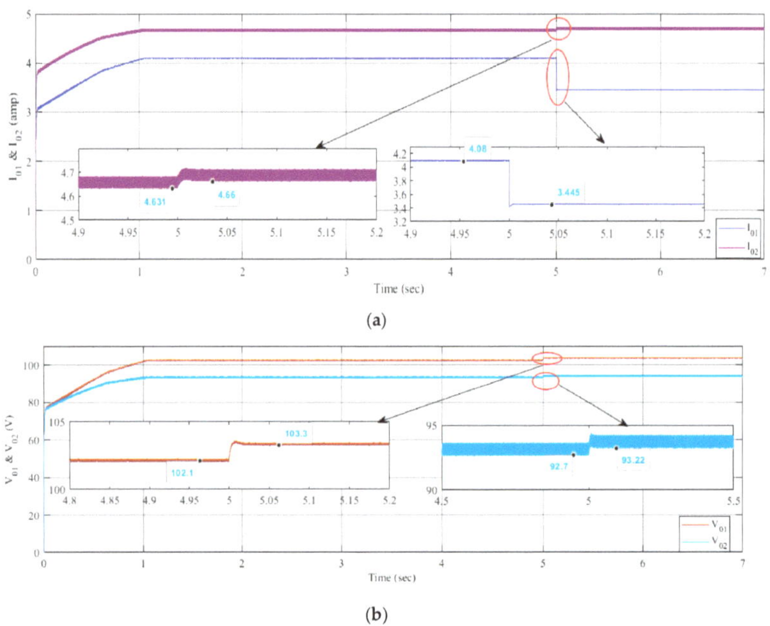

Figure 13. Simulated results of moderate step increase in load change using PID controller (**a**). Output current I_{01} and I_{02} (**b**) Output voltage V_{01} and V_{02}.

Case 3: Moderate step decrease at load output 1: At t = 5 s, the load resistance R_1 is stepped down from 25 Ω to 20 Ω and the results are displayed in Figure 14. It can be seen that the output current I_{01} increases from 4.08 A to 5.021 A as the load decreases. Due to the incremental change in I_{01} current, the other load current I_{02} decreases from 4.63 A to 4.595 A, and the output voltage V_{01} and V_{02} are modulated with respect to change in load. The slight increment in I_{01} does not affect the second load current I_{02}, and it is regulated with the difference of 0.041 A, while the cross-regulation is calculated as 0.006 (or 0.6%). Thus, the controller reduces the cross-regulation for the moderate load change but sudden change produces ripples in the output voltage V_{02}.

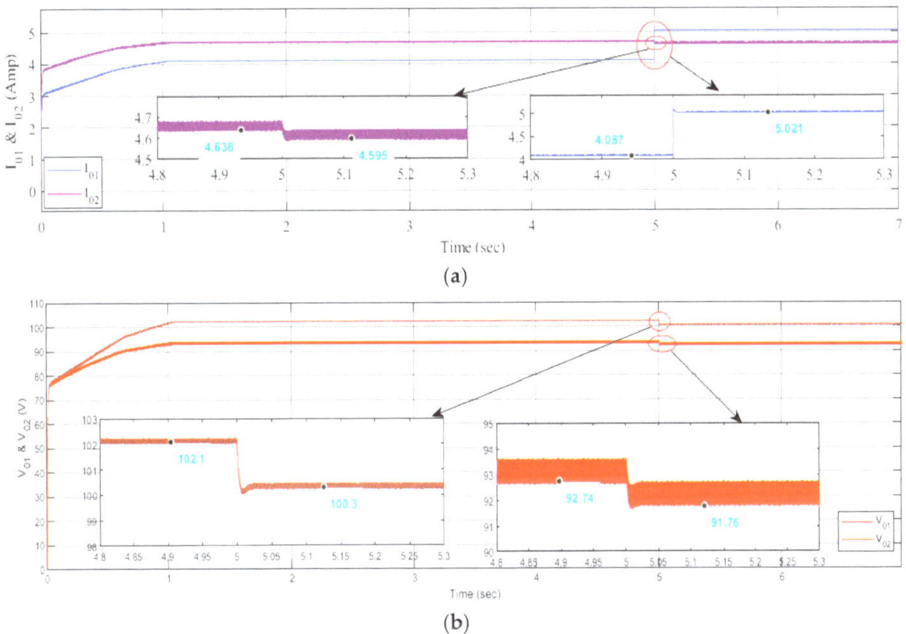

Figure 14. Simulated results of moderate step decrease in load using PID controller. (**a**) Output current I_{01} and I_{02} (**b**) Output voltage V_{01} and V_{02}.

From these results, it is understood that the cross-regulation was reduced to an acceptable level of 0.6 to 2% for various cases of step change in the load. The drawback is the requirement of two PID controllers, to control the switches across the two loads, which produces rippled output. This may increase the complexity of the converter. To overcome this difficulty and to obtain a faster response, a DMPC controller is utilized.

5.2. Simulation of DMPC Controller

The PSLLC was simulated in MATLAB for closed-loop condition with a digital model predictive controller (DMPC). To analyze the DMPC controller and to compare its performance with the PID controller, the same cases are discussed for the evaluation of cross-regulation.

Case 1: Higher step change at load output 1: At t = 0.3 s, the load resistance R_1 is stepped up from 25 Ω to 50 Ω and the results are displayed in Figure 15. It can be seen that the output current I_{01} decreases from 4.401 A to 2.233 A as the load increases. Due to decremental change in I_{01} current, the other load current I_{02} increases from 2.189 A to 2.203 A, and the output voltage V_{01} and V_{02} are controlled to track the reference voltage just before and after the higher load change at any of the output. The load change in I_{01} does not affect the second load current I_{02} and it is regulated with the difference of 0.014 A and the cross-regulation is estimated as 0.006 (or 0.6%). Therefore, the proposed controller effectively suppresses the cross-regulation for the higher load change at any one of the outputs.

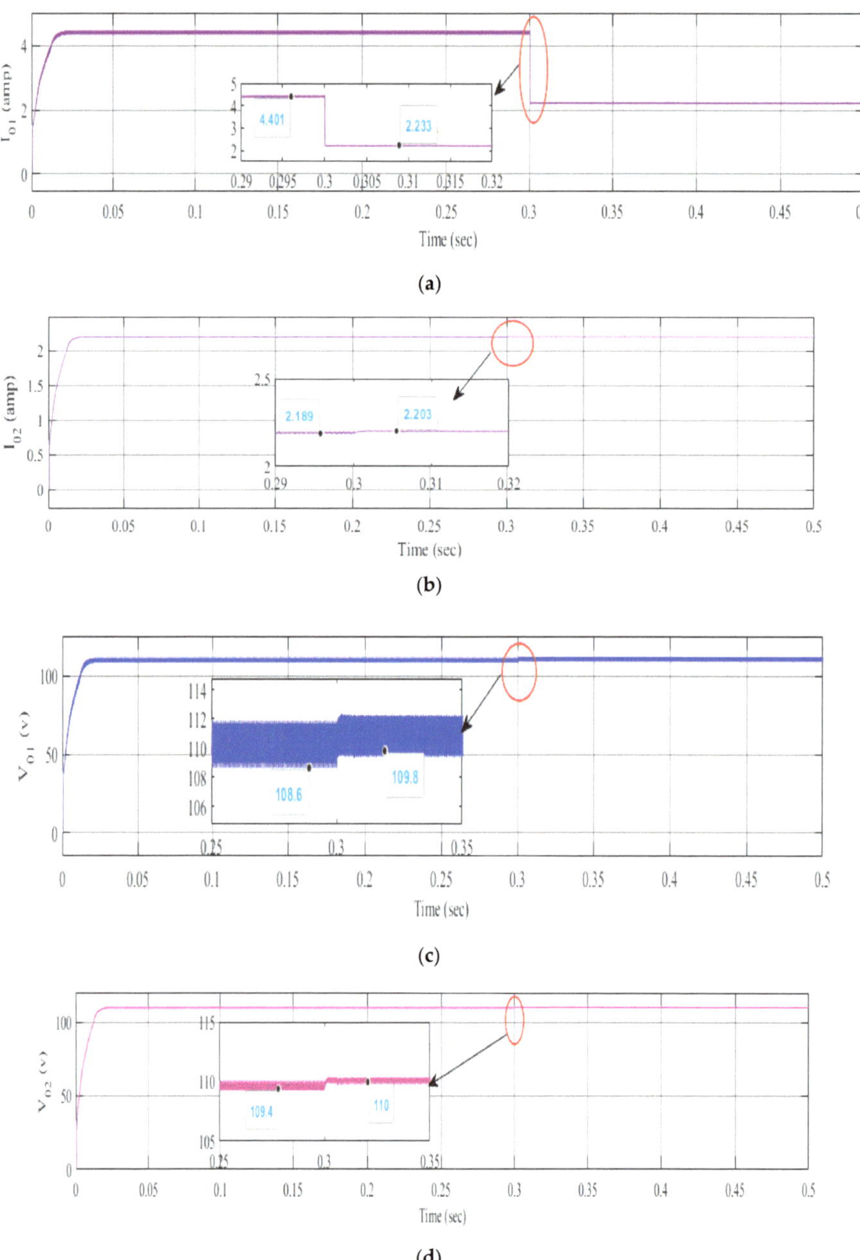

Figure 15. Simulated results of higher step change at load output 1 using DMPC controller (**a**) Output current I_{01} (**b**) Output current I_{02} (**c**) Output voltage V_{01} (**d**) Output voltage V_{02}.

Case 2: Moderate step increase at load output 1: At t = 0.3 s, the load resistance R_1 is stepped up from 25 Ω to 30 Ω and the results are displayed in Figure 16. It can be seen that the output current I_{01} decreases from 4.475 A to 3.671 A as the load increases. Due to decremented change in I_{01} current, the other load current I_{02} increases from 2.195 A to 2.202 A and the output voltage V_{01} and V_{02} are controlled to track the reference voltage nearly before and after the moderate load change at any of the output. The full load change in I_{01} does not affect the second load current I_{02}, and it is regulated with the difference of 0.007 A, while the cross-regulation is calculated as 0.004 (or 0.4%). Thus, the controller effectively minimizes the cross-regulation for the moderate load change at any one of the outputs.

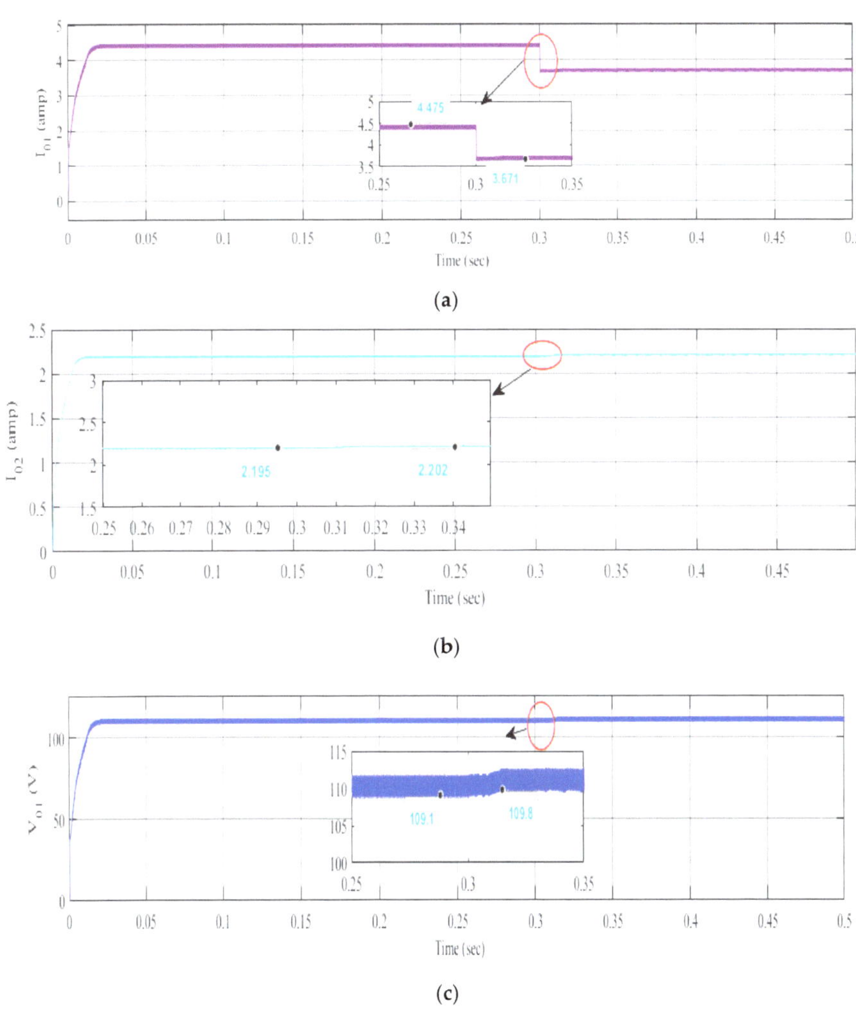

Figure 16. *Cont.*

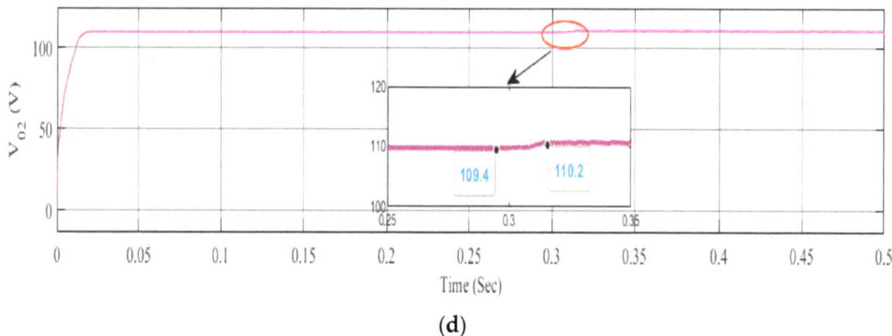

(d)

Figure 16. Simulated results of moderate step increase in load change using DMPC controller (a) Output current I_{01} (b) Output current I_{02} (c) Output voltage V_{01} (d) Output voltage V_{02}.

Case 3: Moderate step decrease at load output 1: At t = 0.3 s, the load resistance R_1 is stepped down from 25 Ω to 20 Ω and the results are displayed in Figure 17. It can be seen that the output current I_{01} increases from 3.89 A to 4.889 A as the load decreases. Due to incremental change in I_{01} current, the other load current I_{02} changes from 3.227 A to 3.243 A, and the output voltage V_{01} and V_{02} are controlled with respect to change in load. The slight increment in I_{01} does not affect the second load current I_{02}, and it is regulated with the difference of 0.016 A, while the cross-regulation is calculated as 0.003 (or 0.3%). Thus, the controller effectively overcomes the cross-regulation effect for the step decrease in load change either at R_{01} or R_{02}.

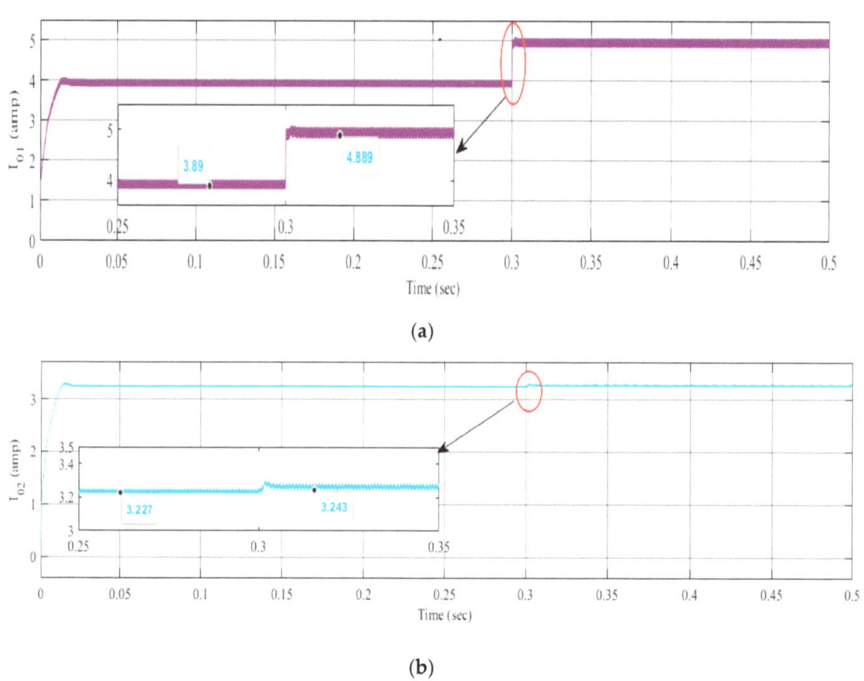

(a)

(b)

Figure 17. *Cont.*

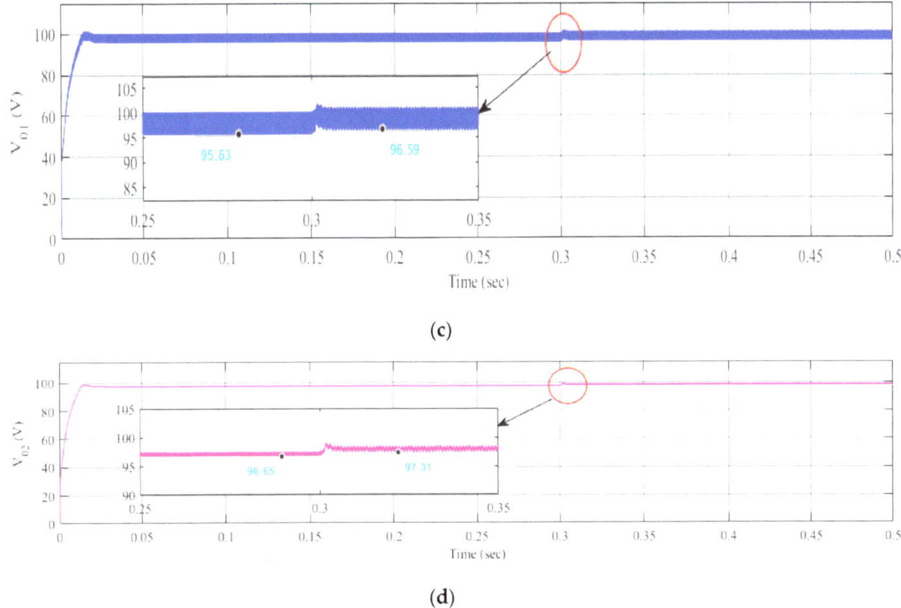

Figure 17. Simulated results of moderate step decrease in load using DMPC controller. (**a**) Output current I_{01} (**b**) Output current I_{02} (**c**) Output voltage V_{01} (**d**) Output voltage V_{02}.

Use of the DMPC shows that the cross-regulation is reduced from 0.3% to 0.6%. The designed DMPC, when compared to a conventional PID controller shows greater performance.

5.3. Comparisons of Various Cases with PID Controller and DMPC Controller

The comparison table shows the various scenarios for analyzing the cross-regulation effect with PID and DMPC controllers. From Table 3, it is observed that the cross-regulation is more reduced in the DIDO system with a DMPC controller than with a PID controller, and the output voltages V_{01} and V_{02} are able to track the desired voltage for any change during load regulation. Additionally, it is proved that for any change in load current I_{01} ranging from 0.1 A to 2 A, the other load current I_{02} is maintained almost constant with the boundary range between 0.007 A to 0.016 A, thus reducing the cross-regulation to an acceptable value. Without any control loop to the converter and by controlling only the load switches, DMPC minimizes cross-regulation in a much more efficient and effective way than a PID controller. Similar conditions were also tested and validated with a second load. However, those results are not included in this paper.

Table 3. Comparisons of various scenarios with PID controller and DMPC controller.

Controller	Change in Load Resistor R_{01} (Ω)	Change in Output Current I_{01} (amp)	Change in Output Current I_{02} (amp)	Change in Output Voltage V_{01} (V)	Change in Output Voltage V_{02} (V)	Cross -Regulation
	25–50	4.08–2.117	4.63–4.728	102.2–105.8	92.69–94.41	0.02
PID	25–30	4.08–3.445	4.631–4.66	102.1–103.3	92.7–93.22	0.008
	25–20	4.08–5.021	4.636–4.595	102.1–100.3	92.74–91.76	0.006
	25–50	4.401–2.233	2.189–2.203	108.6–109.8	109.4–110	0.006
DMPC	25–30	4.475–3.671	2.195–2.202	109.1–109.8	109.4–110	0.004
	25–20	3.89–4.889	3.227–3.243	95.63–96.59	96.65–97.31	0.003

5.4. Comparison with Existing Works

To illustrate the performance of the developed prototype and its controller, comparisons are made with the existing work in the literature [13–15,22,28]. The designed values with the cross-regulation are listed in Table 4. It may be clearly seen that the performance of the designed controller is excellent, by comparing with the key factors such as cross-regulation and the sensitivity index. Cross-regulation is quantified in terms of performance index in [14,15]. The lower the value of cross- regulation, the greater the stability achieved in the converter. Thus, the obtained cross-regulation, in terms of performance index, indicates that the proposed system maintains desirable transient response with the load regulation. Furthermore, in order to highlight the performance of the designed controller, sensitivity indices are estimated. It is proved that the designed controller has a good response by incorporating the step change in load current in one output to other output current variation. The sensitivity (S) index is defined as in Equation (36):

$$S_{I_{out_i}}^{I_{out_j}} = \left(\frac{\Delta I_{out_j}}{\Delta I_{out_i}}\right) \times \frac{I_{out_i}}{I_{out_j}} \tag{36}$$

where ΔI_{out_j} is change in the output current with cross-channel and ΔI_{out_i} is the change in the output current with self-channel. The proposed method has minimum sensitivity indices of 0.012, whereas in [28], it is found to be 0.016, which is calculated based on the step change in voltage variations. It is observed that while calculating the sensitivity index, the value obtained is less than the value mentioned in [13–15,22]. In summary, the DMPC controller used in PSLLC converter is superior to the existing methods.

Table 4. Comparisons of existing papers with the proposed converter.

S.N$_0$	Parameters	[This Paper]	[29]	[23]	[14]	[15]	[15]
1	Input Voltage	**40 V, 20 V**	24, 20 V	12 V	5 V	5 V	4.8 V
2	Output Voltage	**103 V, 96 V**	12, 8 V	1.2 V, 1.5 V	2.5V, 3.3V	1 V, 1.5 V	3.3 V, 1.2 V
3	Output Power	**250 W**	35 W	~0.76 W	~1.5 W	1.25 W	0.78 W
4	Control method	**Digital Model Predictive controller**	Model Predictive control	Multivariable PID and LQR controller	Average current mode and charge control	Decoupling method	Cross -derivative state feedback
5	Switching frequency	**100KHz**	20–100 KHz	10 KHz	500 KHz	500 KHz	100KHz
6	Inductor	**1 mH**	100 µH	1 mH	4 µH	5 µH	10 µH
7	Capacitor	**680 µF**	220 µF	220 µF	10 µF	10 µF	10 µF
8	Step change in Output Current	**4.401-2.233 A @I$_{01}$**	8–10 V@V$_2$	0.5–1.05 A @I$_{02}$	1 A @I$_{01}$	500–250 mA @I$_{01}$	100–200 mA @I$_{01}$
		4.475-3.671 A @I$_{01}$	0.61–0.8 A @I$_1$	2–2.5 A @I$_{02}$	1 A @I$_{02}$	250–500 mA @I$_{01}$	100–200 mA @I$_{02}$
9	Cross-Regulation	**0.006A @I$_{02}$**	0.025	1.2 V @0.01s	0.5 A @I$_{02}$	0.02	0.03
		0.004A @I$_{02}$	-	3.3V @0.007s	0.66 A @I$_{01}$	0.01	0.008
10.	Sensitivity	**0.012**	0.017, 0.016	0.83	0.05	0.01	0.01

5.5. Experimental Results and Discussion

To validate the designed DMPC in an environment where hybrid energy resources are integrated to the loads using PSLLC converter experimentally, a scaled down prototype of 100W capacity is developed, while maintaining the concept of analyzing the cross-regulation effect across the two loads. The dynamic condition is tested and the proposed control strategy is estimated based on Figure 1b with the parameters mentioned in Table 2. Figure 18 shows the hardware model. The prototype model is executed with the aid of FPGA Spartan 6-XC6SLX9 controller board. The components used in the prototype consist of IGBT switch-MG1215H-XBN2MM, PWM driver circuit-ICTLP250, Inductor -IHA 205, Hall Effect sensor-ACS712 for sensing the current, diode rating-ISL9R3060P2, and a 4-inch cooling fan for heat dissipation. Initially, the proposed converter works in the steady state operation. The input voltage for the prototype model is taken as 30V and the obtained output load voltages V_{01} and V_{02} are 50 V with I_L as 1.93 A. The output voltages are maintained at their reference values without any external disturbance by the designed controller.

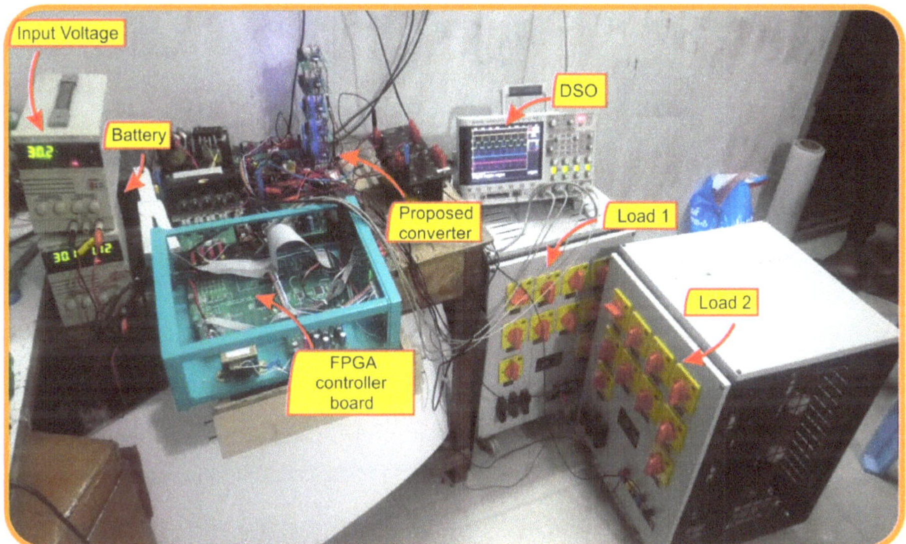

Figure 18. Hardware setup of proposed converter.

Step change is introduced as an external disturbance in the load side and various cases are analyzed which are discussed as follows:

Case 1. Step change at load voltage V_{01}:

To verify the concept of cross-regulation, a step change is introduced at load 1. Initially, the converter works in steady state mode of operation with $I_{01} = 0.1$ mA. At $t_1 = 1$ s, I_{01} is stepped up from 0.1 mA to 0.4 mA and at $t_2 = 4$ s, I_{01} is stepped down from 0.4 mA to 0.1 mA. The respective voltage changes in load 1 (V_{01}), drops from 50 V to 49 V at t_1 and reverts back at t_2. From Figure 19, it is noted that the voltage of second load output V_{02} is constant, maintaining at 50 V for ensuing step changes in the load current I_{01}. The obtained results of Figures 19 and 20 show that the DMPC controller is capable of reducing the cross-regulation effect with the change in dynamic response.

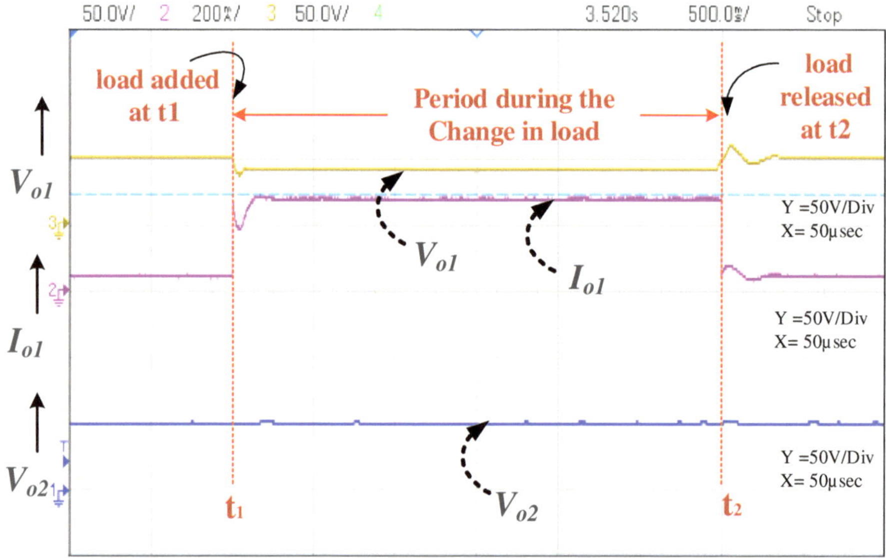

Figure 19. Case 1—Experimental results of step change at load 1.

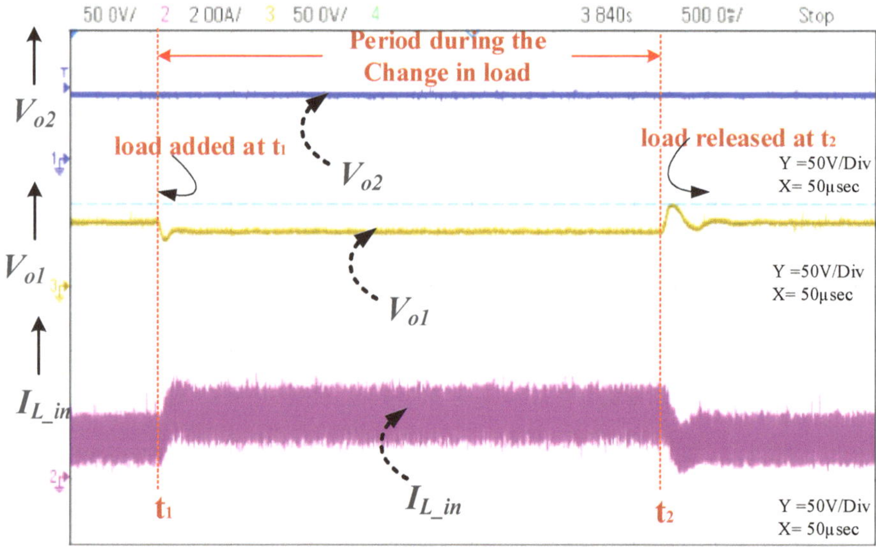

Figure 20. Case 1—Experimental results of step change at load 1.

Case 2. Step change at load voltage V_{02}:

To establish the effect of cross-regulation, a step change is introduced at load 2. Initially, the proposed converter works in steady state mode of operation with $I_{02} = 0.1$ mA. At $t_1 = 1$ s, I_{02} is stepped up from 0.1 mA to 0.4 mA and at $t_2 = 0.4$ s, I_{02} is stepped down from 0.4 mA to 0.1 mA. The voltage changes in load 2 (V_{02}), drops from 50 V to 49 V at t_1, and returns at t_2. From Figure 21, it is noted that the voltage of first load output V_{01} is constant, maintaining a value of 49.8 V for an ensuing step change in the load current I_{02}. The obtained results of Figures 21 and 22 shows that the proposed converter ensures that the output voltages are maintained at their reference values with negligible cross-regulation.

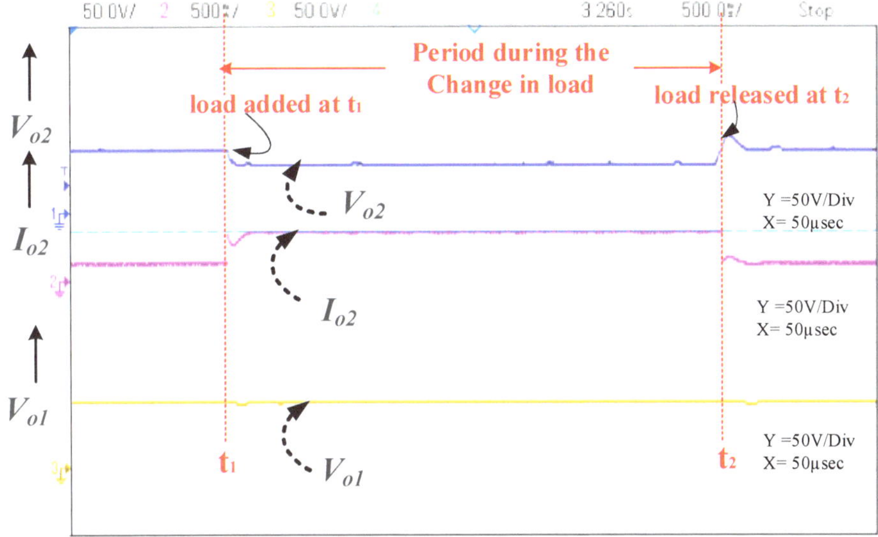

Figure 21. Case 2—Experimental results of step change at load 2.

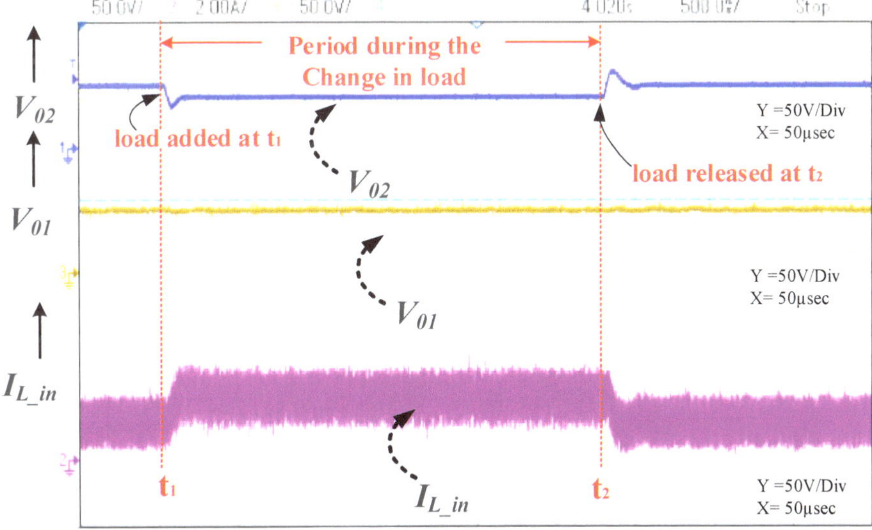

Figure 22. Case 2—Experimental results of step change at load 2.

Case 3. Step change in input voltage variations:

To authenticate the process of line regulation, a step change is introduced in the input voltage while the converter is maintained under steady state operation with I_{01}, I_{02} = 0.1 mA and V_{dc} = 30 V. Figure 23 shows the experimental response to the change in the input voltage. During the time period at t_1 = 1 s, Input voltage V_{in} is stepped down from 30 V to 25 V and at t_2 = 5.8 s, V_{in} is stepped up from 25 V to 30 V. Despite the change in the input voltage, load voltage across V_{01} and V_{02} remains stable at the desired voltage of 50 V. This case proves the proposed DIDO converter remains stable with respect to change in input voltage. Thus, the results indicate that the designed controller is resistant to any change in input voltage.

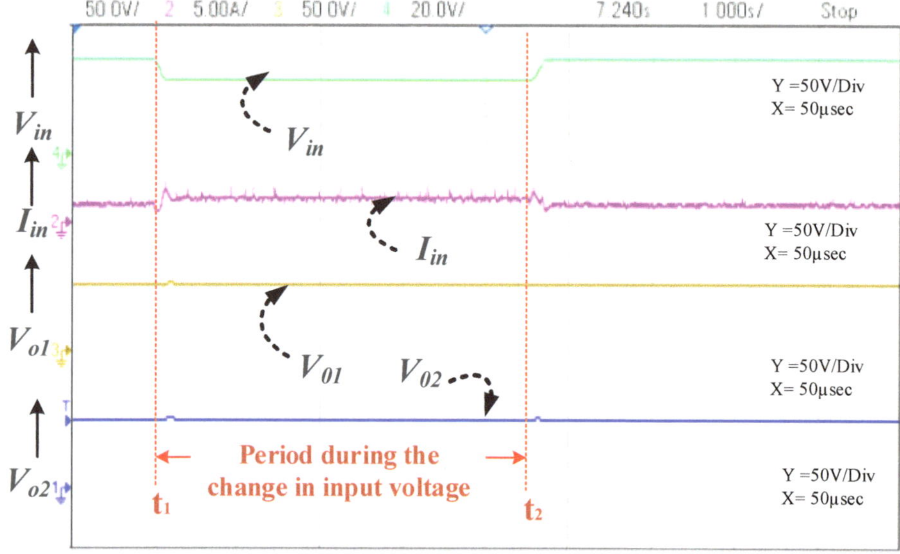

Figure 23. Case 3—Experimental results of step change input voltage.

The comparison Table 5 shows the various cases obtained by the experimental results. It is proved that for a step change in one load does not affect the other loads, which in turn reduces the cross-regulation effectively.

Table 5. Comparisons of different cases of the experimental results.

	t (sec)	I_{01} (mA)	V_{01} (V)	I_{02} (mA)	V_{02} (V)	Cross-regulation
Step Change in load 1	$t_1 = 1$	0.1–0.4	50–49	~0.1	~50	0.001
	$t_2 = 4$	0.4–0.1	49–50	~0.1	~50	
	t (sec)	I_{02} (mA)	V_{02} (V)	I_{01} (mA)	V_{01} (V)	
Step Change in load 2	$t_1 = 1$	0.1–0.4	50–49	~0.1	~49.8	0.002
	$t_2 = 4$	0.4–0.1	49–50	~0.1	~49.8	
	t (sec)	V_{in} (V)	I_{in} (mA)	V_{01} (V)	V_{02} (V)	
Input Voltage Variation	$t_1 = 1$	30–25	1–2	~50	~50	0.001
	$t_2 = 5.8$	25–30	2–1	~50	~50	

To summarize the results obtained from Figures 18–22, it is clearly shown that the proposed controller has accomplished the job of suppressing the cross-regulation and maintaining the line regulation for various cases of incremental and decremented step change in the resistive loads. The sensitivity index and cross-regulation are estimated in the hardware result and observed to be satisfied with the theoretical values.

6. Conclusions

This paper proposed a digital model predictive controller (DMPC) for DIDO-PSLLC converter to subdue the effect of cross-regulation. At first, the conventional PID controller was designed for the proposed converter and compared with the proposed digital model predictive controller.

This paper proposed a MIMO controller for DIDO hybrid energy system to subdue the effect of cross-regulation. The conclusions are summarized as follows:

- This paper proposes the PV and battery-connected MIMO positive Super-Lift Luo Converter with high-voltage transfer gain, high power density, high efficiency, reduced ripple voltage and current.
- The proposed DMPC controller has the advantage of fast dynamic response and suppression of cross-regulation by controlling the load switches. The decoupling method is preferred to overcome the interaction of the proposed hybrid DIDO system with renewable energy resources.
- The DMPC controller shows greater performance in cross-regulation and sensitivity index when compared with existing literature.
- Thus, MIMO controller is implemented in a DIDO hybrid energy system, which can be interfaced for electric vehicle application.

Theoretical analysis followed by comparison results proves the outstanding performance of DMPC method in minimizing the cross-regulation in proposed DIDO system. By considering the architecture of the proposed structure, this work can be flexibly extended to an arbitrary number of inputs and outputs. In addition, the proposed converter with the cross-regulation examination can find a place in electric vehicle lighting applications.

Author Contributions: All authors were involved in developing the concept, simulation, and experimental validation and making the article error free and getting a technical outcome for the set investigation work. All authors have read and agreed to the published version of the manuscript.

Funding: This research received no external funding.

Conflicts of Interest: The authors declare no conflict of interest.

References

1. Rehmani, M.H.; Reisslein, M.; Rachedi, A.; Erol-Kantarci, M.G.; Radenkovic, M.G. Integrating Renewable Energy Resources into the Smart Grid: Recent Developments in Information and Communication Technologies. *IEEE Trans. Ind. Inform.* **2018**, *14*, 2814–2825. [CrossRef]
2. Suresh, K.; Chellammal, N.; Bharatiraja, C.; Sanjeevikumar, P.; Blaabjerg, F.; Nielsen, J.B.H. Cost-efficient nonisolated three-port DC-DC converter for EV/HEV applications with energy storage. *Int. Trans. Electr. Energy Syst.* **2019**, *29*, 1. [CrossRef]
3. Xiang, W.; Lin, W.; Miao, L.; Wen, J. Power balancing control of a multi-terminal DC constructed by multiport front-to-front DC–DC converters. *IET Gener. Transm. Distrib.* **2017**, *11*, 363–371. [CrossRef]
4. Sidorov, D.; Panasetsky, D.; Tomin, N.; Karamov, D.N.; Zhukov, A.; Muftahov, I.; Dreglea, A.; Liu, F.; Li, Y. Toward Zero-Emission Hybrid AC/DC Power Systems with Renewable Energy Sources and Storages: A Case Study from Lake Baikal Region. *Energies* **2020**, *13*, 1226. [CrossRef]
5. Anuradha, C.; Chellammal, N.; Maqsood, S.; Vijayalakshmi, S. Design and Analysis of Non-Isolated Three-Port SEPIC Converter for Integrating Renewable Energy Sources. *Energies* **2019**, *12*, 221. [CrossRef]
6. Karthikeyan, M.; Elavarasu, R.; Ramesh, P.; Bharatiraja, C.; Padmanaban, S.; Mihet-Popa, L.; Mitolo, M. A Hybridization of Cuk and Boost Converter Using Single Switch with Higher Voltage Gain Compatibility. *Energies* **2020**, *13*, 2312. [CrossRef]
7. Chen, H.; Zhang, Y.; Ma, D. A SIMO Parallel-String Driver IC for Dimmable LED Backlighting with Local Bus Voltage Optimization and Single Time-Shared Regulation Loop. *IEEE Trans. Power Electron.* **2011**, *27*, 452–462. [CrossRef]
8. Ma, N.D.; Ki, N.W.-H.; Tsui, N.C.-Y.; Mok, P.K.T. Single-inductor multiple-output switching converters with time-multiplexing control in discontinuous conduction mode. *IEEE J. Solid State Circuits* **2003**, *38*, 89–100. [CrossRef]

9.   Ma, D.; Ki, W.H.; Tusi, C.Y. A pesudo-CCM/DCM SIMO switching converter with freewheel switching. *IEEE J. Solid State Circuits* **2003**, *38*, 1007–1014.

10.   Le, H.-P.; Chae, C.-S.; Lee, K.-C.; Wang, S.-W.; Cho, G.-H. A Single-Inductor Switching DC–DC Converter with Five Outputs and Ordered Power-Distributive Control. *IEEE J. Solid State Circuits* **2007**, *42*, 2706–2714. [CrossRef]

11.   Shao, H.; Li, X.; Tsui, C.-Y.; Ki, W.-H. A Novel Single-Inductor Dual-Input Dual-Output DC–DC Converter with PWM Control for Solar Energy Harvesting System. *IEEE Trans. VLSI Syst.* **2013**, *22*, 1693–1704. [CrossRef]

12.   Qian, Y.; Zhang, H.; Chen, Y.; Qin, Y.; Lu, D.; Hong, Z. A SI-DIDO DC–DC converter with dual-mode and programmable-capacitor-array MPPT control for thermoelectric energy harvesting. *IEEE Trans. Circuits Syst. II Express Briefs* **2017**, *64*, 952–956.

13.   Trevisan, D.; Mattavelli, P.; Tenti, P. Digital Control of Single-Inductor Multiple-Output Step-Down DC–DC Converters in CCM. *IEEE Trans. Ind. Electron.* **2008**, *55*, 3476–3483. [CrossRef]

14.   Pizzutelli, A.; Ghioni, M. Novel control technique for single inductor multiple output converters operating in CCM with reduced cross-regulation. *Appl. Power Electron. Conf. Expo.* **2008**, 1502–1507. [CrossRef]

15.   Patra, P.; Ghosh, J.; Patra, A. Control Scheme for Reduced Cross-Regulation in Single-Inductor Multiple-Output DC–DC Converters. *IEEE Trans. Ind. Electron.* **2013**, *60*, 5095–5104. [CrossRef]

16.   Dasika, J.D.; Bahrani, B.; Saeedifard, M.; Karimi, A.; Rufer, A. Multivariable Control of Single-Inductor Dual-Output Buck Converters. *IEEE Trans. Power Electron.* **2014**, *29*, 2061–2070. [CrossRef]

17.   Wang, Y.; Xu, J.; Zhou, G. A Cross Regulation Analysis for Single-Inductor Dual-Output CCM Buck Converters. *J. Power Electron.* **2016**, *16*, 1802–1812. [CrossRef]

18.   Huang, M.-H.; Chen, K.-H. Single-Inductor Multi-Output (SIMO) DC-DC Converters with High Light-Load Efficiency and Minimized Cross-Regulation for Portable Devices. *IEEE J. Solid State Circuits* **2009**, *44*, 1099–1111. [CrossRef]

19.   Behjati, H.; Davoudi, A. A Multiple-Input Multiple-Output DC–DC Converter. *IEEE Trans. Ind. Appl.* **2013**, *49*, 1464–1479. [CrossRef]

20.   Park, Y.-J.; Khan, Z.H.N.; Oh, S.J.; Jang, B.G.; Ahmad, N.; Khan, D.; Abbasizadeh, H.; Shah, S.A.A.; Pu, Y.; Hwang, K.C.; et al. Single Inductor-Multiple Output DPWM DC-DC Boost Converter with a High Efficiency and Small Area. *Energies* **2018**, *11*, 725. [CrossRef]

21.   Wang, B.; Zhang, X.; Ye, J.; Gooi, H.B. Deadbeat Control for a Single-Inductor Multiple-Input Multiple-Output DC–DC Converter. *IEEE Trans. Power Electron.* **2019**, *34*, 1914–1924. [CrossRef]

22.   Soman, D.E.; Leijon, M. Cross-Regulation Assessment of DIDO Buck-Boost Converter for Renewable Energy Application. *Energies* **2017**, *10*, 846. [CrossRef]

23.   Lindiya, A.; Subashini, N.; Karuppaiyan, V. Cross Regulation Reduced Optimal Multivariable Controller Design for Single Inductor DC-DC Converters. *Energies* **2019**, *12*, 477. [CrossRef]

24.   Kwak, S.; Park, J.-C. Model-Predictive Direct Power Control with Vector Preselection Technique for Highly Efficient Active Rectifiers. *IEEE Trans. Ind. Informatics* **2015**, *11*, 44–52. [CrossRef]

25.   Cortes, P.; Ortiz, G.; Yuz, J.I.; Rodriguez, J.; Vazquez, S.; Franquelo, L.G. Model Predictive Control of an Inverter with Output *LC* Filter for UPS Applications. *IEEE Trans. Ind. Electron.* **2009**, *56*, 1875–1883. [CrossRef]

26.   Karamanakos, P.; Geyer, T.; Manias, S. Direct Voltage Control of DC–DC Boost Converters Using Enumeration-Based Model Predictive Control. *IEEE Trans. Power Electron.* **2014**, *29*, 968–978. [CrossRef]

27.   Cheng, L.; Acuna, P.; Aguilera, R.P.; Jiang, J.; Wei, S.; Fletcher, J.E.; Lu, D.D.-C. Model Predictive Control for DC–DC Boost Converters with Reduced-Prediction Horizon and Constant Switching Frequency. *IEEE Trans. Power Electron.* **2018**, *33*, 9064–9075. [CrossRef]

28.   Kamalesh, M.; Senthilnathan, N.; Bharatiraja, C. Design of a Novel Boomerang Trajectory for Sliding Mode Controller. *Int. J. Control. Autom. Syst.* **2020**, *18*, 2917–2928. [CrossRef]

29.   Wang, B.; Kanamarlapudi, V.R.K.; Xian, L.; Peng, X.; Tan, K.T.; So, P.L. Model Predictive Voltage Control for Single-Inductor Multiple-Output DC–DC Converter with Reduced Cross Regulation. *IEEE Trans. Ind. Electron.* **2016**, *63*, 4187–4197. [CrossRef]

30.   Luo, F.L.; Ye, H. Hybrid split capacitors and split inductors applied in positive output super-lift Luo-converters. *IET Power Electron.* **2013**, *6*, 1759–1768. [CrossRef]

31. Balaji, C.; Chellammal, N.; Sanjeevikumar, P.; Subramaniam, U.; Holm-Nielsen, J.B.; Leonowicz, Z.; Masebinu, S.O. Non-Isolated High-Gain Triple Port DC-DC Buck-Boost Converter with Positive Output Voltage for Photovoltaic Application. *IEEE Access* **2020**, *8*, 1. [CrossRef]
32. Kamaraj, V.; Nallaperumal, C. Modified multiport Luo converter integrated with renewable energy sources for electric vehicle applications. *Circuit World* **2020**, *46*, 125–135. [CrossRef]

Publisher's Note: MDPI stays neutral with regard to jurisdictional claims in published maps and institutional affiliations.

Article

Voltage-Balancing Strategy for Three-Level Neutral-Point-Clamped Cascade Converter under Sequence Smooth Modulation

Le Yu [1], Xu Peng [2,*], Chao Zhou [2] and Shibin Gao [1]

[1] School of Electrical Engineering, Jiaotong University Chengdu, Chengdu 611756, China;
   njspeed@163.com (L.Y.); gao_shi_bin@126.com (S.G.)
[2] Aviation Engineering Institute Civil, Aviation Flight University, Guanghan 618307, China; zc_cafuc@163.com
* Correspondence: pengxuswjtu@foxmail.com; Tel.: +86-0838-5188137

Received: 31 July 2020; Accepted: 8 September 2020; Published: 22 September 2020

Abstract: Three-level neutral-point clamped cascaded converters (3LNPC-CC) are widely used in high power nigh-voltage applications. This paper mainly discusses the open-circuit fault in DC-side of the 3LNPC-CC. Optimized by the sequence pulse modulation, a sequence smooth modulation (SSM) is proposed to keep the DC-side voltage balance while the 3LNPC-CC suffers open-circuit fault from DC-side. The SSM found efficient switch-state path through a 3-D cube model and simplified the path from thousands of switch state. The SSM avoids the complex calculation in the voltage-balancing modulation, while the dynamic character of it was less influenced. At the same time, the modulation changes the voltage level smoothly and balances the fault DC-side voltage effectively. The characters of the proposed modulation are verified by the simulation and the experiment.

Keywords: 3LNPC-CC; voltage-balancing; sequence smooth modulation open-circuit fault

1. Introduction

The traction power supply system is one of the four core systems of a railway and the only source of train power [1]. In addition, the traction power supply system is also an important load of the three-phase power system [2,3]. Because the transformer production technology is mature and simple relatively, the railway mainly power supply is three-phase to single-phase. In the traction power supply system, the power realized the conversion of three-phase 110 kV to single-phase 27.5 kV. However, with the development of industrialization and the full use of renewable energy, many disadvantages of this power supply mode have gradually emerged. Therefore, the traction power supply method based on the application of multilevel converters was widely researched and applied [4,5].

An inverter is the core piece of equipment in the multilevel converter [2]. In this topology, photovoltaic and wind-generator energy is connected to the system through the inverter to provide energy for traction power supply and railway load. The cascaded multilevel inverter is a typical inverter, which was proposed in 1981 [6]. The advantages of the multilevel inverter include:

(1)   High-voltage and high-power output can be realized by low-voltage power-switching devices;
(2)   The switching frequency of a single power-switching device is low;
(3)   Low loss of power-switching devices;
(4)   The filter device is physically small in size.

H-bridge topology, the neutral-point clamped (NPC) topology and modular multilevel converters (MMC) topology are the most widely used structures in multilevel converters. The H-bridge topology is simple in structure, but weak in voltage-bearing capacity, while the MMC topology is strong in voltage-bearing capacity, but complex in structure. In this way, the H-bridge topology and the MMC topologies are not applicable to the railway power supply system due to their limitations. The NPC cascaded converter has been widely used due to its simple structure relatively and high voltage-bearing capacity [7]. The NPC topology consists of several IGBT and clamp diodes. However, with the increase of the number of output levels, the topology of the converter is relatively complex, and a large number of sensors is needed to control the converter. Once charging and discharging time of the supporting capacitor is inconsistent, the voltage of the series capacitor will be unbalanced [8]. In addition, when the DC-side capacitance or internal switching devices of the cascaded multilevel inverter develop an open-circuit fault, the control performance of the system will be lost and even seriously affect the operation of the system. Therefore, the study in DC-side open-circuit fault of the cascaded inverter has become an urgent problem in academic and industry [9].

For the direct current-side (DC-side) open-circuit fault of the system, the direct measurement method of voltage sensor is adopted. However, the measurement method of the voltage sensor is limited by the DC-side support capacitance, which reduces the open-circuit-fault diagnosis accuracy and greatly increases the open-circuit-fault diagnosis time. In [10], the control strategy used the proportional-integral controller (PI) controller is adjusted. In this paper, the equivalent duty cycle is adjusted by PI controller and the power flow of each module is changed to realize the DC-side voltage balanced. However, the DC-side voltage balanced stability with the PI controller has a contradiction between the mean voltage velocity and stability. In addition, the regulation speed of the DC-side voltage is slow.

In contrast, a modulation voltage-balancing strategy could change the turn-on and turn-off sequence of the switches directly, so the fault tolerance effect is stronger—and the speed is faster. In [11], the redundant vector-selection modulation strategy is used to realize the stability control after DC-side voltage unbalanced of each module in cascaded H-bridge converter (CHBC). The DC-side voltage control is integrated into the modulation strategy in this strategy, which can achieve a more rapid, wide range of fault tolerance effects. The strategy proposed in [12] indicates that the opposite vectors could be a fantastic way to ensure the stability of the DC-side voltage, which has been demonstrated by a large number of simulations and experiments in this paper. This method can greatly widen the stable range of the fault tolerance strategy. However, there is no intermediate transition vector when the opposite vector is inserted, it will lead to the sudden change of port voltage level, which will lead to the increase of switching frequency, switching loss. In order to solve the above problems, a smooth voltage-balancing strategy is proposed in [13]. By sacrificing parts of the dynamic response capabilities, this strategy selects the redundancy and smooth voltage vector to control the DC-side voltage after open-circuit fault, which eliminates the mutation of port voltage level. In [14], a sequence pulse modulation (SPM) is proposed, which has strong fault tolerance capability and smooth change of voltage level. However, it is complicated to calculate the switching state. Generally speaking, the [12–14] could balance the DC-link voltage under modulation index among 0.8. However, [12] balances the voltage with the price of switch state jump, [14] works with drastic fluctuation, the calculation of [13] was too complex to become engineered. A fault reconfiguration strategy with strong control ability and good performance need to be studied [15–17]. Additionally, the voltage-balancing capability is difficult to realize the quantification [18,19]. In [20,21], the range of voltage-balancing capability is described by energy flow between DC-side and AC-side. Among these strategies, the DC-link voltage could remain balance while the modulation index is at least below 0.8. However, a qualitative description is not suitable for different voltage level systems and has some inevitable errors [22–25]. Thus, a precise voltage-balancing capability mathematical model is of great importance to establish for different 3LNPC-CC systems.

In this paper, an SSM strategy is proposed for the 3LNPC-CC system. The rest of this paper is organized as follows: Section 2 presents the 3LNPC-CC and its control system. Then, the proposed strategy is discussed in detail in Section 3. Section 4 verifies the theoretical analysis by simulation and experimental results, respectively.

2. The Configuration and Control Strategy of 3LNPC-CC

2.1. The Configuration of the Multimodule 3LNPC-CC

The basic structure of 3LNPC-CC applied in the hybrid AC–DC–AC smart grid is shown in Figure 1. The nigh-voltage input is connected with the distributed network. The module voltage input is a bidirectional port connecting with the battery group, super capacitor or flywheel energy-storage system. The low-voltage output provides three-phase 380-V AC-voltage for industry applications, urban lighting and residential electrical equipment. Additionally, the 3LNPC-CC plays an important role in the realization of high and low-voltage power conversion. The transformer is mainly used for electrical isolation while the filter is applied to improve power quality. The output of this system is used for the daily life.

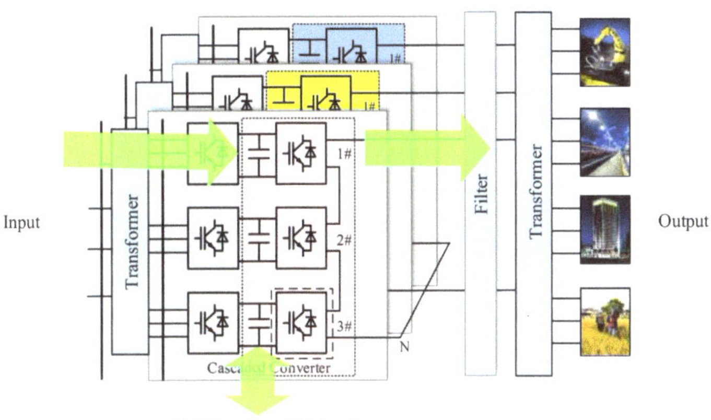

Figure 1. Basic structure of the hybrid AC–DC–AC smart grid.

As is shown in Figure 2, the topology of 3LNPC-CC is composed of a DC side, several NPC modules and an AC side. In each NPC module, there are one isolated switch, two series capacitors, eight IGBTs and four diodes. Each module is cascaded to establish one phase's voltage; the other two phases have similar topologies. Thus, the input of 3LNPC-CC is composed of 3n DC ports, while the output of 3LNPC-CC is three-phase AC port and one neutral point.

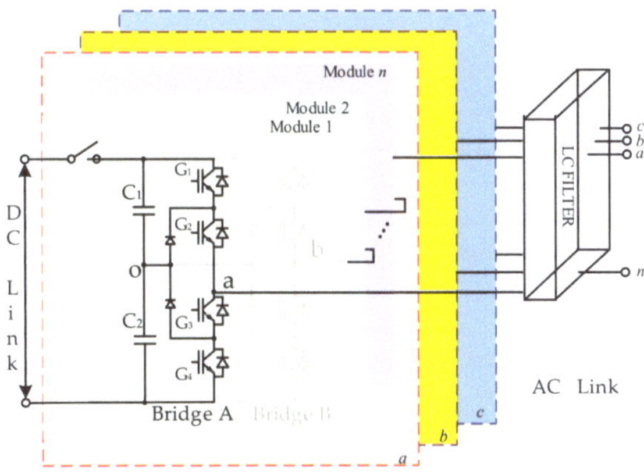

Figure 2. Topology of 3LNPC-CC.

2.2. The Control Strategy of the Multimodule 3LNPC-CC

In multimodule 3LNPC-CC, a voltage–current closed loop control is applied to regulate the output voltage and current, respectively. The control strategy aims to provide the AC voltage according to the reference value and improve the power quality. The control strategy is mainly composed of a voltage loop and a current loop based on the PI controller. The voltage and current are decoupled with the active component d and passive component q. Once the output AC voltage develops fluctuation, the control variable v_d and v_q will be regulated under the d–q axis to maintain the output voltage. Moreover, once the output AC current contains a reactive component, the control variable i_d and i_q will be regulated to maintain the unit power factor. Thus, the mathematical model of the double closed loop control strategy can be defined as Formulas (1) and (2).

$$\begin{cases} u_d^* = k_{ip}(i_d^* - i_d) + k_{ii}\int (i_d^* - i_d)dt + \omega L i_q \\ u_q^* = k_{ip}(i_q^* - i_q) + k_{ii}\int (i_q^* - i_q)dt + \omega L i_d \end{cases} \tag{1}$$

$$\begin{cases} i_d^* = i_{Ld} + k_{vp}(v_d^* - v_d) + k_{vi}\int (v_d^* - v_d)dt + \omega C v_{Lq} \\ i_q^* = i_{Lq} + k_{vp}(v_q^* - v_q) + k_{vi}\int (v_q^* - v_q)dt + \omega C v_{Ld} \end{cases} \tag{2}$$

where u_d, u_q, i_d, i_q are control variables under the d–q axis while the u^*_d, u^*_q, i^*_d, i^*_q are reference values under the d–q axis. The PI controller has two pairs parameters which are k_{vp}, k_{vi}, k_{ip} and k_{ii}. Moreover, the L is the filter value of the inductor, the C is the filter value of capacitor and ω is the frequency of 3LNPC-CC. According to the mathematical model, the control scheme of 3LNPC-CC can be illustrated in Figure 3. v_{an}, v_{bn} and v_{cn} are the three phase voltage. 1 is the normalized reference value of v_d. 0 is the normalized zero value of v_q. i_{La}, i_{Lb} and i_{Lc} are the three phase current. ω is the angle of the three phase current, which is locked by PLL (phase-locked loop). The voltage signal sampled from voltage sensor is the input of the voltage loop, which can produce the reference of d–q current. Similarly, the current signal from current sensor is transformed to current loop to produce the reference of d–q voltage. Thus, the d–q voltage can be decoupled with the three-phase voltage signal as the output signal of the control strategy. In Figure 3, the modulation strategy and the proposed voltage-balancing strategy are illustrated, which are thoroughly introduced in Sections 3 and 4. The final control results in Figure 3 will be transferred as IGBT signal ($G_1,G_2,G_3 \dots G_n$), for controlling the gate of IGBT in Figure 2.

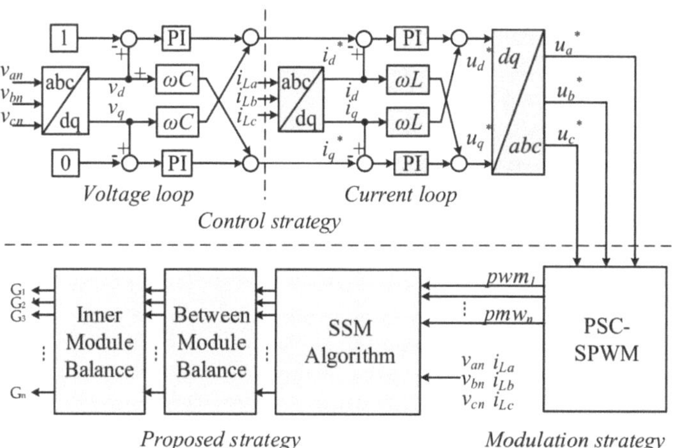

Figure 3. Control scheme of 3LNPC-CC.

3. The Proposed Strategy of 3LNPC-CC

3.1. The Proposed Strategy

A three-module 3LNPC-CC is used to illustrate the theory of voltage balancing. In the DC voltage V_{dc1}, V_{dc2} and V_{dc3} of Module 1, Module 2 and Module 3 are sampled, respectively. While one module suffers from the open-circuit fault of DC-side, the DC-side voltage of the module will drop drastically. In fact, the function of voltage-balancing strategy is to stop the DC-side voltage dropping and bring DC-side voltage to normal. When the output current is positive, the open-circuit fault module should try it best to synthesize −2, −1 and 0 voltage level, so that the voltage level could balance the DC-side voltage by charging the DC-side capacitor. Similarly, When the output current is negative, the fault module should try its best to synthesize 2, 1 and 0 voltage levels.

To achieve the voltage-balancing strategy, [13] proposed an SPM as shown in Table 1—in which all modules synthesize the voltage level based on table according to the voltage level. The relationship DC-side voltage of Module S_1, Module S_2 and Module S_3 are $V_{dc1}> V_{dc2}> V_{dc3}$. However, as it is shown in Table 1, when the total voltage-level changes from 2 to 1, at the same time, the rank number of the DC-side voltage changes as $V_{dc1}> V_{dc3}> V_{dc2}$, the voltage level of cell 2 will change from 2 to −2, namely the voltage-level jump appears. The voltage-level jump will lead an equivalent increase in switching frequency. If the voltage-level jump needs to be avoided, the rank of the DC-side voltage could change while the modulation level M equals 6, 0, −6. However, the limitation will cause dynamic performance loss. In addition—to keep the voltage-level change smooth—the −2 voltage level is banned while $M = 1$, as does $M = 2$ while $M = -1$. Therefore, the voltage-balancing capability of SPM could still be increasing.

Table 1. Voltage level allocation.

M	S_{1st}	S_{2nd}	S_{3th}
6	2	2	2
5	2	2	1
4	2	2	0
3	2	2	−1
2	2	2	−2
1	1	1	−1
0	0	0	0
−1	−1	−1	1
−2	−2	−2	2
−3	−2	−2	1
−4	−2	−2	0
−5	−2	−2	−1
−6	−2	−2	−2

An ideal 3-D cube model voltage-balancing strategy is shown in Figure 4, three-module consist of $5^3 = 125$ switch states, which makes up the number of total voltage levels is 13. Each module is shown on an axis in Figure 4, a cube appears. Each color represents one kind of the total voltage level. The full line shows the path when the total voltage-level changes. For example, when the total voltage level is 13, which is on the point of (2,2,2), the red line shows the path of the total voltage between 13 and 12, the point changes from (2,2,2) to (2,2,1), (2,1,2) or (1,2,2). Thus, the voltage-level changes smoothly. To realize this strategy, the modulation needs to calculate voltage level of each module which is related to its DC-side voltage and the initial switch state. However, this also means that once a bug occurs in the calculation, all following calculations will error in succession.

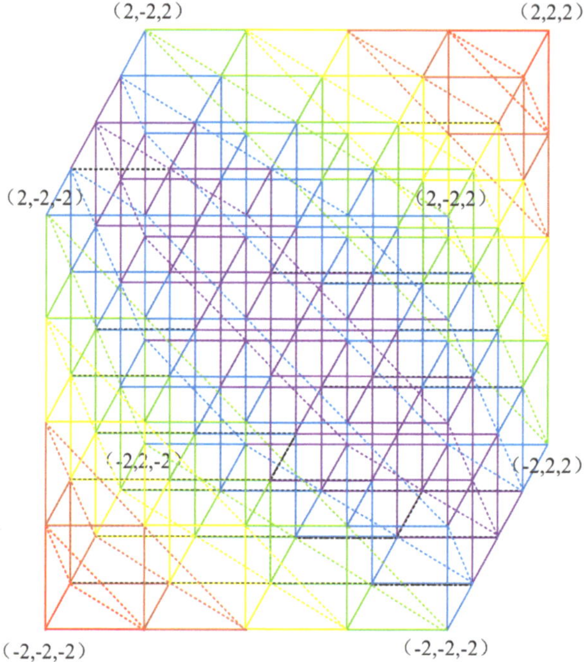

Figure 4. 3-D cube model voltage-balancing strategy.

Considering the advantage and disadvantages of the SPM and the calculation modulation above, the proposed SSM combines the advantage of them. The basic frame of the SSM is to calculate all the possibility of the switch-state changes, then list the changes in Table 1. Thus, a 3 Module 3LNPC-CC has 125 kinds of switch state and 6 kinds of the DC-side voltage rank. Each switch state has 2 kinds of voltage-level change tail, which is the voltage level plus and voltage level minus. Thus, the number of the total switch-state changes is about 1500. The 1500 kinds of switch states could be classified as follows:

(a)　DC-side voltage rank: $V_{dc1} > V_{dc2} > V_{dc3}$, voltage level: positive plus, the number of module which the switch state is 2:0, initial example: (1,1,1), result example (2,1,1);

In this situation, the switch state of the module (Module 1) which DC-side voltage rank is the highest of all has rising space, thus the switch state of Module 1 increases from 1 to 2.

(b)　DC-side voltage rank: $V_{dc3} > V_{dc2} > V_{dc1}$, voltage level: positive plus, the switch state of Module 3 is 2, initial example:(1,1,2), result example (1,2,2);

In this situation, the switch state of the module (Module 3) which DC-side voltage rank is the highest of all, but it has no rising space, thus the switch state of Module 2 increases from 1 to 2, which has the second-highest of the DC-side voltage rank.

(c)　DC- side voltage rank: $V_{dc2} > V_{dc3} > V_{dc1}$, voltage level: positive plus, the switch state of Module 3 is 2 and the switch state of Module 2 is 2, initial example:(−1,2,2), result example (0,2,2);

In this situation, the switch state of the module (Module 2) which DC- side voltage rank is the highest of all, but it has no rising space, neither does Module 3, thus the switch state of Module 1 increases from −1 to 0, which has the second-highest of the DC-side voltage rank.

(d)   DC- side voltage rank: $V_{dc1} > V_{dc2} > V_{dc3}$, voltage level: positive minus, the number of module which the switch state is −2:0, initial example:(1,1,1), result example (1,1,0);

In this situation, the switch state of the module (Module 3) which DC- side voltage rank is the lowest of all has falling space, thus the switch state of Module 3 decreases from 1 to 0.

(e)   DC- side voltage rank: $V_{dc3} > V_{dc2} > V_{dc1}$, voltage level: positive minus, the switch state of Module 1 is −2, initial example:(1,2, −2), result example (1,1, −2);

In this situation, the switch state of the module (Module 3) which DC- side voltage rank is the lowest of all, but it has no falling space, thus the switch state of Module 2 increases from 2 to 1, which has the second-highest of the DC-side voltage rank.

(f)   DC- side voltage rank: $V_{dc1} > V_{dc2} > V_{dc3}$, voltage level: negative plus, the number of module which the switch state is 2:0, initial example:(−1, −1, −1), result example (−1, −1,0);

In this situation, the switch state of the module (Module 3) which DC- side voltage rank is the lowest of all has rising space, thus the switch state of Module 3 increases from −1 to 0.

(g)   DC- side voltage rank: $V_{dc3} > V_{dc2} > V_{dc1}$, voltage level: negative plus, the switch state of Module 1 is 2, initial example:(−1, −1, −2), result example (−1,0, −2);

In this situation, the switch state of the module (Module 1) which DC- side voltage rank is the lowest of all, but it has no rising space, thus the switch state of Module 2 increases from −1 to 0, which has the second-highest of the DC-side voltage rank.

(h)   DC- side voltage rank: $V_{dc1} > V_{dc2} > V_{dc3}$, voltage level: negative minus, the number of module which the switch state is -2:0, initial example:(−1, −1, −1), result example (−1, −1, −2);

In this situation, the switch state of the module (Module 3) which DC- side voltage rank is the highest of all has rising space, thus the switch state of Module 3 decreases from 1 to 0.

(i)   DC- side voltage rank: $V_{dc3} > V_{dc2} > V_{dc1}$, voltage level: negative minus, the switch state of Module 1 is −2, initial example: −2, −1, −1), result example (−2, −2, −1);

In this situation, the switch state of the module (Module 3) which DC- side voltage rank is the highest of all, but it has no falling space, thus the switch state of Module 2 increases from −1 to −2, which has the second-highest of the DC-side voltage rank.

(j)   DC- side voltage rank: $V_{dc2} > V_{dc3} > V_{dc1}$, voltage level: negative minus, the switch state of Module 3 and Module 3 are −2, initial example:(−2, −2,1), result example (−2, −2,0);

In this situation, the switch state of the module (Module 2) which DC- side voltage rank is the highest of all, but it has no falling space, neither does Module 3, thus the switch state of Module 1 decreases from 1 to 0, which has the second-highest of the DC-side voltage rank.

(k)   The special situation; (2,2,2) cannot be added anymore, and the (−2, −2, −2) cannot be reduced any more.

The changes of switch states are calculated and ranked, to make a list and saved in the controller. Then the modulation will work as the list arranging. In addition, the result of the modulation will operate like Figure 4 shows. First, the SSM chooses the proper switch state to balance the voltage smoothly. Second, the SSM could choose the off-line switch state—the final result of the switch state could be generated as a table, avoiding the complex calculations while 3LNPC-CC is working. Finally, the SSM is easy to extend as the off-line calculating. Compared to traditional strategies, the proposed SSM is optimized, simplified and extendable. Although the expression of the SSM is complex, the calculation while the SSM working is much easier than that of [13] and [20].

3.2. The Balance Range of the Proposed Strategy

Although the proposed voltage-balancing strategy provides a strong balance capability for 3LNPC-CC, the balance range has its limits. In order to find the boundary of the balance capability, the mathematical model of balance capability is established in this section, based on the conservation of energy theorem.

First, the balance index ζ must be defined in Equation (3), which can quantitatively measure the degree of voltage imbalance. In the traditional voltage-balancing algorithm, the definition of balance index is always dependent on power. However, it is difficult for traditional definitions to maintain uniformity at different power levels. Thus, based on the normalization technique, the balance index ζ is defined by DC-side voltage reference and DC-side voltage difference. If all DC-side inputs are equal to the rated voltage, the balance index ζ is equal to zero. Similarly, if one module develops an open-circuit fault of DC side, the balance index ζ is equal to 0. Thus, full range of load can be represented regardless of the power level of system:

$$\zeta = \frac{n \cdot (v_{dc.\max} - v_{dc.\min})}{\sum v_{dc}} \tag{3}$$

Second, the relationship between the balance index ζ and energy flow can be established in Equation (4). Due to the structure of 3LNPC-CC, all modules are cascaded with a similar current flow even if one module is faulty. Thus, the balanced index is proportional to DC-voltage, DC-power and energy flow of each module. In Equation (4), the Δv_{dc}, ΔW and ΔE are different from DC-voltage, power and energy flow between faulty modules and other normal modules. Moreover, k_1 and k_2 are the proportional factor in the process, which have no dimensions:

$$\zeta = \frac{\Delta v_{dc}}{\sum v_{dc}} = k_1 \frac{\Delta W}{\sum W} = k_2 \frac{\Delta E}{\sum E} \tag{4}$$

Next, the mathematical model of the proposed voltage-balancing strategy is established in Equation (5). According to Equation (5), it indicates that the range of voltage-balancing is determined by the difference in energy flow, including input energy and output energy. The input energy can be changed by the value of DC-side voltage while the output energy can be controlled by the proposed voltage-balancing strategy. Thus, in Equation (5), the energy flow difference can be described where $PWM_{fault}(m, \omega t)$ is the voltage level of the faulty module at the moment of t and ω is the angular frequency of 3LNPC-CC. v_{dc} and i_s is the DC-side voltage and AC-side current of 3LNPC-CC, respectively:

$$\zeta = \Delta v_{dc} / \sum v_{dc} = \Delta W / \sum W = \frac{\int_0^{2\pi} v_{dc} i_s PWM_{fault}(m, \omega t)\, dt}{\int_0^{2\pi} v_{dc} i_s \sin \omega t\, dt} \tag{5}$$

Finally, the relationship between balance index ζ, modulation index m, and module number n can be illustrated in Figure 5. In respect to the fixed module number, the boundary of voltage-balancing is composed of two parts. Taken the 3 modules system as an example, when modulation index is lower than 0.82, the voltage-balancing can be acquired no matter how serious the input situation is. However, when modulation index is higher than 0.82, the voltage-balancing capability is deceased. Voltage-balancing can only be achieved under specific load conditions. When modulation index is close to 1, the voltage-balancing capability hardly exists. Additionally, with the increasing of module numbers, the voltage-balancing capability is stronger. However, in consideration of the engineering requirements, the module number needs to be as little as possible. Thus, one compromise must be made between voltage-balancing capability and module number. The calculation result is shown in Figure 5.

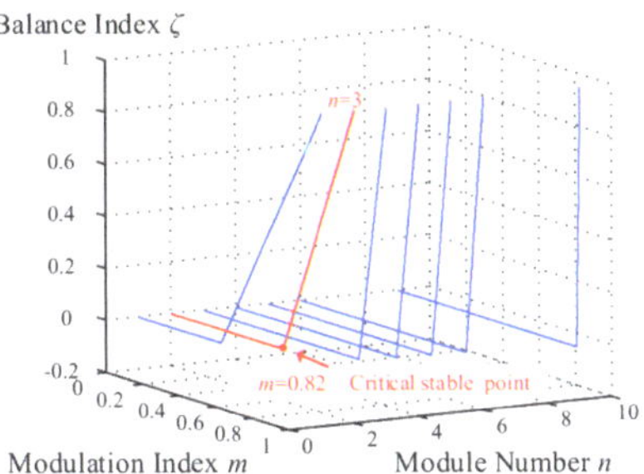

Figure 5. Balance boundary of 3LNPC-CC.

4. Simulation and Experiment

4.1. Simulation

The simulation of the proposed SSM is built in Matlab/Simulink, a three-Module 3LNPC-CC to verified it. The detailed parameters of the simulation are shown in Table 2, where the DC-side voltage is about 48 V in each module.

In order to verify the unit power factor of the AC-side achieved by the control strategy of 3LNPC-CC, the output voltage and current of the 3LNPC-CC is simulated and shown in Figure 6. In the initial stage, the 3LNPC-CC operates under the modulation index of 0.6, the AC-side voltage and current keep the same phase and the frequency. When the sudden change of modulation index comes at 0.445 s, the amplitude of the AC-side voltage and current changes to 120 V and 2.5 A, respectively and the system enters a new steady-state rapidly. After the sudden change, the 3LNPC-CC operates under the modulation index of 0.8 and still keeps the unit power factor operation.

As shown in Figure 7, the DC-side source of Module 3 develops an open-circuit fault. After the open-circuit fault of the DC side, the whole output port voltage u_{ab} still remain stable and the number of voltage level is eleven under the modulation index of 0.8. The whole output port voltage u_{ab} is shaped like a regular staircase wave, which indicates the good performance of the control strategy. However, the output port voltages of each module are no longer the staircase waves. As is obviously seen in Figure 7, Module 1 and Module 2 transfer their switching frequency to Module 3. Module 1 and Module 2 remain in switching states 2 and −2 most of the time while Module 3 changes its switching states rapidly for the purpose of balancing the DC-side voltage. More important, the voltage level of three modules is changed only 1 or −1 at one time, the short circuit and overvoltage which may happen in bridge arm are prevented. Thus, the voltage level of Module 3 is operated as the strategy needs.

In order to verify the effectiveness of the proposed strategy, the DC-side source of Module 3 is set to have an open-circuit fault before the 3LNPC-CC starts. The DC-side voltages of three modules are shown in Figure 8. In the beginning, V_{dc1} and V_{dc2} remain stable at 48 V and V_{dc3} is 0 because of the open-circuit fault. After the response time of 0.055 s, V_{dc3} increases rapidly to 48 V due to the proposed strategy. When the 3LNPC-CC enters the stable period, DC-side voltages of each module keep the same track at 48 V with a low amplitude ripple. In this way, the validity of the proposed strategy is proved.

With reference to Figure 9a–c, each DC-side voltage with different modulation index is illustrated based on the proposed strategy. The DC-side source of Module 3 is cut off at 0.15 s due to the open-circuit fault in all these three figures. In Figure 9a, the modulation index is 0.78 which is lower than its critical value 0.82, the DC-side voltage-balancing will be maintained after the open-circuit fault. In Figure 9b, the modulation index is the critical value 0.82, V_{dc3} is parallel to V_{dc1} and V_{dc2} and the system is in a critical stable state. However, in Figure 9c, the modulation index is 0.88, and the system works out of the balanced area. Therefore, V_{dc3} cannot be balanced at 48 V, which verifies the correctness of the proposed reconfiguration capability calculation in Section 4.

Table 2. Simulation Parameter.

Parameter Name	Value
Number of the modules	3
DC-side voltage of each module	48 V
Modulation index	0.8
Carrier wave frequency	2 kHz
Switches	IGBT
Filter inductance	1 mH
Filter capacitor	10 μF
DC-side capacitor	470 μF
Load	50 Ω

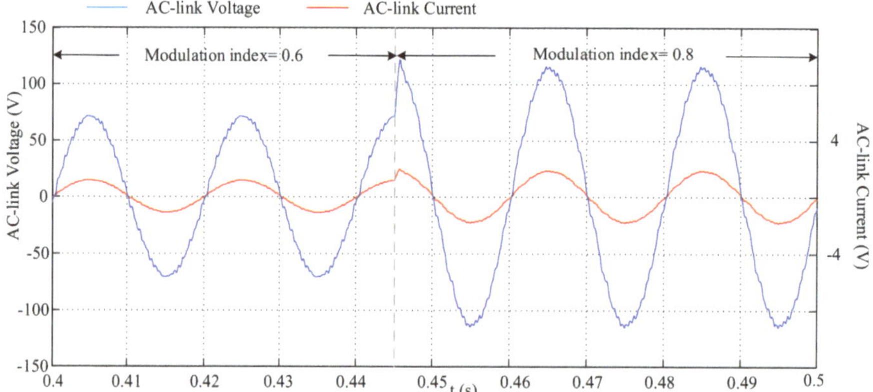

Figure 6. AC-side voltage and current for a step in the modulation index.

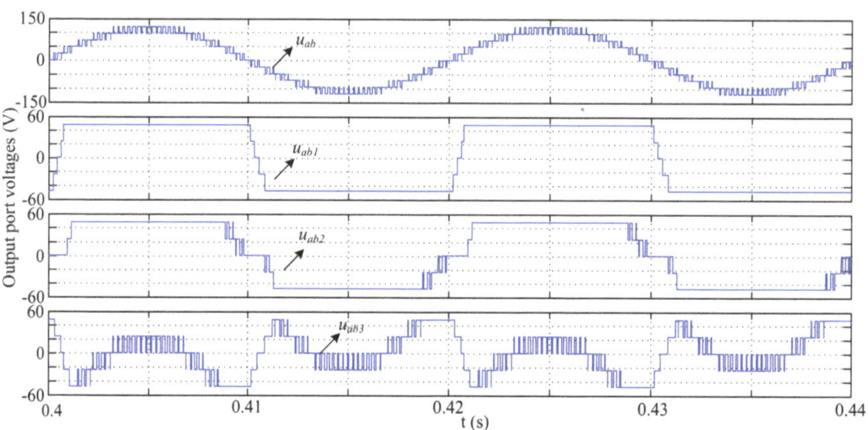

Figure 7. Output port voltage of the proposed strategy.

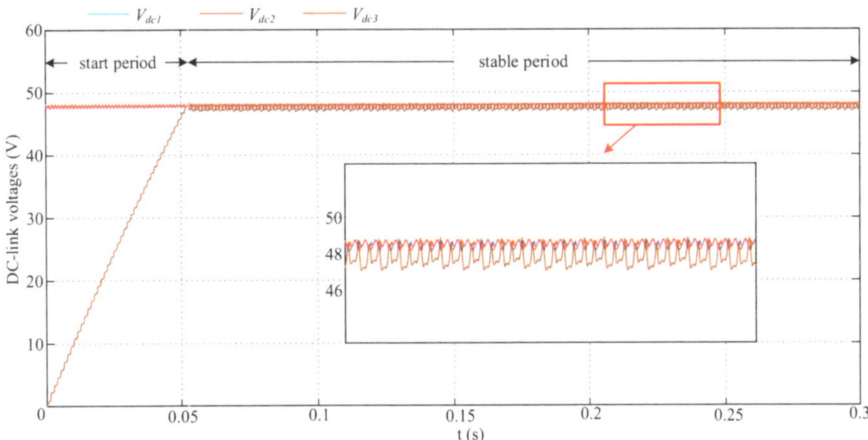

Figure 8. DC-side voltages of each module.

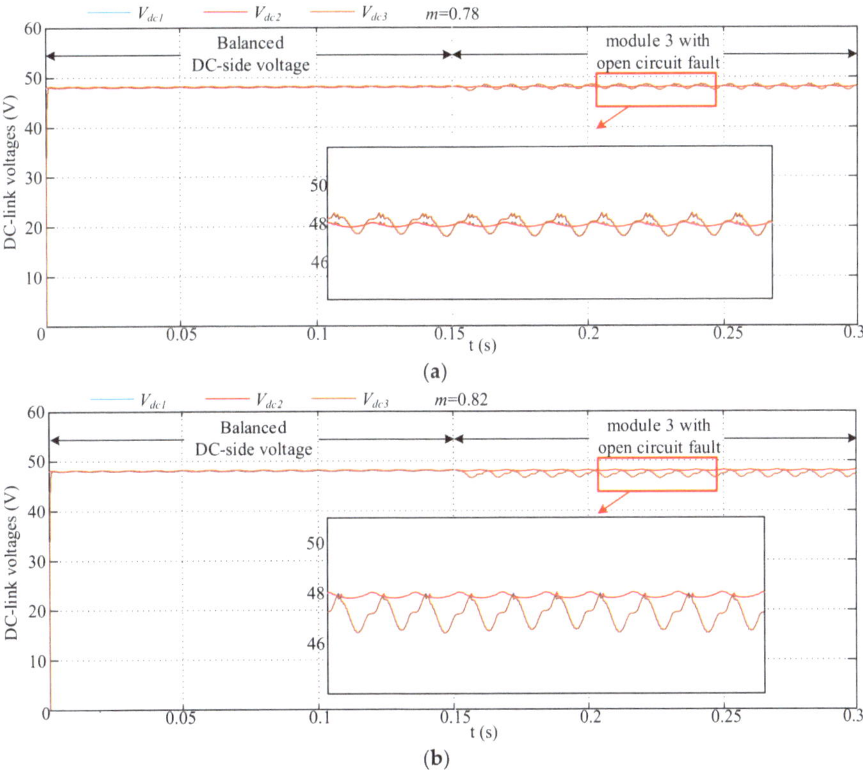

Figure 9. *Cont.*

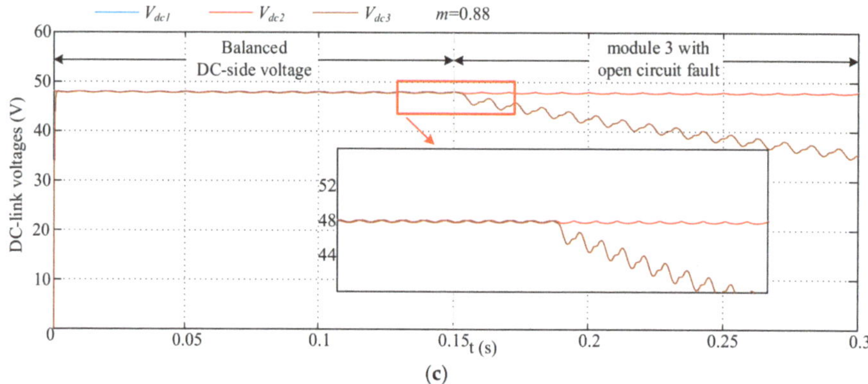

Figure 9. DC-side voltages of each module. (**a**) Each DC-side voltage of the proposed strategy ($m = 0.78$); (**b**) each DC-side voltage of the proposed strategy ($m = 0.82$); (**c**) each DC-side voltage of the proposed strategy ($m = 0.88$).

4.2. Experiment

The low power experiment of a three-Module 3LNPC-CC prototype was designed to illustrate the dynamic performance of the proposed strategy. The parameter and the prototype are shown in Table 3 and Figure 10, respectively. A prototype was designed and two experiments are conducted to illustrate the dynamic performance of the proposed strategy. The FPGA EP4CE10F17C8N is used as a core controller, the process of FPGA is mainly made up of PLL, d–q decoupling (dq–abc), d—q coupling (abc–dq), PI controller, PSC modulation and SSM proposed, as it is shown in Figure 11.

In order to verify the topology and the modulation of the prototype, the experiment of using phase-shift carrier (PSC) modulation and the proposed strategy is completed, the static waveform of the 3LNPC-CC is shown in Figure 12. In Figure 12a, an eleven-level output voltage is synthesized by the PSC-SPWM, while the voltage levels of each module are evenly distributed. Figure 12b shows the static wave of output voltage when the proposed strategy works, while Module 1 is regarded as the one who suffers from open-circuit fault. Based on the proposed strategy, the voltage level of Module 1 changes rapidly, while the voltage level of the other two modules remains at switching state ± 2 for most of the time. Moreover, the total output voltage U_{ab} of the 3LNPC-CC remains normally, although the output voltage of each module is rearranged by the proposed strategy.

Figure 13 shows the start period of the 3LNPC-CC when Module 1 suffering from open-circuit fault. Because of the DC-side voltage loss, the input and output voltage of Module 1 is 0 at the beginning. Figure 13a is the output voltage of the 3 modules. Because of the open-circuit fault of Module 1, the total output voltage U_{ab} deformed during the start period. Gradually, with the DC-side voltage of Module 1 recovering, the U_{ab} become normal. However, due to the proposed strategy, the switching times of the open-circuit fault module increases to keep its DC-side voltage steady. Although the switch states of all modules are rearranged, the total voltage of 3LNPC-CC remains unchanged. Figure 13b illustrates the DC-side voltage of each module and the AC-side current. Similarly, the DC-side voltage of Module 1 is zero at the beginning. After the proposed strategy is activated, the DC-side voltage of Module 1 increases and traces the DC-side voltage of the other two modules within 0.4 s. This means that the SSM could ensure the 3LNPC-CC working while the DC-port of one module of a module is fault while it. Figure 13b also shows the proposed SSM could balance the DC-link voltage under the modulation index among 0.8 while one module of 3LNPC-CC under 0.8, this keeps the average level such as [12–14] and [20].

Table 3. Experiment parameters.

Parameter Name	Value
Number of the cascade module	3
DC-side voltage	48 V
Output voltage	85 V
Controller (FPGA)	EP4CE10F17C8 N
Switch (IGBT)	FGA25N120ANTD
DC-side capacitance	470 μF
Filter inductance	1 mH
Filter capacitance	10 μF
Load	50 Ω
Carrier wave frequency	2 kHz

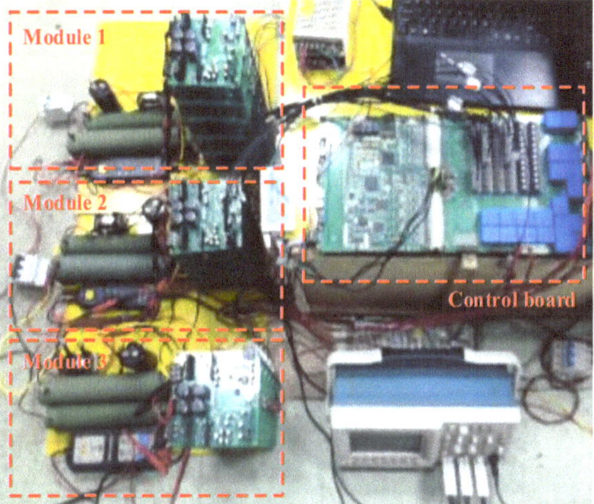

Figure 10. Prototype of the three-module 3LNPC-CC.

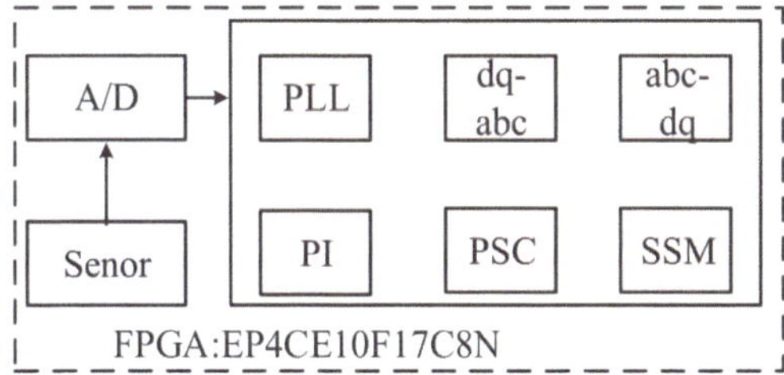

Figure 11. Controller.

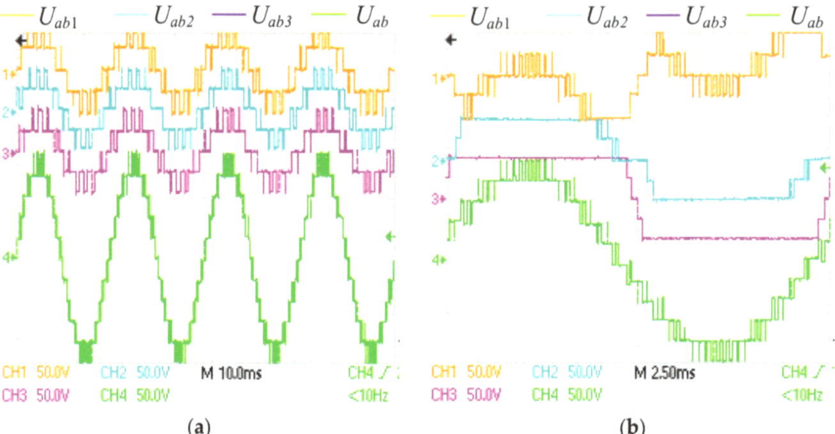

Figure 12. The static waveforms of three-module 3LNPC-CC. (**a**) The PSC modulation of 3LNPC-CC; (**b**) proposed strategy of 3LNPC-CC.

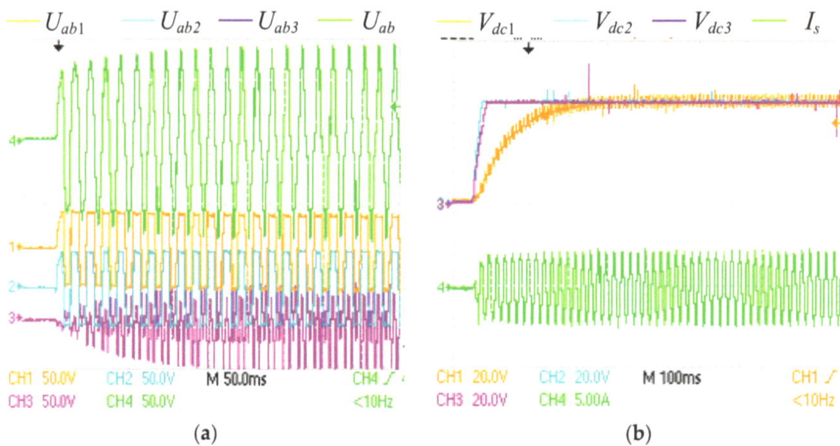

Figure 13. Experiment start with Module 1 suffering from an open-circuit fault. (**a**) Output voltages of 3LNPC-CC; (**b**) DC-side voltages and AC-side current.

Figure 14 shows the DC-side voltages and AC-side current using the proposed strategy of three-module 3LNPC-CC. When the system enters the steady-state operation, the open-circuit fault of DC-side source occurs in Module 1 and the DC-side source is completely cut off. Therefore, the DC-side voltage of Module 1 dropped greatly about 28 V. However, under the influence of the proposed strategy, the DC-side voltage of Module 1 does not drop continuously and increases to 50 V at the end. The process of recovering is less than 0.04 s, which is faster than the start experiment. In this way, the validity of the proposed strategy is proved. Furthermore, the fluctuation is much smaller than [13], similar with others article.

Since the modulation index could have an effect on the voltage-balancing capability, Figure 15 shows the DC-side voltages and AC-side current with different modulation index and the DC-side source of Module 1 is cut off in Figure 15a–c. In Figure 15a, the system works at a limit condition with the modulation index of 0.82 and the DC-side voltage of Module 1 fluctuates in a large range. As shown in Figure 15b,c, the proposed strategy is verified with a modulation index of 0.8 and 0.78. The fluctuation of DC-side voltages of Module 1 decreases and the DC-side voltage of three modules follow the trails of each other because the applying of the proposed strategy. These experimental results are in accordance with the theoretical and simulation analysis.

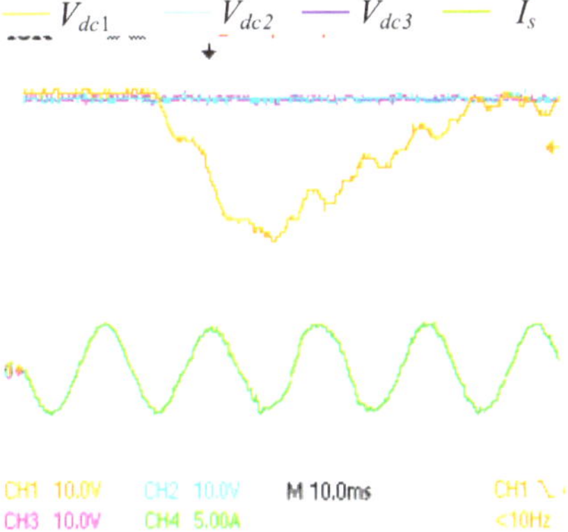

Figure 14. DC-side voltages and AC-side current during DC-side source open-circuit fault.

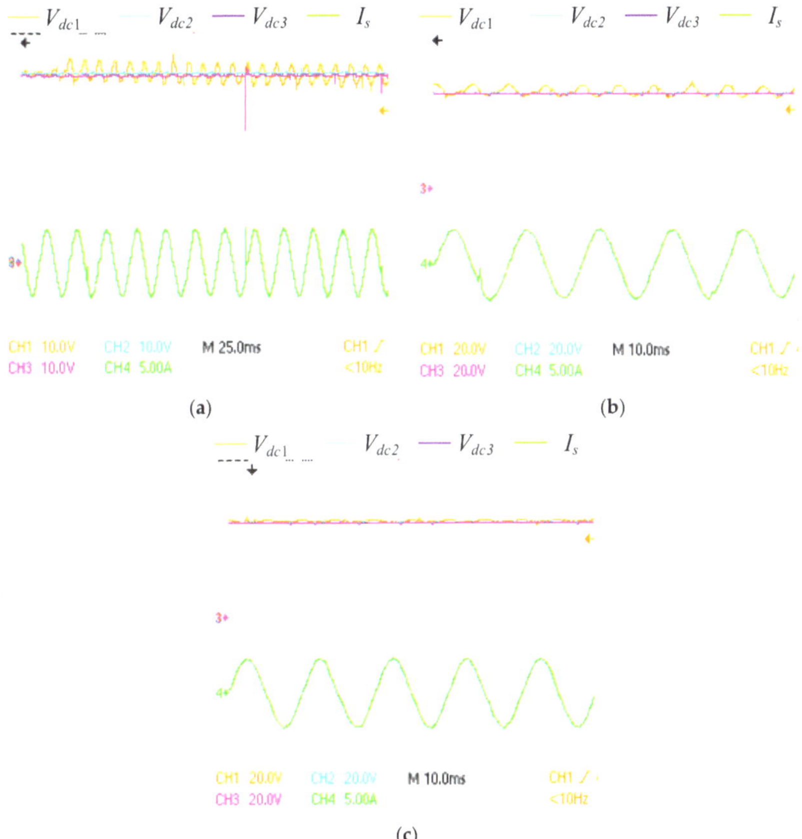

Figure 15. DC-side voltages and AC-side current under different modulation index. (**a**) m = 0.82; (**b**) m = 0.8; (**c**) m = 0.78.

5. Conclusions

In this paper, a novel voltage-balancing strategy applied in a three-level neutral-point clamped cascaded converter is proposed to improve the balance of DC-side voltage. The basic configuration of 3LNPC-CC was designed, and the normal control strategy is illustrated, which achieved the regulation of AC-side voltage and current. A SSM of voltage-balancing strategy and the optimal path of voltage level transformation is also proposed. The input DC-side voltage can be regulated—even if one module develops an open-circuit fault of DC-side. The voltage level can be changed smoothly in this strategy. The recovery time of the input DC-side source from full power to no power is approximately 0.28 s. The main contributions of this paper are as follows:

(a) An SSM structure of 3LNPC-CC is proposed, which represents the change of different working states. When the input of 3LNPC-CC is unbalanced, the normal path of working states is changed and an improved path of working states is assigned that can recover the balance of DC-side balance. The advantage of the improved path is a smooth switch of voltage level during the whole working cycle;

(b) A coordinated control strategy is designed, which is composed of double-closed loop and sequence smooth modulation, based on AC-side voltage, AC-side current and DC-side voltage. The performance of AC-side and DC-side can be assured at the same time;

(c) An experiment platform is established in this paper. The dynamic performance is illustrated when DC-side input develops an open-circuit fault and must be cut off. The DC-side voltage can be recovered even if the extreme imbalance occurs. The ripple of the DC-side voltage is improved with the decreasing of modulation index. Additionally, the 3LNPC-CC has an excellent dynamic performance.

In conclusion, the advantages of proposed SSM are as follows:

(a) The proposed SSM could balance the DC-link voltage of the DC-port fault module, while the modulation index is under 0.8. The voltage-balancing ability is fairly high-level among voltage-balancing strategies;

(b) The proposed SSM is able to find a proper switch-state path for voltage balancing that is suitable for the voltage balancing strategy. In addition, the proper switch-state path ensures the minimum switch-state changes and minimum frequency;

(c) The proposed SSM takes the advantages from the smooth modulation and SPM by using a calculated table—not only for finding the proper switch state, but also for avoiding complex calculations while the 3LNPC-CC is working.

Author Contributions: L.Y. conceived the strategy and the experiments; X.P. performed the experiments; S.G. analyzed the data; C.Z. acquihired the financial support for the project leading to this publication. S.G. contributed experiment prototype and L.Y. wrote the paper. All authors have read and agreed to the published version of the manuscript.

Funding: This research received no external funding.

Acknowledgments: This research was funded by Program of the CAFUC Grant Number XM3736 and the National Key R&D Program of China Grant Number F2015KF02.

References

1. Ronanki, D.; Singh, S.; Williamson, S. Comprehensive Topological Overview of Rolling Stock Architectures and Recent Trends in Electric Railway Traction Systems. *IEEE Trans. Transp. Electron.* **2017**, *3*, 724–738. [CrossRef]

2. Chung, Y.; Lee, C.; Kim, D.; Kang, H.; Park, Y.; Yoon, Y. Conceptual Design and Operating Characteristics of Multi-Resonance Antennas in the Wireless Power Charging System for Superconducting MAGLEV Train. *IEEE Trans. Appl. Supercon.* **2017**, *27*, 1–5. [CrossRef]

3.	Meng, T.; Song, Y.; Wang, Z.; Ben, H.; Li, C. Investigation and Implementation of an Input-Series Auxiliary Power Supply Scheme for High-Input-Voltage Low-Power Applications. *IEEE Trans. Power Electron.* **2018**, *33*, 437–447. [CrossRef]

4.	Yang, S.Y.; Bryant, A.; Mawby, P.; Xiang, D.W.; Ran, L.; Tavner, P. An industry-based survey of reliability in power electronic converters. In Proceedings of the IEEE Energy Conversion Congress and Exposition, San Jose, CA, USA, 20–24 September 2009. [CrossRef]

5.	An, F.; Song, W.; Yang, K.; Yang, S.; Ma, L. A Simple Power Estimation with Triple Phase-Shift Control for the Output Parallel DAB DC-DC Converters in Power Electronic Traction Transformer for Railway Locomotive Application. *IEEE Trans. Transp. Electron.* **2019**, *5*, 299–310. [CrossRef]

6.	Nabae, A.; Takahashi, I.; Akagi, H. A New Neutral-Point-Clamped PWM Inverter. *IEEE Trans. Ind. Appl.* **1981**, *14*, 518–522. [CrossRef]

7.	Kouro, S.; Malinowski, M.; Gopakumar, K.; Franquelo, J.; Wu, B.; Rodriguez, J. Recent Advances and Industrial Applications of Multilevel Converters. *IEEE Trans. Ind. Electron.* **2010**, *57*, 2553–2580. [CrossRef]

8.	Tu, P.; Yang, S.; Wang, P. Reliability and Cost-Based Redundancy Design for Modular Multilevel Converter. *IEEE Trans. Ind. Electron.* **2019**, *66*, 2333–2342. [CrossRef]

9.	Qi, C.; Chen, X.; Tu, P.; Wang, P. Cell-by-Cell-Based Finite-Control-Set Model Predictive Control for a Single-Phase Cascaded H-Bridge Rectifier. *IEEE Trans. Power Electron.* **2018**, *33*, 1654–1665. [CrossRef]

10.	He, X.; Lin, X.; Peng, X.; Han, P.; Shu, Z.; Gao, S. Control Strategy of Single-Phase Three Level Neutral Point Clamped Cascaded Rectifier. *Energies* **2017**, *10*. [CrossRef]

11.	She, X.; Huang, A.Q. Current sensorless power balance strategy for DC/DC converters in a cascaded multilevel converter based solid state transformer. *IEEE Trans. Ind. Electron.* **2014**, *29*, 17–22. [CrossRef]

12.	Peng, X.; He, X.; Han, P.; Lin, H.; Gao, S.; Shu, Z. Opposite Vector Based Phase Shift Carrier Space Vector Pulse Width Modulation for Extending the Voltage Balance Region in Single-Phase 3LNPC Cascaded Rectifier. *IEEE Trans. Power Electron.* **2016**, *32*, 7381–7393. [CrossRef]

13.	Peng, X.; He, X.; Han, P.; Guo, A.; Shu, Z.; Gao, S. Smooth Switching Technique for Voltage Balance Management Based on Three-Level Neutral Point Clamped Cascaded Rectifier. *Energies* **2016**, *9*, 803. [CrossRef]

14.	Peng, X.; He, X.; Han, P.; Lin, X.; Shu, Z.; Gao, S. Sequence Pulse Modulation for Voltage Balance in a Cascaded H-Bridge Rectifier. *J. Power Electron.* **2017**, *17*, 664–673. [CrossRef]

15.	Shu, Z.L.; He, X.Q.; Wang, Z.Y.; Qiu, D.Q.; Jing, Y.Z. Voltage Balancing Approaches for Diode-Clamped Multilevel Converters Using Auxiliary Capacitor-Based Circuits. *IEEE Trans. Power Electron.* **2012**, *28*, 2111–2124. [CrossRef]

16.	Shu, Z.L.; Zhu, H.F.; He, X.Q.; Ding, N.; Jing, Y.Z. One-inductor-based auxiliary circuit for dc-link capacitor voltage equalisation of diode-clamped multilevel converter. *IET Power Electron.* **2013**, *6*, 1339–1349. [CrossRef]

17.	Han, P.; He, X.; Zhao, Z.; Yu, H.; Wang, Y.; Peng, X.; Shu, Z. DC-Link Capacitor Voltage Balanced Modulation Strategy Based on Three Level Neutral Point Clamped Cascaded Rectifier. *J. Power Electron.* **2019**, *19*, 99–107. [CrossRef]

18.	Kwon, Y.A.; Kim, S.K. A High-Performance Strategy for Sensorless Induction Motor Drive Using Variable Link Voltage. *IEEE Trans. Power Electron.* **2007**, *22*, 329–332. [CrossRef]

19.	Tallam, R.M.; Naik, R.; Nondahl, T.A. A Carrier-Based Pwm Scheme for Neutral-Point Voltage Balancing in Three-Level Inverters. *IEEE Trans. Ind. Appl.* **2005**, *41*, 1734–1743. [CrossRef]

20.	Lin, H.; Shu, Z.; He, X.; Liu, M. N-D SVPWM with DC Voltage Balancing and Vector Smooth transition Algorithm for Cascaded Multilevel Converter. *IEEE Trans. Ind. Electron.* **2018**, *65*, 3837–3847. [CrossRef]

21.	He, X.; Yu, H.; Han, P.; Zhao, Z.; Peng, X.; Shu, Z.; Koh, L.; Wang, P. Fixed and Smooth-Switch-Sequence Modulation for Voltage Balancing Based on Single-Phase Three-Level Neutral Point Clamped Cascaded Rectifier. *IEEE Trans. Ind. Appl.* **2020**. [CrossRef]

22.	Han, P.; He, X.; Ren, H.; Wang, Y.; Peng, X.; Shu, Z.; Gao, S.; Wang, Y.; Chen, Z. Fault Diagnosis and System Reconfiguration Strategy of a Single-Phase Three-Level Neutral-Point-Clamped Cascaded Inverter. *IEEE Trans. Ind. Appl.* **2020**, *55*, 3863–3876. [CrossRef]

23.	Busquets-Monge, S.; Griñó, R.; Nicolas-Apruzzese, J.; Bordonau, J. Decoupled DC-Link Capacitor Voltage Control of DC—AC Multilevel Multileg Converters. *IEEE Trans. Ind. Electron.* **2016**, *63*, 1344–1349. [CrossRef]

24. Zhao, C.; Dujic, D.; Mester, A.; Steinke, J.K.; Weiss, M.; Lewdeni-Schmid, S.; Chaudhuri, T.; Stefanutti, P. Power Electronic Traction Transformer—Medium Voltage Prototype. *IEEE Trans. Ind. Electron.* **2013**, *61*, 3257–3268. [CrossRef]
25. Dujic, D.; Kieferndorf, F.; Canales, F.; Drofenik, U. Power Electronic Traction Transformer Technology. In Proceedings of the IEEE 7th International Power Electronics and Motion Control Conference, Harbin, China, 2–5 June 2012. [CrossRef]

 *energies*

Article

Trusted Simulation Using Proteus Model for a PV System: Test Case of an Improved HC MPPT Algorithm

Abdelilah Chalh [1,*], Aboubakr El Hammoumi [1], Saad Motahhir [2], Abdelaziz El Ghzizal [1], Umashankar Subramaniam [3] and Aziz Derouich [1]

[1] Innovative Technologies Laboratory, EST, SMBA University, Fez 30000, Morocco; aboubakr.elhammoumi@usmba.ac.ma (A.E.H.); abdelaziz.elghzizal@usmba.ac.ma (A.E.G.); aziz.derouich@usmba.ac.ma (A.D.)

[2] Engineering, Systems and Applications Laboratory, ENSA, SMBA University, Fez 30000, Morocco; saad.motahhir@usmba.ac.ma

[3] Renewable Energy Lab, College of Engineering, Prince Sultan University; Riyadh 12435, Saudi Arabia; shankarums@gmail.com

* Correspondence: abdelilah.chalh@usmba.ac.ma

Received: 9 February 2020; Accepted: 10 March 2020; Published: 15 April 2020

Abstract: The real implementation of the maximum power point tracking (MPPT) controllers for the photovoltaic (PV) systems is still a big challenge for researchers working in this field. Often, they use simulation tools to assess the performance of their MPPT algorithms before actual implementation. In this context, this paper aims to propose a trusted simulation of a PV system designed under Proteus software. The proposed PV simulator can be used to verify and evaluate the performance of MPPT algorithms with a closer approximation to the real implementation. The main advantage of this model that it contains a real microcontroller, as can be found in reality, so that same code for the MPPT algorithm used in the simulation will be used in real implementation. In contrast, when using (Powersim Software) PSIM or Matlab/Simulink, the code of the algorithm must be rewritten once the real experiment begins, because these tools don't provide a microcontroller or an electronic board in which our algorithm can be implemented and tested in the same way as the real experiment. After this section, a modified Hill-Climbing (HC) algorithm is introduced. The proposed algorithm can avoid the drift problem posed by conventional HC under a fast variation in insolation. The simulation results show that this method presents good performance in terms of efficiency (99.21%) and response time (10 ms), which improved by 1.2% and 70 ms respectively compared to the conventional HC algorithm.

Keywords: Proteus; MPPT algorithm; PV Simulator; Drift problem

---

1. Introduction

Today, solar energy currently has taken a large part of the market compared with other sources of renewable energy [1–3]. This development has pushed many researchers to search for important solutions to increase the PV energy extracted from the PV panels. Among these solutions, we find MPPT techniques [4–7]. These techniques are used to control a DC-DC converter to extract the maximum power from the PV system. The role of the DC-DC converter is to provide impedance matching between the PV array and load. In addition, many works have investigated the effect of partial shading conditions (PSC) on the extracted PV energy, always with the aim of improving the PV system efficiency. As a result, several GMPPT techniques have been proposed in the literature to make PV systems working at the actual best efficiency under PSC [8–13]. On the other hand, to validate the

71

performance of these techniques, it is necessary to test them before implementation. For this reason, researchers often use simulation tools such as Matlab-Simulink or PSIM to model PV systems and then test the performance of their MPPT algorithms [14–16]. However, the main disadvantage of these tools is that they do not contain microcontrollers or embedded boards (such as Digital Signal Processor (DSP), Field-Programmable Gate Array FPGA, Arduino, Programmable Interface Controllers (PIC) ...) in which the MPPT algorithm can be implemented and tested as it is done in an actual prototype. Therefore, once we start the actual implementation of the MPPT algorithm for a real PV system, several problems may occur due to the disparities that can appear between the software development and the requirements during the development process. Moreover, as all the components used (DC/DC converter, sensors, actuators and microcontroller) are real, unlike those designed in the simulation tool, it is difficult to know the component responsible in the event of a bug. That can increase the time spent in debugging runtime errors. Against, these problems can be avoided by using the Proteus tool as an alternative for modelization and simulation.

Proteus is an electronic circuit design software that includes diagrammatic capture, PROSPICE simulation, and printed circuit board (PCB) layout modules. It provides in its libraries embedded boards such as PIC and Arduino, where we can implement our controller by uploading the hex code to the microcontroller as it actually happens in reality. In addition, this tool has been developed to contain components or models that are close to reality, which can give results that are more accurate. Figure 1 shows a comparison between the steps of developing a circuit board in Proteus and in traditional design tools. From this figure, it can be noticed that when using traditional design tools (such as Matlab/Simulink or PSIM), the software development and testing system cannot begin until a PCB or physical prototype is available. Thereafter, if something is wrong with the hardware design, the entire process must be repeated. However, using Proteus, software development can begin as soon as the schematic is drawn, and the combination of hardware and software can be thoroughly tested before physical prototyping. In addition, the same algorithm's code implemented in the microcontroller on Proteus will be used in the real experiment. Whereas, using PSIM or Matlab/Simulink, we have to rewrite the code of the algorithm once we begin the real experiment. For these reasons, researchers and engineers can really benefit from Proteus to test and evaluate the performances of their MPPT algorithm before the real implementation.

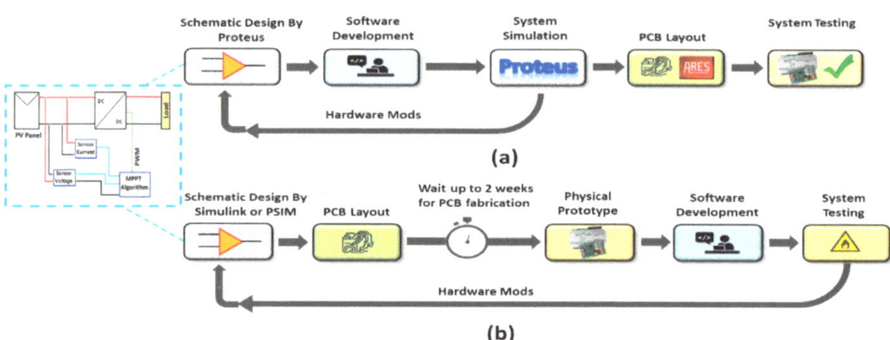

Figure 1. Steps to develop a circuit board in Proteus tool: (**a**) compared to those for developing it in traditional design tools, (**b**) such as Matlab/Simulink or PSIM.

However, to the authors' best knowledge, there is no simulation found in the literature of an entire PV system under Proteus software. While recently, some works have addressed the simulation of PV panel in Proteus as the first attempts in this subject. A single-diode Proteus model of PV module has been proposed and validated in our previous work [17–20]. Authors in [21] have been proposed a two-diode model of PV panel in Proteus. Against this background, this paper aims to present an entire design of a PV system based on Proteus, which can be used as a simulator to test the performances of MPPT algorithms. This simulator is composed of a PV panel model, DC-DC converter, voltage and current sensors, graph analysis, a Liquid-Crystal Display (LCD) screen, and the Arduino UNO board in which the MPPT algorithm is implemented. The design aspects of the PV system under Proteus are extensively described in this paper. Moreover, a modified Hill-Climbing (HC) MPPT algorithm is proposed in this paper to avoid the drift problem posed by the conventional HC algorithm when under a fast variation in insolation.

The remainder of this paper is organized as follows. Section 2 presents the PV system design in Proteus and describes the analysis of the effect of the drift problem under a fast variation in insolation with the conventional and improved HC MPPT method. Section 3 discusses the simulation results obtained. Finally, Section 4 concludes with a summary and discussion of future works.

2. Materials and Methods

2.1. Description of a PV System

2.1.1. Description of the Entire PV System

Figure 2 shows the schematic diagram of the entire PV system on Proteus. The latter is composed of a PV panel, a DC-DC converter, an MPPT controller, and a load. The PV panel is connected to the load via the boost converter. The MPPT controller implemented in the Arduino board uses the voltage and PV current data to control the boost converter to reach the MPP. The code of this controller is developed firstly in the Arduino software (IDE), and then its hexadecimal code is uploaded to the board on Proteus.

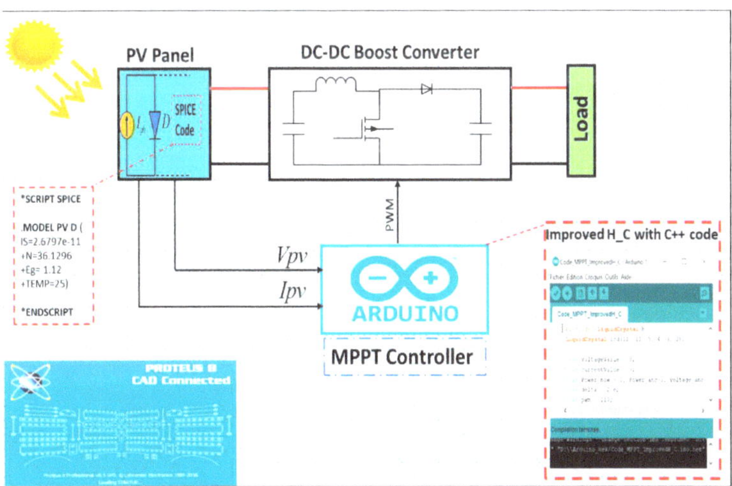

Figure 2. Schematic diagram of the entire photovoltaic (PV) system.

2.1.2. Proteus PV Panel Model

The equivalent circuit for the most used solar cell model consists of a diode and a current source connected in parallel, as well as a shunt resistor R_{sh} and a series resistor R_s, as shown in Figure 3. Based on this circuit diagram, the output current is given by the following expression [22]:

$$I = I_{ph} - I_{s,0}\left(\exp\left(\frac{V + IR_s}{aV_t}\right) - 1\right) - \frac{V + IR_s}{R_{sh}} \tag{1}$$

where: I_{ph}: Represents the photocurrent of the solar cell (A); $I_{s,0}$: Represents the reverse saturation current of the solar cell (A); V: Represents the output cell voltage of the solar cell (V); a: Represents the diode ideality factor of the solar cell; V_t: Represents the thermal voltage of the solar cell.

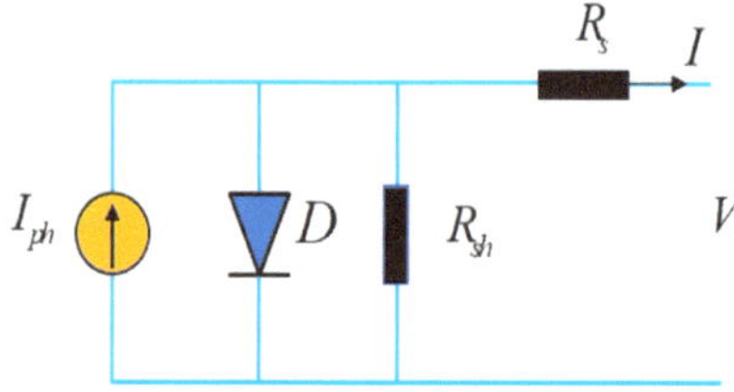

Figure 3. Equivalent circuit for single diode model of solar cell.

In this study, the TDC-M20-36 panel is used, and Table 1 presents its specifications. Figure 4 presents the Proteus PV panel model with the Spice code. In this latter, the parameters of the used diode (saturation current, number of cells, ideality factor, and bandgap energy) according to the TDC-M20-36 panel's specification, must be introduced. Note that presented the PV panel model is already validated through simulation/ experimental results in our previous paper [17].

Table 1. Characteristics of the used PV model at Standard Test Conditions (STC).

TDC-M20-36	
Pmax of PV panel	20 W
Vmpp at Pmax of PV panel	18.76 V
Impp at Pmax of the PV panel	1.07 A
Current Isc at Short-circuit (SC)	1.17 A
Voltage Voc at Open-circuit (OC)	22.70 V
Temperature coefficient Kv at OC	$-0.35\%/°C$
Temperature coefficient Ki at SC	$-0.043\%/°C$
Number of cells	36

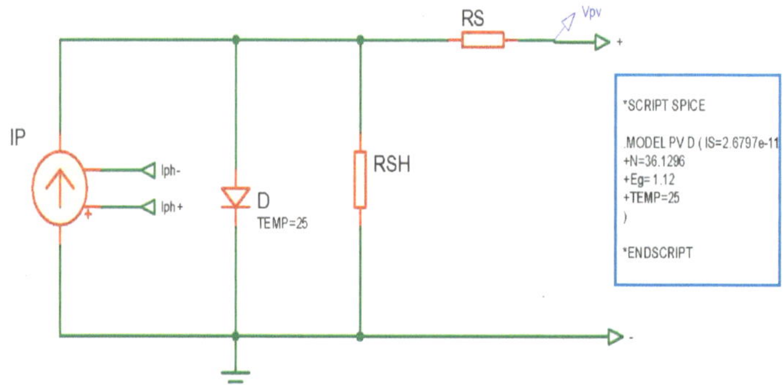

Figure 4. The equivalent circuit of panel model under Proteus.

2.1.3. DC-DC Boost Converter

The DC-DC converter is a principle element of the PV systems with an MPPT controller, which provides impedance matching between the PV array and load. Figure 5 presents the circuit diagram of the used DC-DC converter. Table 2 presents its principal parameters, which are the inductor

(L), input capacitor (Cin), output capacitor (Cout), and the switching frequency (fs). In addition, Equations (2) and (3) define the relationship between its inputs and outputs.

$$V_0 = \frac{V}{1 - \alpha} \tag{2}$$

$$I_0 = I(1 - \alpha) \tag{3}$$

$$R_{eq} = \eta(1 - \alpha)^2 R_{load} \tag{4}$$

where: I_0: Represent the output current of the Boost converter (A); I: Represent the output current of the PV model (A); V_0: Represent the output voltage of the boost converter (V); V: Represent the output voltage of the PV model (V); R_{eq}: The equivalent resistance as seen by the PV module (Ω); R_{load}: The load resistance of the PV system (Ω); η: The efficiency of the boost converter (%); α: The duty cycle.

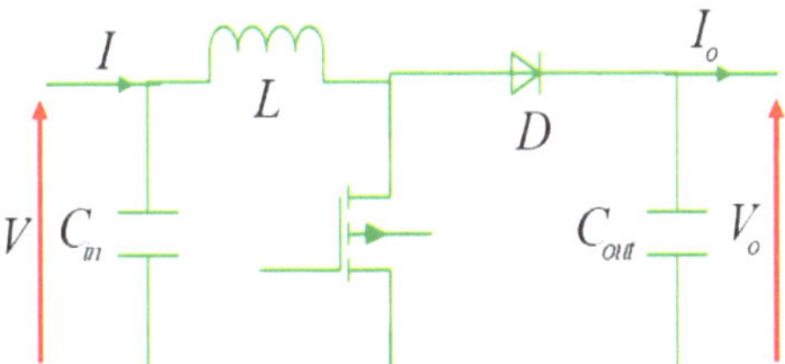

Figure 5. Circuit diagram of the DC-DC boost converter.

Table 2. The values used of the parameters of the DC-DC boost converter in our work.

Parameters	Value
L	20 mH
Cin	220 μF
Cout	470 μF
fs	1 kHz

2.1.4. MPPT algorithm

The MPPT technique is a controller that varies the duty cycle of the DC-DC converter in order to extract the maximum power from PV panels. In this paper, a classical algorithm named the HC algorithm is used to make the MPPT control of the PV system. But, as reported in the literature, this algorithm suffers from drift problem when the fast variation in insolation [23,24]. Therefore, a modified HC algorithm is proposed to solve this problem.

Conventional HC MPPT algorithm

The conventional HC algorithm is based on the variation of PV power (dP) and PV voltage (dV) by considering the P-V characteristics curve. However, this MPPT algorithm suffers from drift problems that occur with the fast variation in insolation (rapid increase in insolation). Figure 6 presents the PV curves under a rapid increase in insolation. From this figure, it can be noted that when an increase in insolation at point 3, the operating point is set to a new point 4 (dP = P4 (kTa) – P3 (kTa) > 0 and dV = V4 (kTa) – V3 (kTa) > 0). Moreover, as shown in the flowchart of the conventional HC MPPT algorithm presented in Figure 7, the duty cycle decreases with increasing insolation, which involves moving point 4 to point 5 of the maximum power point (MPP) in the new curve, called the drift problem. Similarly, thing occurs for an increase of insolation on other points. So, A drift problem occurs when there is a rapid change in insolation due to cloudy days. This problem is due to the incorrect decision of the conventional HC MPPT algorithm when dP/dP > 0.

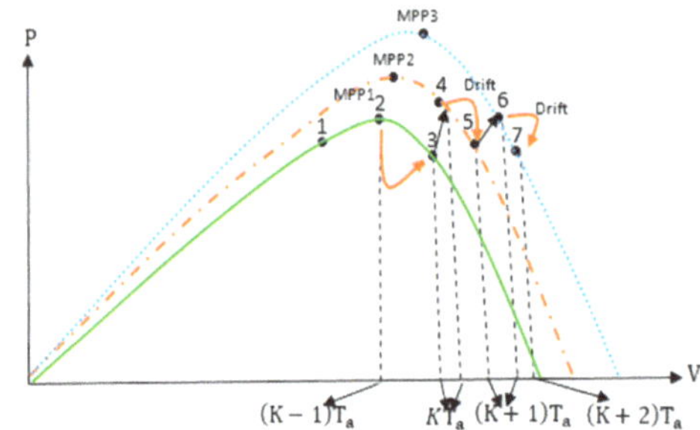

Figure 6. The P-V curves of the PV module under a rapid increase in insolation.

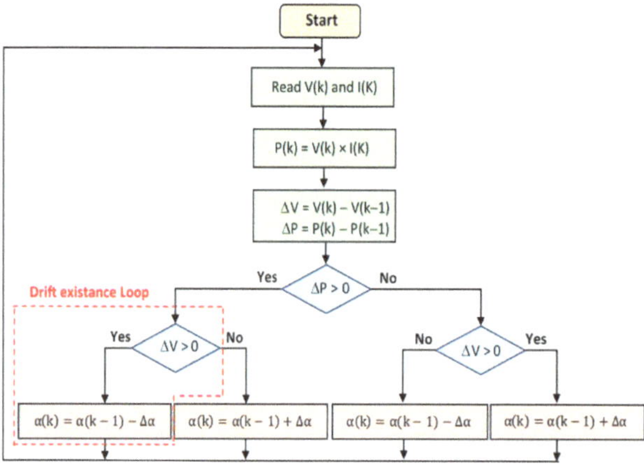

Figure 7. The flowchart of the conventional HC MPPT algorithm.

Improved HC MPPT algorithm

After detecting the loop in which the conventional HC algorithm makes a bad decision, a simple solution is proposed to overcome the obvious problem posed by HC algorithm.

The relationship between the PV current and voltage can be written as follows:

$$I = \frac{V}{R_{eq}} \tag{5}$$

With the value of the equivalent resistance is defined by Equation (4).

This relationship can also be expressed based on the single-diode model of the PV module as follows:

$$I = I_{ph} - I_0 \left(exp\left(\frac{q(V + IR_s)}{\partial KTN_s}\right) - 1\right) - \frac{(V + IR_s)}{R_{sh}} \tag{6}$$

By substituting Equation (5) into Equation (6), and by considering Taylor's series expansion up to first order, Equation (7) can be expressed as follows:

$$\frac{V}{\eta(1-\alpha)^2 R_{load}} = I_{ph} - \frac{I_0}{\partial KTN_s}\left(qV + \frac{R_s}{\eta(1-\alpha)^2}V\right) - \frac{V}{R_{sh}} - \frac{R_s}{R_{sh}}\frac{V}{\eta(1-\alpha)^2 R_{load}} \tag{7}$$

By simplifying Equation (7), the V can be expressed in terms of I_{ph} for an insolation G and the slope of the load line as follows:

$$V/_G = \frac{I_{ph}/G}{\frac{1}{\eta R_{load}(1-\alpha)^2}\left(1 + \frac{R_s}{R_{sh}}\right) + \frac{I_0}{aKTN_s}\left(q + \frac{R_s}{\eta R(1-\alpha)^2}\right) + \frac{1}{R_{sh}}} \tag{8}$$

By substituting (8) into (5) yields

$$I/_G = \frac{1}{\eta R_{load}(1-\alpha)^2} \frac{I_{ph}/G}{\frac{1}{\eta R_{load}(1-\alpha)^2}\left(1 + \frac{R_s}{R_{sh}}\right) + \frac{I_0}{aKTN_s}\left(q + \frac{R_s}{\eta R(1-\alpha)^2}\right) + \frac{1}{R_{sh}}} \tag{9}$$

At an insolation of G, I_{ph} can be expressed in terms of $I_{sc,n}$ as follows:

$$I_{ph}/_G = (I_{sc,n} + K_1\Delta T)\frac{G}{G_n} \tag{10}$$

With $I_{sc,n}$ is the short-circuit current at nominal conditions, K_1 is the short circuit current/temperature coefficient, and $\Delta T = T - T_n$ (T and T_n are the present and nominal temperatures respectively). From Equations (8) and (9), we get the Equation (10). In addition, Equations (11) and (12) can be obtained to derive the expressions V and I according to insolation.

$$\frac{dV}{dG} = \frac{(I_{sc,n} + K_1\Delta T)\frac{1}{G_n} + K_1\frac{G}{G_n}\frac{dT}{dG}}{\frac{1}{\eta R_{load}(1-\alpha)^2}\left(1 + \frac{R_s}{R_{sh}}\right) + \frac{I_0}{aKTN_s}\left(q + \frac{R_s}{\eta R_{load}(1-\alpha)^2}\right) + \frac{1}{R_{sh}}} > 0 \tag{11}$$

$$\frac{dI}{dG} = \frac{1}{\eta R_{load}(1-\alpha)^2} \frac{(I_{sc,n} + K_1\Delta T)\frac{1}{G_n} + K_1\frac{G}{G_n}\frac{dT}{dG}}{\frac{1}{\eta R_{load}(1-\alpha)^2}\left(1 + \frac{R_s}{R_{sh}}\right) + \frac{I_0}{aKTN_s}\left(q + \frac{R_s}{\eta R_{load}(1-\alpha)^2}\right) + \frac{1}{R_{sh}}} > 0 \tag{12}$$

The conditions $\frac{dV}{dG} > 0$ and $\frac{dI}{dG} > 0$ are validated in the case of a fast increase in insolation. Because in equations (11) and (12) the numerator and the denominator are positive values. Hence, from the I-V characteristics curve of the PV module, which are shown in Figure 8, it can be noticed that $dV > 0$ and $dI > 0$ in the case of a fast increase in insolation. Therefore, to avoid the drift problem, we rely on the information of dV and dI to make a good decision.

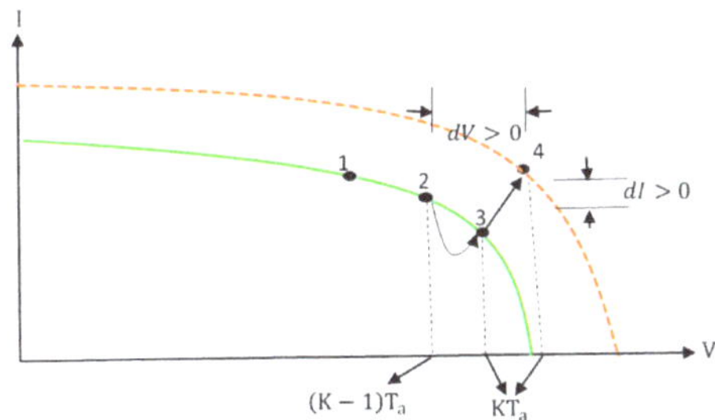

Figure 8. The I-V curves for analysis of the drift problem in case of a rapid increase in insolation with the observation of change in current (*dI*).

As shown in Figure 8, for two irradiation difference values, the dV and dI indicators are always positive when we have a fast increase in insolation. Therefore, to resolve the drift problem, the modified HC MPPT algorithm take a decision present in the Flowchart, as shown in Figure 9.

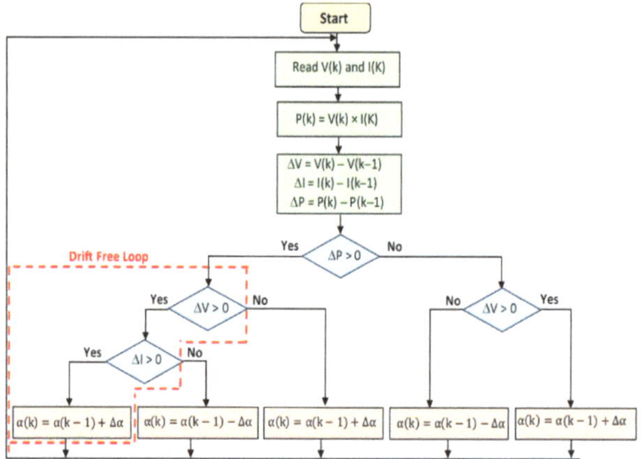

Figure 9. The flowchart of the modified HC MPPT algorithm.

2.2. PV System Design in Proteus Software

Figure 10 presents the PV system design, made in Proteus. As shown, this PV system can be divided into eight blocks as follows:

- Block (1): presents the subcircuit of the PV panel model.
- Block (2): presents the boost converter.
- Block (3): presents the embedded board (Arduino Uno).
- Block (4): presents the LCD screen, which is used to display the values of PV voltage, current and power.
- Block (5): presents the driver (TC4420), which is used to control the metal–oxide–semiconductor field-effect transistor (MOSFET) transistor of the Boost converter.
- Block (6): presents the current sensor (INA169) used for measuring the PV current. The modelization of this sensor is based on the INA168, which is available in the Proteus Tool. In order to model this sensor, you will need to follow the next steps:

   ✓ Launch the Proteus tool application.
   ✓ Open the Pick Devices.
   ✓ Select the INA168 component.
   ✓ Add two resistances R_5 (0.1 Ω) and R_6 (50 KΩ). Where Rs is a shunt resistor placed in series between the output of the PV module and the Boost converter, and RL is a load resistor connected between the Pin 1 of the INA168 and the ground.
   ✓ Set the power-supply voltage to 6 V in order to adapt it with that of INA169.

Then, the PV panel output current can be defined by the Equation (13) [25].

$$(I_S = \frac{V_{out1}1K\Omega}{R_5 R_6})$$
(13)

- Block (7): presents the module of the voltage sensor (B25 Voltage Sensor Module) used for measuring the PV voltage [26]. It is basically a voltage divider using two series resistances. The PV panel output voltage is defined by Equation (14):

$$(V_{out2} = \frac{R_3}{R_3 + R_4}V_{in})$$
(14)

where ($R_3 = 25\text{K}\Omega$) and ($R_4 = 100\text{K}\Omega$)

- Block (8): presents the graph analysis, it is used to display the simulation results.

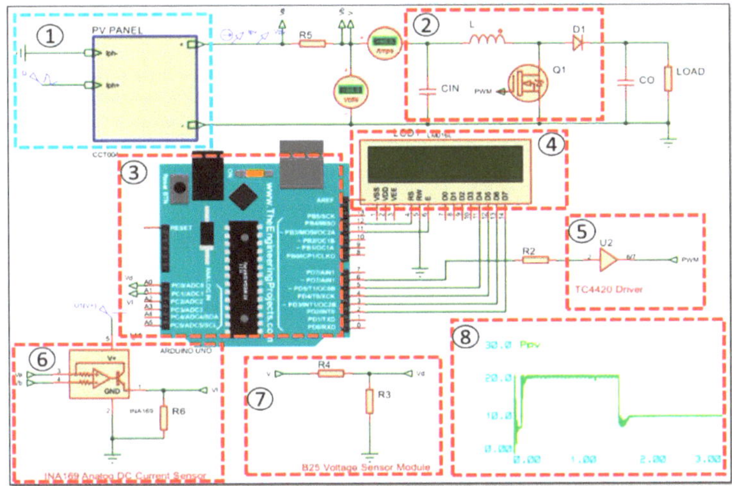

Figure 10. Schematic of PV system design in Proteus software.

3. Results and Discussion

In order to examine the effectiveness of the conventional and modified HC MPPT algorithms to solve the drift problem, a test case of a fast variation in insolation from 500 W/m^2 to 750 W/m^2 and from 750 W/m^2 to 1000 W/m was performed. The corresponding results are shown in Figure 11. From this figure, it can be seen that the conventional HC algorithm cannot avoid the drift problem when the insolation is increased, while the modified HC algorithm tracks the MPP without the drift problem. In addition, Table 3 presents a comparison between conventional and modified HC MPPT under different uniform irradiance. The test case of a fast variation in insolation from 500 W/m^2 to 750 W/m^2 presents regarding efficiency (99.15%) and response time (10 ms). Further, the test case from 750 W/m^2 to 1000 W/m^2 in insolation presents regarding efficiency (99.21%) and response time (10 ms). In these two test cases, the efficiency and response time are improved by 1.2% and 70 ms respectively as compared to conventional HC. Therefore, the modified HC algorithm can improve the efficiency of the PV system by gaining additional power when drift occur, compared to the conventional HC algorithm. Hence, the present method allows obtaining a significant energy gain throughout the life cycle of the PV panel.

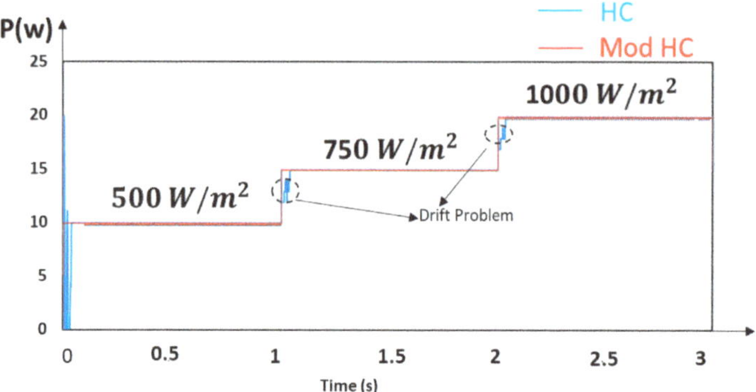

Figure 11. Output power with conventional and modified HC MPPT algorithm under a fast variation in insolation.

Table 3. Comparison between conventional and modified HC MPPT under different uniform irradiance.

MPPT Algorithms	G = 500 W/m²		G = 750 W/m²		G = 1000 W/m²	
	Efficiency	Response Time	Efficiency	Response Time	Efficiency	Response Time
Conventional HC	98.39%	10 ms	98.55%	80 ms	98.85%	80 ms
Modified HC	99.11%	5 ms	99.15%	10 ms	99.21%	10 ms

4. Conclusions

In this paper, a simple PV system developed under Proteus software is proposed. This system can be used as a simulator to test the performances of MPPT algorithms before real implementation. The present simulator is composed of a PV panel model, DC-DC converter, voltage and current sensors, graph analysis, LCD screen, and an Arduino UNO board. It is demonstrated in this work that the choice of the Proteus software to test MPPT algorithms improves the PV system realization time. Because the same C MPPT code used in the simulation will be used for the real physical prototype. Further, the simulation results of the conventional and Modified HC MPPT algorithms under a fast variation in insolation shown that the modified method can avoid the drift problem. As well as addition, the efficiency and response time are improved by 1.2% and 70 ms respectively as compared to conventional HC.

As a perspective, future studies will be focused on the following two aspects: (i) Design of a simple method to find the GMPP under PSC with simple instructions; And (ii) the implementation of this method using the proposed PV Simulator.

Author Contributions: A.C., S.M. and A.E.G. performed the study of a simple PV simulator. In addition, A.C. performed the simulation examination. A.E.H., U.S. and A.D. revised the manuscript. All authors have read and agreed to the published version of the manuscript.

Funding: The authors declare that they have no funding for the research.

Acknowledgments: The authors like to express their sincere gratitude to the Renewable Energy research lab, College of Engineering, Prince Sultan University, Riyadh, Saudi Arabia for providing technical support.

Conflicts of Interest: The authors declare that they have no competing interests.

Abbreviations

PV	photovoltaic
CAs	conventional algorithms
MPPT	maximum power point tracking
HC	Hill-Climbing
GMPP	global maximum power point
PCB	printed circuit board

Nomenclatures

a	The diode ideality factor of the solar cell
I_{ph}	The photocurrent of the solar cell [A]
I_0	The output current of the Boost converter [A]
I_{mpp}	The current at MPP [A]
G	The solar irradiance level [W/m2]
G_n	The solar irradiance nominal [W/m2]
K	The constant of Boltzmann [J. K-1]
N_s	The number of cells connected in series
R_{load}	The load resistance [Ω]
R_{eq}	The equivalent input resistance of the converter Boost [Ω]
R_s	The series resistance of the solar cell [Ω]
R_{sh}	The shunt resistance of the solar cell [Ω]
T	The junction temperature [K]
T_n	The nominal temperatures [K]
V	The PV panel output voltage [V]
V_{mpp}	The voltage at MPP [V]
V_0	The Boost output voltage [V]
η	The efficiency of the DC-DC converter Boost [%]
α	The duty cycle.

References

1. Kousksou, T.; Allouhi, A.; Belattar, M.; Jamil, A.; El Rhafiki, T.; Arid, A.; Zeraouli, Y. Renewable energy potential and national policy directions for sustainable development in Morocco. *Renew. Sustain. Energy Rev.* **2015**, *47*, 46–57. [CrossRef]
2. Kabir, E.; Kumar, P.; Kumar, S.; Adelodun, A.A.; Kim, K.H. Solar energy: Potential and future prospects. *Renew. Sustain. Energy Rev.* **2018**, *82*, 894–900. [CrossRef]
3. Subramaniam, U.; Vavilapalli, S.; Padmanaban, S.; Blaabjerg, F.; Holm-Nielsen, J.B.; Almakhles, D. A Hybrid PV-Battery System for ON-Grid and OFF-Grid Applications—Controller-In-Loop Simulation Validation. *Energies* **2020**, *13*, 755. [CrossRef]
4. Subudhi, B.; Pradhan, R. A comparative study on maximum power point tracking techniques for photovoltaic power systems. *IEEE Trans. Sustain. Energy* **2013**, *4*, 89–98. [CrossRef]
5. Ferrero Bermejo, J.; Gómez Fernández, J.F.; Pino, R.; Crespo Márquez, A.; Guillén López, A.J. Review and Comparison of Intelligent Optimization Modelling Techniques for Energy Forecasting and Condition-Based Maintenance in PV Plants. *Energies* **2019**, *12*, 4163. [CrossRef]
6. Motahhir, S.; El Hammoumi, A.; El Ghzizal, A. The most used MPPT algorithms: Review and the suitable low-cost embedded board for each algorithm. *J. Clean. Prod.* **2019**, *246*, 118983. [CrossRef]
7. Lee, H.S.; Yun, J.J. Advanced MPPT Algorithm for Distributed Photovoltaic Systems. *Energies* **2019**, *12*, 3576. [CrossRef]
8. Pathy, S.; Subramani, C.; Sridhar, R.; Thentral, T.; Padmanaban, S. Nature-inspired MPPT algorithms for partially shaded PV systems: A comparative study. *Energies* **2019**, *12*, 1451. [CrossRef]
9. Bahrami, M.; Gavagsaz-Ghoachani, R.; Zandi, M.; Phattanasak, M.; Maranzanaa, G.; Nahid-Mobarakeh, B.; Pierfederici, S.; Meibody-Tabar, F. Hybrid maximum power point tracking algorithm with improved dynamic performance. *Renew. Energy* **2019**, *130*, 982–991. [CrossRef]

10. Huang, C.; Wang, L.; Yeung, R.S.C.; Zhang, Z.; Chung, H.S.H.; Bensoussan, A. A prediction model-guided Jaya algorithm for the PV system maximum power point tracking. *IEEE Trans. Sustain. Energy* **2017**, *9*, 45–55. [CrossRef]

11. Bingöl, O.; Özkaya, B. Analysis and comparison of different PV array configurations under partial shading conditions. *Solar Energy* **2018**, *160*, 336–343. [CrossRef]

12. Belhachat, F.; Larbes, C. Comprehensive review on global maximum power point tracking techniques for PV systems subjected to partial shading conditions. *Solar Energy* **2019**, *183*, 476–500. [CrossRef]

13. Huang, C.; Wang, L.; Zhang, Z.; Yeung, R.S.C.; Bensoussan, A.; Chung, H.S.H. A Novel Spline Model Guided Maximum Power Point Tracking Method for Photovoltaic Systems. *IEEE Trans. Sustain. Energy* **2019**. [CrossRef]

14. Ishaque, K.; Salam, Z. A comprehensive MATLAB Simulink PV system simulator with partial shading capability based on two-diode model. *Solar Energy* **2011**, *9*, 2217–2227. [CrossRef]

15. Bagyaveereswaran, V.; Umashankar, S.; Arulmozhivarman, P. Simulation Tool for Tuning and Performance Analysis of Robust, Tracking, Disturbance Rejection and Aggressiveness Controller. *Algorithms* **2019**, *12*, 144. [CrossRef]

16. Motahhir, S.; El Ghzizal, A.; Sebti, S.; Derouich, A. Proposal and Implementation of a novel perturb and observe algorithm using embedded software. In Proceedings of the 2015 3rd International Renewable and Sustainable Energy Conference (IRSEC), Marrakech, Morocco, 10–13 December 2015; pp. 1–5.

17. Motahhir, S.; Chalh, A.; El Ghzizal, A.; Sebti, S.; Derouich, A. Modeling of Photovoltaic Panel by using Proteus. *J. Eng. Sci. Technol. Rev.* **2017**, *10*, 8–13. [CrossRef]

18. Motahhir, S.; Chalh, A.; El Ghzizal, A.; Derouich, A. Development of a low-cost PV system using an improved INC algorithm and a PV panel Proteus model. *J. Clean. Prod.* **2018**, *204*, 355–365. [CrossRef]

19. Motahhir, S. Data for: Development of a low-cost PV system using an improved INC algorithm and a PV panel Proteus model. *Mendeley Data* **2020**. [CrossRef]

20. Motahhir, S. PV Panel in Proteus. *Mendeley Data* **2017**. [CrossRef]

21. Yaqoob, S.J.; Obed, A.A. Modeling, Simulation and Implementation of PV System by Proteus Based on Two-diode Model. *J. Tech.* **2019**, *1*, 39–51.

22. Chaibi, Y.; Salhi, M.; El-Jouni, A.; Essadki, A. A new method to extract the equivalent circuit parameters of a photovoltaic panel. *Solar Energy* **2018**, *163*, 376–386. [CrossRef]

23. Killi, M.; Samanta, S. Modified perturb and observe MPPT algorithm for drift avoidance in photovoltaic systems. *IEEE Trans. Ind. Electron.* **2015**, *62*, 5549–5559. [CrossRef]

24. Xiao, W.; Dunford, W.G. A modified adaptive hill climbing MPPT method for photovoltaic power systems. In Proceedings of the 2004 IEEE 35th Annual Power Electronics Specialists Conference (IEEE Cat. No. 04CH37551), Aachen, Germany, 20–25 June 2004; Volume 3, pp. 1957–1963.

25. INA1x9 Datasheet [WWW Document]. 2017. Available online: www.ti.com (accessed on 8 January 2018).

26. Voltage Sensor Module-Arduino Compatible [WWW Document], n.d. Available online: https://www.emartee. com/product/42082/VoltageSensor (accessed on 8 January 2018).

 *energies*

Article

Microcontroller-Based Strategies for the Incorporation of Solar to Domestic Electricity

Nkateko E. Mabunda [1,*] **and Meera K. Joseph** [2]

[1] Faculty of Engineering and the Built Environment, Department of Electrical and Electronic Engineering Technology, University of Johannesburg, 2000 Johannesburg, South Africa
[2] IEEE Computer Society, 2059 Johannesburg, South Africa
* Correspondence: nkatekom@uj.ac.za; Tel.: +27-84-503-5868

Received: 4 June 2019; Accepted: 16 July 2019; Published: 22 July 2019

Abstract: Microcontrollers have been largely used in applications that include reducing power consumption. Microcontroller development tools are now readily available. Many countries are faced with energy challenges such as lack of enough power capacity and growth in energy demand. It is therefore important to introduce innovative methods to reduce reliance on national grid energy and to supplement this source of energy with alternative methods. In this study, the microcontroller is used to monitor the energy consumed by household equipment and then decide, based on the power demand and available solar energy, the type of energy source to be used. In this research, a special circuit was also designed to control geyser power and align it to the capacity of the renewable energy source. This geyser control circuit includes a Dallas temperature sensor and a triode for alternating current (TRIAC) circuit that is included to control output current drawn from a low power, renewable energy source. Alternatively, two heating elements may be used instead of the TRIAC circuit. The first heating element is powered by solar to maintain the water temperature and to save energy. The second heating element is powered by national grid power and is used for the initial heating, and therefore saves water heating time. The strategy used was by adding a programmed microcontroller-based control circuit and a low power element or one current controlled element to a geyser whereby Photovoltaic (PV) energy was used to save the energy geysers consume from the domestic electricity source when they are not in use. A microcontroller, current sensor, battery level sensor, and relay board was used to incorporate solar-based renewable energy to the commercial energy supply.

Keywords: incorporation of solar; microcontrollers for solar; solar to domestic electricity; photovoltaics

1. Introduction

Challenges can arise when photovoltaic (PV) energy is used as a primary source of electricity, such as instability, which results from this energy source and it is very difficult to predict when the sun [1,2] will give out enough energy. Atmospheric conditions such as dust, fog, cloud cover, etc. can change spontaneously and therefore affect the availability of solar energy [3]. Consumers can only increase the capacity of PV energy by purchasing high power solar panels in large numbers and batteries to be prepared for longer periods with less sun irradiance, thus increasing the system cost. It is for these reasons many people are still not using solar power as a source of electricity [4]. Solar energy is described as one of the cheapest, cleanest and easily obtainable energy sources [5]. However, the initial cost of PV energy systems is much higher when compared to that of non-renewable sources [6]. PV systems can be categorized as grid-connected and standalone systems. Grid-connected PV systems are characterized by a connection to the local grid, and their loss of power in the absence of the sun unless they are used with a backup source [7]. The continued monitoring of several parameters in photovoltaic devices will lead to effective usage of solar energy [8]. The microcontroller has been used to acquire various signals for monitoring purposes or to control activities such a battery charging and solar tracking [9]. PV energy was used to optimize the performance of auxiliary power source by employing an automatic transfer switch [10]. A study to employ PV energy for landing crafts and therefore reduce their gas emission and fuel consumption was conducted in [11]. Microcontrollers were also used to interface PV-derived energy with the grid supply as a way of providing a reference point for synchronization [12]. A multi-control single-input switch was used to select between diesel, wind and photovoltaic sources in [13]. PV energy was integrated to the smart grid system by using energy management systems that shifted load connections from one energy system to the other during certain time slots [14]. Fuzzy logic was also used to manage energy consumed by loads as means of reducing electrical bills [15]. The application of solar-derived electrical energy directly to DC water heating element has been studied in [16]. There are ongoing studies for the continued improvement of photovoltaic/thermal (PV/T) in water heating applications as in [17–19]. Although energy management techniques are available and several strategies are used to reduce energy consumption in water heating, a research gap is available to study methods of incorporating off-grid PV systems into the domestic supply irrespective of the PV system size or the load capacity. The developments in sensors and microcontrollers formed a good support base and their broader application was demonstrated [20].

In our research, we have used sensors to detect the available solar resources, water temperature and load current. The microcontroller selects utility supply for cold-water temperature and/or higher load currents. Energy produced by a standalone photovoltaic system is selected when its capacity matches the load condition. As a result, any PV system size can be incorporated to a domestic supply where high power applications (where energy is consumed at a faster rate) are reserved for a non-renewable source. A renewable energy source will be then used to supply low power devices or operate when energy consumption is at a slow rate. The microcontroller also chooses the geyser current profile to be used based on the selected energy source. High current is used for initial heating and low current to increase the water temperature at a slower rate to ultimately keep the water warm. Irrespective of the PV system size and load capacity, the research still offers benefits that includes reduction of carbon emissions, prevention of complete blackouts during the absence of utility supply and partial powering of electric water heaters by using photovoltaic energy.

2. Similar Studies

Previous researchers studied electrical energy consumption from commercial sources to support the regulations of consumptions. A survey to study the consumption of electricity by small and medium enterprises was conducted [21]. The aim in this study was to enforce a mindful electricity usage by using campains, elevated energy prices and rebates for lower consumptions. In [22], other authors presented on how law can be used to impose the regulation of energy consumed from the national grid. ISO 50001 is the standard that governs the use of supplied energy. Ineffectiveness of electricity usage was studied for a municipality in [23]. In this study, various legislation used for the regulation of energy consumptions are listed.

This research is aimed to encourage a reduction in energy consumption from the national grid; however, a technical solution is used whereby the solar energy is incorporated to supplement grid supply as opposed to legislation.

The study has attempted to unify several aspects that can bring along one solution to enable simultaneous use of both (renewable energy source) RES and domestic electricity supply source. The microcontroller continuously monitors the battery level, water temperature and load current. The monitored parameters influence a choice of electrical energy supply source. For these reasons, individual studies discussed in this section can only cover fewer aspects; however, collectively, they will cover most of them.

2.1. Microcontrollers and Energy Management in PV Systems

Other researchers designed an embedded low-cost microcontroller-based logger for the PV system in [24]. In their research, an Arduino microcontroller is used to control activities of the modules such as current sensors, temperature sensors and analog-to-digital convertors that were interfaced by using a variety of protocols. This prototype has 18-bit resolution and provides eight analog inputs that can be used for measurement of up to eight photovoltaic modules within an array or string. Measured current can be either DC or AC, and shunt resistors were used for the detection.

Other researchers have accomplished remote monitoring of PV currents by using a hall current sensor that is interfaced to the Arduino board [25]. Zigbee-based communication is a protocol used for the transmission of data to the receiving system that has a LabVIEW-based application. In this application, data is received via the USB port connected to the wireless signal circuit. A DC current of 1.5 amps has been detected and transmitted for monitoring.

A system that uses a microcontroller to disconnect a load that consumes power that exceeds the set limit of 500 W was prototyped in [26]. Here, the system is also able to protect generators against excessive load currents. In [26], the relays that are driven by NPN transistors are employed to do the switching and a current transformer is used to perform the current detection.

A unit that can alter the load connections between the generator, PV energy source and domestic utility was designed in [27] and ATMEGA16L uses the relay board to select the appropriate supply based on the availability. The prototype also includes a LCD screen to display the status.

A home energy management system was used to switch DC and AC load connections between the installed PV system and utility supply [28], and a mathematical approach was used to formulate the prediction algorithm.

A home automation based on DC load-matching technique was studied [29]. A strategy to remove DC–AC converters and AC–DC converters was investigated in order to reduce losses.

A battery management system based on the Internet of Things (IOT), where a Raspberry Pi model 2 acquires battery information as well as PV information and sends it to a cloud for distribution is discussed in [30]. The Raspberry Pi uses various sensors to acquire temperature, voltage and currents. Computers and mobile devices can be used to access the acquired data by connecting to the cloud database. The PV system consisted of battery bank (8 × 100 Ah), PV array (20 modules × 50 Wp), grid tied inverter and the load.

A switching system between the grid and PV system was implemented by using the PIC16F877 microcontroller [31]. Automatic switching is done in order to avoid over discharging of the battery. The switching circuit was made of a Darlington pair transistor configuration and a relay. The prototype has been tested by using a 3-W photovoltaic panel, which produced an output voltage up to 10.5 V.

2.2. Strategies of Supplementing Grid Power with PV Energy and to Control Electric Water Heater (Geyser) Power Consumptions

A microcontroller was used to implement a water heater that uses both solar and the domestic energy supply [32]. An electric heater is used when the solar radiations become insignificant and the installed thermostat sets the highest water temperature. The microcontroller automates the process and controls the activities of the circulation pump.

AC–DC hybrid and PV generation with a battery backup using a smart grid system was designed by other experts to ensure continuous supply of electricity in a cost-effective manner [33]. The system's algorithm checks the availability of solar energy and then connects to it whenever it can provide a required capacity, if not, it utilizes the local supply. The system also predicts weather conditions and reschedules high power consumption tasks to the period with high-energy production. Backup batteries are kept at charge level, which is above 50 percent.

The demand response strategy used in [34] clarifies that the thermal loads account to huge energy consumptions from the grid. If the thermal loads are not properly controlled, they create overloading when operating in great numbers at the same time. In this study, information and communication technologies are used to support the potential of renewable energy source.

2.3. Novelty and Contribution

Although microcontrollers were used for entirely different purposes, for instance, for switching between grid and PV as in [31], battery management system as in [30] and for developing solar trackers as in [9] by other experts, this research is novel. In addition to switching between a domestic energy supply and PV energy, a special circuit was designed to control geyser power and align it to the capacity of the renewable energy source in this research. Two options were used to control the geyser current; the first option uses the microcontroller select between one of the two types of heating elements as influenced by the available energy source. In the second option, the geyser control circuit uses a Triode for Alternating Current (TRIAC) circuit, which is included to control output current drawn from a low power, a renewable energy source when is in use. Therefore, our contribution further enabled the development of a programmed microcontroller-based control circuit for a geyser high current and low current heating profiles. In this research, PV energy was used to save energy when entire domestic power requirements match the available solar resources.

3. Objectives

The objectives of this research are:

- To prototype and test the microcontroller-based electric water heater (geyser) that utilizes both PV and domestic energy supplies. We propose two ways to do this; the first option is to use two elements where the calculation for the size of each element is based on capacity of the corresponding input and the second option is to include a current-limiting circuit, which will regulate current accordingly.
- To prototype and test the microcontroller-based circuit that changes the connection between PV-derived energy and domestic supplied energy based on energy supply availability and load consumptions.

4. Methods

4.1. Hardware Components

In Figure 1, a solar array that produces 200 W output is located at the rooftop. The solar panels output current charges the 12 V, 100 Ah battery via a 40 A solar charge controller designated here as MPPT (maximum power point tracker). The charge controller manages the charging process. Three sensors are connected at the input of the microcontroller, namely: current sensor (used for load management), battery level sensor (used to check when to connect and disconnect from RES) and a Dallas temperature sensor (for geyser water temperature monitoring). Several microcontroller outputs are connected to the relay circuit as illustrated on Figure A1. As explained in Section 1, the software in Appendix B is used to control the relay circuit outputs.

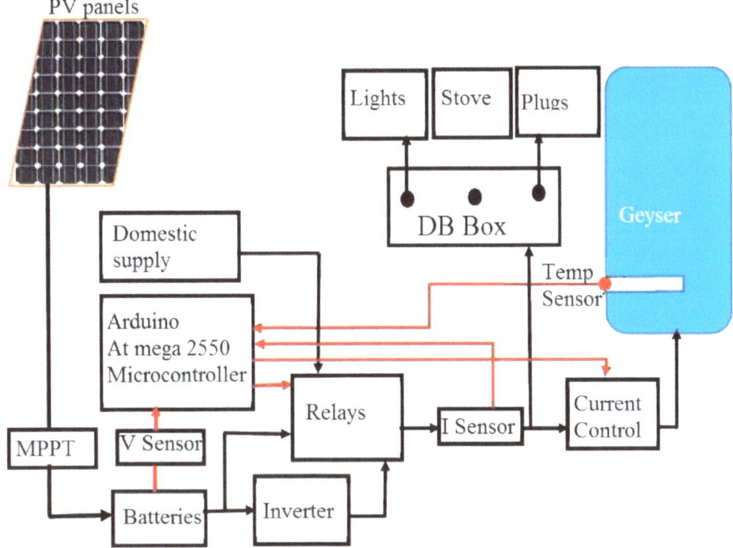

Figure 1. Hardware layout.

The system has three major sensors: battery level sensor, inverter-output current sensor and the Dallas temperature sensor [35] for the monitoring of water temperatures. A maximum charge current of 16 A has been measured. If the software never allows the battery level to drop below 50% of its maximum capacity, i.e., 50 Ah, it will therefore take a minimum of about 3 h to fully restore the battery charge to its maximum level. To create about 50 Ah from 16 A, you need only about 3 h of sunshine from a day with a potential to give six peak-sun hours. At the same time, when ignoring losses, where a maximum load power of 600 W is permissible for a PV source—a fully charged 12 V battery can supply the load for a duration of about 1 h (50 Ah). For a maximum connected load, this system will therefore charge battery for 3 h to supply continuous energy for 1 h and have a potential to provide domestic electricity savings of about 1 kWh per day, as influenced by the available sunshine duration. To produce more savings, one will have to increase the size of solar panels, battery and a charge controller. If the load requires about 120 W to operate and the system in Figure 1 is used to give backup power, it will work for about 10 h and still save 1 kWh of energy per day.

Figure A1 is a schematic diagram that represents the overall circuit that has been discussed. In this schematic, sensors discussed on the previous sections were excluded. The LCD used to monitor the load controller status and activities is shown in Figure 2.

Figure 2. LCD display.

From the LCD, it can be seen that the battery level capacity is at 55.9%, the equipment consumes only 110 W of solar power and that the status is "SOLAR", meaning solar energy is available. On the LCD, three statuses can be displayed: charging, idle, and solar (PV usage).

4.1.1. Battery Level Sensing

Upon powering on, the microcontroller will check the battery level by reading the voltage across the current limiting resistor (R_1) illustrated in Figure 3, where a 1 W, 9.1 V Zener diode has been used. The Arduino ATmega2550 microcontroller analog to digital converter supports a maximum of 5 V at a 10-bit resolution. Even though a fully charged battery gives about 12.7 V, by using this sensor the microcontroller will withstand up to a maximum 14.1 V (9.1 V + 5 V) across the battery. A Cotek SK1000-212 inverter with specifications outlined in [36] was used. Therefore, based on its specifications, this inverter will operate properly over the voltage range between 10.5 V and 15 V. To avoid excessive battery discharging, the battery minimum level is kept at 12.1 V, which is a represents 50% of its maximum charge. The microcontroller never allows battery usage by the load when the level at its analog to digital converter input becomes lower than 3 V (12.1 V − 9.1 V) or less. When this occurs, the load is diverted to the domestic electricity supply.

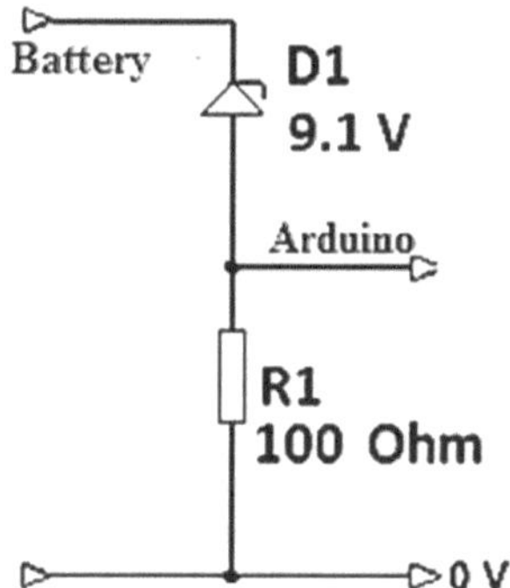

Figure 3. A simple voltage sensor circuit.

4.1.2. Output Current Sensing

A circuit shown in Figure 4 was used to detect a high load current by using the current transformer (CT) illustrated. The current transformer used can deliver up to a maximum of 141 mA peak to peak output current through the 33 Ω resistor.

By using ohm's law; this will translate to a maximum peak to peak voltage of 4.65 V applied to the input of' the microcontroller. From the 1000 W invertor used and excluding any form of calibrations; the maximum output power was restricted to 600 W at 230 V output, which then produced the peak-to-peak current of 7.379 A. The CT current ratio is 100 A: 50 mA.

Therefore a 7.379 A peak-to-peak current will produce 3.689 mA through the load. By using Ohm's law, a peak-to-peak voltage of 3.689 mA × 33 Ω = 121.764 mV results. The Microcontroller will suspend any connections to PV energy source and connect all loads to the commercial electricity supply after reading 121.764 mV or higher.

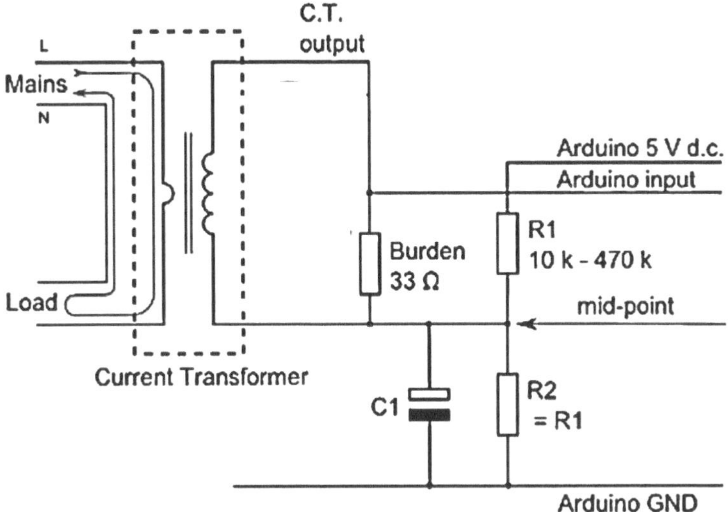

Figure 4. Current sensor circuit [37].

4.1.3. Geyser Water Temperature Sensing

A temperature probe and connection diagram as in [38] was used as a guide for this research for the detection of water temperature. Before using it, one must download software library from an open source. The library software is then included as a header file in the source code (see Appendix B). In source code, the sensor is initialized to prepare it for temperature reading. The existing geyser thermostat can also be used for extra protection, but it will have to be set to the value, which is slightly higher than maximum preset cut-off temperature of the sensor in [38].

4.1.4. Controlling Geyser Element Current

Special attention was given to geysers since they consume huge amounts of energy. In this research, it is proved that geysers can be configured to draw high current only to heat up water when the temperature is less than the pre-defined value and thereafter uses less current to bring water temperatures to even higher values. It was already proved that a significant amount of energy savings would result in reducing geyser connection time to the domestic electricity supply source. Instead of fully disconnecting geysers to the power source, in this research, we bring down their power consumption and thereafter connect it to low power PV energy source. By doing this, the domestic energy savings is further improved as the water temperature will be increased slowly to even higher temperatures. Two methods of doing this were evaluated. It may be done by using two elements or a TRIAC-based circuit that operates like a high-power light dimmer. Various low power light dimmers are discussed in [39].

4.1.4.1. Using Two Water-Heating Elements

In Figure 5, two elements are used. The low power element can be connected directly to the batteries or DC to AC inverter output but not directly to the solar panels: The high-power element will raise water from its lowest temperature to 50 °C, thereafter, the low power element ensures that the water stays warm by raising its temperature further to 72 °C. To do this: The microcontroller reads the temperature sensor. If the water temperature is below 50 °C, it then connects the high-power element to the domestic electricity supply to raise the water temperature to 50 °C. Once the temperature of 50 °C has been reached, the microcontroller checks the level of the battery shown in Figure 1. If the battery level is more than 50%, it then keeps the water warm by connecting the low power element to renewable source until the battery level is just below 50% or the water temperature reaches about 72 °C. When the battery level becomes less than 50%; the low power element is kept off until the battery recovers to a level of 75%. The battery charging and water heating process will be repeated until the water temperature of 72 °C is reached. At the water temperature of 72 °C, all geyser elements are disconnected from their corresponding energy source to avoid geyser overflow. If it happened during the battery charging phase whereby water temperature drops to a temperature below 50 °C, the high-power element takes over.

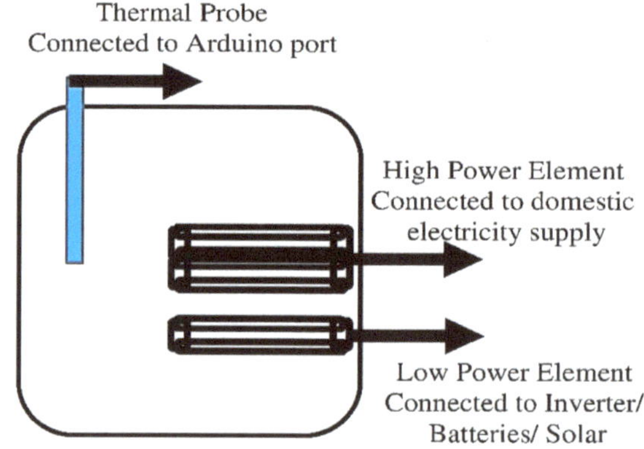

Figure 5. Two-element water heating system.

4.1.4.2. Using TRIAC Circuit to Control Geyser Current

In Figure 6, a special switched TRIAC circuit is used to control geyser current. A TRIAC is a device that can be used to regulate output current by controlling its on/off periods.

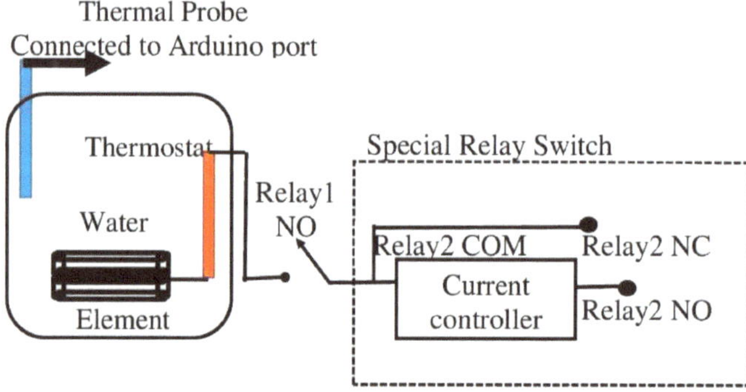

Figure 6. One current controlled element.

This setup is well suited for conventional geysers since it requires less modification to them. The difference between the operations explained in Section 4.1.4.1 and the one for the setup in Figure 6 is as follows:

- One element is used instead of two.
- Current control circuit regulates current from source.
- Relay1 is used to switch off all elements.
- Relay2 normally closed (NC) contact connects to the national grid whilst the normally open contact (NO) connects to renewable energy source.

4.2. Software Component

The Geyser and a load controller (a circuit that switches between solar energy and the national grid supply) were tested independently using separate software as represented by the flowcharts that will be discussed later. Appendix B gives a complete source code that was used to control geyser operations. Since there is greater similarity between the software used to control the load and that of the geyser, it is unnecessary to attach both sets of source codes. Overall, the source code can be organized to have a main program that can either call the geyser control or the load controller's software algorithm. As a result, the source code in Appendix B can be rewritten to become a function.

4.2.1. Geyser Controlling Software

The flowchart in Figure 7 uses national grid energy to raise the initial water temperature up to 50 °C and thereafter uses solar energy to keep water temperature between 50 °C and 72 °C.

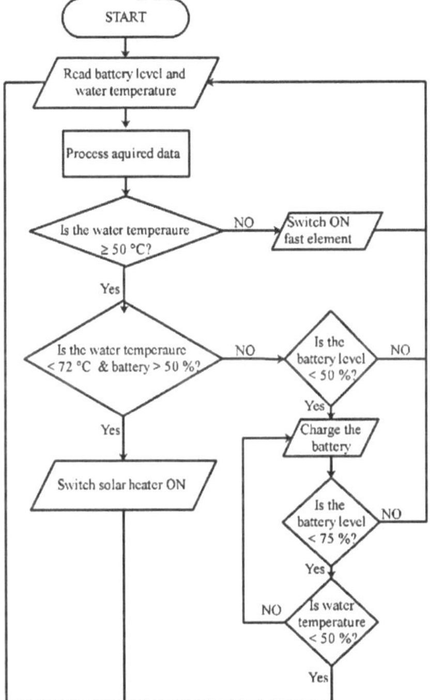

Figure 7. Geyser control software algorithm.

Upon powering on, the software reads outputs from both the battery and temperature sensors. These acquired values are then translated into meaningful data that is later used for decision-making. If the water temperature is below 50 °C, domestic electricity is applied to heat the water until the temperature of 50 °C is reached, thereafter the renewable energy is applied to heat the water further if the battery level is still above 50% and water temperature is below 72 °C. If the battery level drops to a value below 50% whilst the water temperature is still above 50 °C, renewable energy is disconnected to allow battery recovery by charging it up to 75% capacity. If the water temperature drops to a temperature below 50 °C whilst the battery is recovering, the mains will connect to the element and raise the water temperature back to 50 °C. A geyser thermostat was also incorporated to protect against overflow. Connecting the mains to the element during the charging phase will never affect the battery recovery process because the batteries are always connected to the PVs via the maximum power point trackers (MPPT).

4.2.2. Load Controlling Software

Figure 8 shows the software algorithm that was used to switch load connection between the renewable energy source and the national grid power. Here, the underlying fact is that the battery must be above 50% and the connected load should not exceed 600 W for the renewable energy source to be used. The national grid power is given the lowest priority. In other words, renewable energy here is a primary source of energy that uses the national grid power as a backup source.

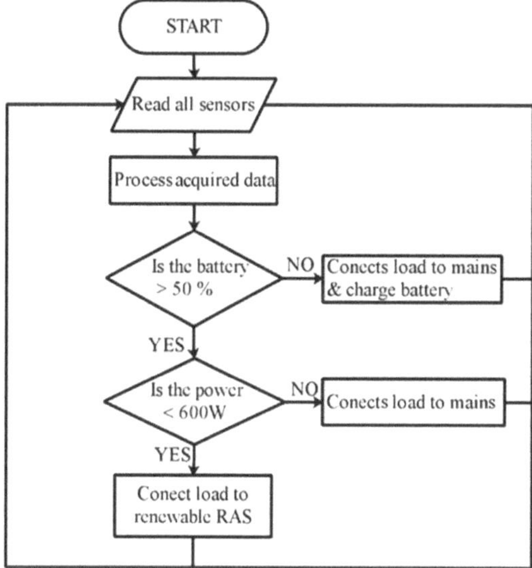

Figure 8. Load control software algorithm.

5. Results

5.1. Performance of the Geyser Prototype

The tests were conducted using a geyser that was prototyped as illustrated in Figure 9 and followed the connections shown in Figure 6. If the low-power element is not available, one high-power element can be used. The circuit that employs a TRIAC moderates the current whilst the microcontroller controlled the geyser operation. The software algorithm illustrated in Figure 7 can also be used for a two-element system as per the connections shown in Figure 6.

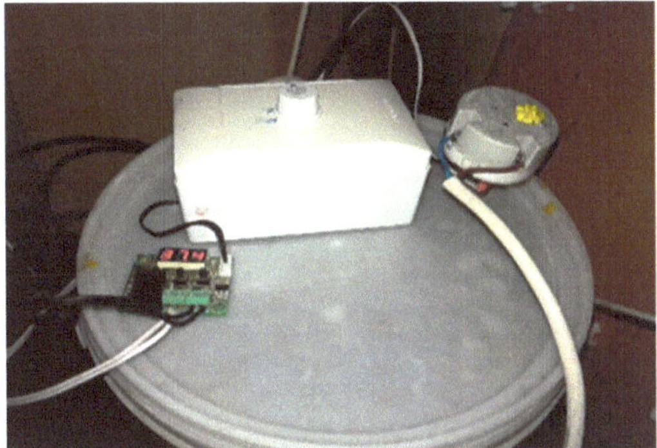

Figure 9. Hybrid geyser that was prototyped using 10-L water bucket, light dimmer, thermostat and the temperature sensor.

5.1.1. Results Obtained When Both PV Energy and Grid Energy are Used

The values of energy consumption presented in this section were logged in by a custom-built energy logger that employs NI myRIO data acquisition card that was programmed by using LabVIEW development platform. The details about logging and tracking of renewable energy using this tool are outlined in [40]. As illustrated by Table 1, 140 Wh of energy from the domestic electricity supply source was used to raise the initial geyser water temperature of 27 °C to 50 °C in a duration of 7 min. The solar energy maintained the water temperature above 50 °C in a time period between 08h07 until 18h48. In some instances, the low power heating overlapped with the a battery levels that are below 50%, that occurred spontaneously due to high DC currents of up to 30 A, which were drawn from this battery, but never last longer than the software delays. Another challenge is to get accurate battery voltage readings (used to derive battery charge level) whilst the load is actively connected. However, in any case, a battery level of lower than 50% is not permitted. The geyser water temperature is maintained by using PV power to gradually raise the water temperature to 72 °C. At a water temperature of 72 °C, all power sources were disconnected. Table 1 also indicates that PV energy was only able to preserve the water temperature of about 72 °C during the period that ranged from 11h00 until 16h33. After 16h33, in the absence of enough sunlight, the water temperature has decayed from the temperature of 72 °C to a value below 50 °C. This is what has retriggered connections to the national grid power at 18h48. Five-hour power saving can be calculated by looking at a time span from 09h05 to 14h06 in Table 1. The saving will therefore equal to about 10 Wh (160 Wh − 150 Wh). A battery level of a percentage lower than 50% occurred around 16h33; note that even though the geyser stopped drawing current from the battery, a DC to AC inverter is still on to power the instruments that continued to reduce the battery charge further.

Table 1. Sample values that highlight the geyser performance during the presence of both national grid energy and solar energy.

Time	Temperature	Battery (%)	Energy (Wh)	Status
07:59:41	27	71.69	0	High Power Heating
08:07:17	50	87.39	140	Low Power Heating
08:07:19	50	85.94	140	Idle
08:07:21	50	83.44	140	Low Power Heating
09:05:02	65	29.99	150	Low Power Heating
11:04:40	73	39.5	150	Battery Charging
11:12:08	70	65.11	150	Idle
11:12:09	70	69.51	150	Low Power Heating
11:40:06	73	46.57	150	Battery Charging
11:41:09	72	65.28	150	Idle
1143:52	73	42.88	150	Low Power Heating
11:45:43	72	65.11	150	Battery Charging
11:45:46	72	65.45	150	Idle
11:50:59	73	51.2	150	Low Power Heating
11:55:02	71	65.62	150	Battery Charging
12:09:26	72	52.45	150	Low Power Heating
12:11:33	72	65.45	150	Battery Charging
12:12:57	73	44.34	150	Low Power Heating
12:16:31	71	66.97	150	Battery Charging
12:26:57	72	46.21	150	Low Power Heating
12:28:14	72	65.28	150	Battery Charging
14:06:01	72	51.2	160	Low Power Heating
16:33:00	72	41.69	160	Battery Charging
18:48:00	50	39.5	160	Battery Charging
18:48:45	49	39.66	160	High Power Heating
18:48:55	49	39.5	170	High Power Heating
18:49:09	50	39.33	170	Battery Charging

5.1.2. Results Obtained When PV Energy Is Not Available

As illustrated in Table 2, the national grid power was connected to raise water temperature from 28 °C to 50 °C and maintain the water temperature of 50 °C for the period ranging from 09h13 until 14h06. Water temperature has been observed to drop by about 1 °C in every 20 min. This has caused the geyser to be switched on every time when the temperature drops to 49 °C. As a result, the energy consumption was raised regularly. As illustrated in Table 2, a total energy of 130 Wh (230 Wh − 100 Wh) was drawn from the national grid source in order to preserve water temperature at 50 °C for a period that is between 09h13 and 14h06. The test process has been terminated earlier to avoid excessive energy consumption.

Table 2. Sample values that highlight the geyser performance during the absence of solar energy.

Time	Water Temperature (°C)	Energy (Wh)
09:05:25	28	10
09:13:15	50	100
09:30:01	50	110
09:30:03	50	110
10:00:03	50	120
10:30:00	49	140
11:00:03	50	150
11:30:05	50	160
12:00:00	50	170
12:30:01	50	190
13:00:03	50	200
13:30:03	49	210
14:06:01	50	230

5.2. Load Controller Performance

Figure 10 shows a PV system consisting of a DC to AC inverter, real time solar power consumption monitoring unit, solar charge controller, energy logging device, Arduino-based load controller, relay board, two 3-W indicating lamps and a solar current/voltage sensor. Three sets of wiring cables were used, i.e., wires carrying large DC current from the battery and solar panels, CAT 5 network cable for the signals from sensors and AC output current cables.

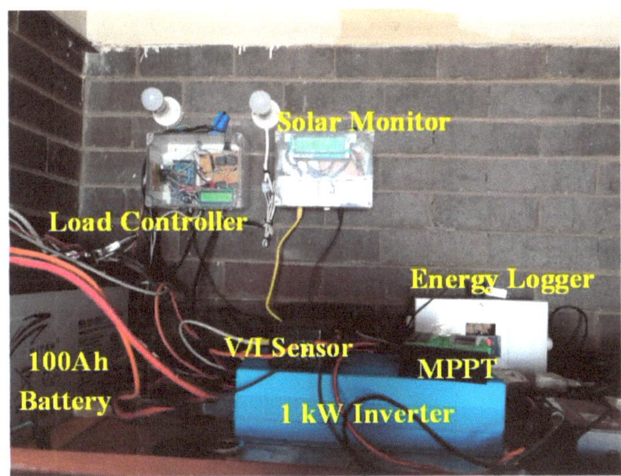

Figure 10. A hybrid renewable energy system.

5.2.1. Results Obtained When PV Energy Is Incorporated to Domestic Electricity Supply

As already described, the solar renewable energy system consists of: 2×100 W PVs mounted on the roof top, 100 Ah battery, 1 kW Inverter, 30 A power point tracker (MPPT), custom-built solar monitor, which shows real time power produced by solar panels, as well as the roof ambient temperature and custom-built energy logger. During a sunny day, the system illustrated on Figure 10 is seen to be active between 11h30 until 14h30. The input solar power was recorded up to the maximum of 200 W and the output AC power ranged between 150 W and 500 W. It should be remembered that the system never allows any power levels above 600 W, as previously discussed. If that happens, it connects the load to the national grid power. In Figure 11, we illustrate that most points of consumed electrical energy do not overlap with a virtual line called "reference line". A reference line represents the power that can be accumulated continuously without any form of disconnection. The actual energy curve goes below it. The space between the two lines therefore signifies the amount of energy savings.

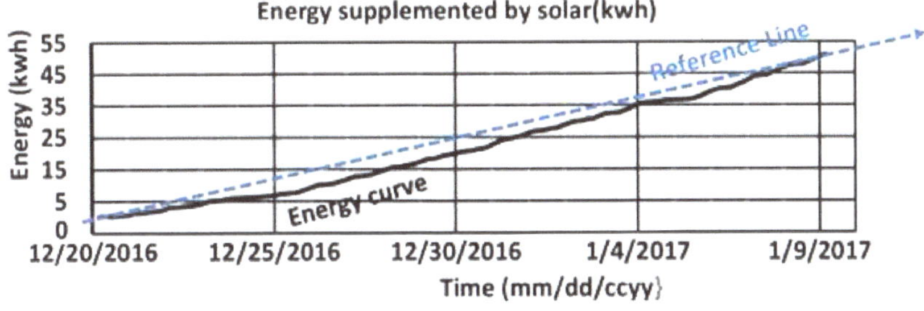

Figure 11. Accumulative energy that is consumed from the national grid source, when solar produced energy is used as supplement.

5.2.2. Results Obtained When Solar Is Not Included

Figure 12 illustrates that most points of energy consumed from domestic source overlap with a virtual line called the "reference line". This is because the grid power is always active.

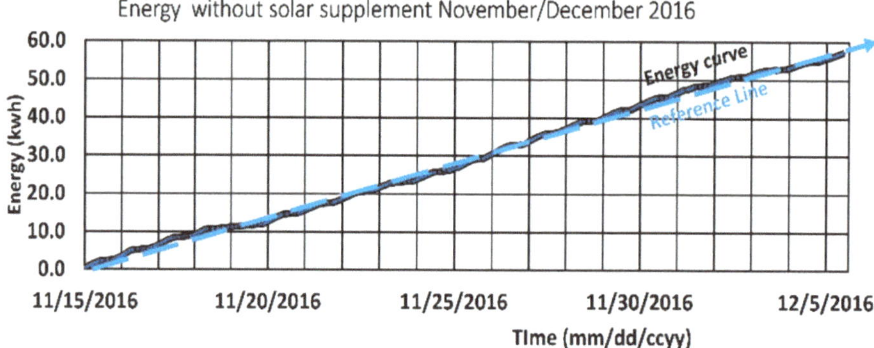

Figure 12. Accumulative energy consumed from commercial source that is not supplemented by solar energy.

6. Discussion

6.1. Geyser Control Circuitry

Please refer to Tables 1 and 2. The geyser control circuitry has managed to reduce the daylight geyser domestic energy consumption from 130 Wh to 10 Wh as calculated in Section 5.1.1. This is a good performance; however, it should be noted that the prototype geyser system used is not thermally protected has less water capacity (10 L). Also note here that the geyser energy consumption here, refers to the energy it consumes when the water is not being used. Using this control circuit with normal geysers will never affect the output that much, since both energy sources will be affected a similar manner, but the geyser control circuit will have to be upgraded to support high current levels and the renewable energy system capacity will have to be increased slightly. On a test that was conducted for two days, from 07h59 until 14h06, a geyser control system has showed to save power given by (1 − 10/130) × 100% = 92%—a percentage that can be improved by changing the capacity of the renewable system. Besides saving energy, the system also takes away the notion of switching off geysers when hot water from it is not scheduled for usage.

6.2. Load Control Circuitry

Please refer to Figures 11 and 12. Unlike values explained in Tables 1 and 2, where energy was presented in Wh over a time domain given as hh:mm:ss, this section presents energy in kWh over a time domain that is given in days. As a result, the performance of solar-based renewable energy system is expected to depreciate since it is not capable of giving high power for longer periods as some periods will produce insignificant sun light. Refer to the components that were listed under Figure 10. On a test conducted for 20 days from each power source, Figure 11 shows that when the renewable system is incorporated to the commercial energy supply, a total energy of 50.9 kWh has been consumed. On the contrary, Figure 12 shows that when renewable energy is not incorporated, the energy consumption has risen to 57.4 kWh. The total energy savings here can be calculated as follows: (57.4 − 50.9) × 100% ÷ 57.4 = 11.3%. The calculated percentage depends on the capacity of a PV-based renewable system.

7. Conclusions

In this paper, we focus on the strategies that can be used to incorporate solar-based renewable energy to the domestic supply in order to reduce energy consumption.

- We developed a prototype and tested the microcontroller-based electric water heater (geyser) that utilizes both PV and domestic energy supplies. A programmed microcontroller-based control circuit with a high-power and a low-power element or one current controlled element was connected to the geyser. That enabled a direct application of PV-derived AC voltage to geysers as a supplement of domestic energy supply, as a result, domestic energy consumption is reduced. A circuit to divert the low power load to a PV source was also included.
- Collectively, a microcontroller, current sensor, battery level sensor and relay board were used to construct a circuit to reduce electricity usage. The significance of the findings is that consumers can therefore only use the domestic electricity to supply high-power devices. This might only occur during peak consumption hours, when they do cooking, laundry, ironing, etc. However, things like low-power lighting and entertainment will automatically be diverted to PV energy sources. Geyser power is also aligned to the PV supply for connections during low-power consumption times. Consumers might not have to worry about switching off geysers when not using water from it.

As opposed to a conventional thermostat, a Dallas temperature supplies continuous temperature readings that enable the choice of geyser heating profile. In this way, both the upper and lower temperature limits are considered.

Author Contributions: Conceptualization, N.E.M.; Software: N.E.M.; Methodology, N.E.M. and M.K.J.; Validation, N.E.M.; Formal Analysis, N.E.M. and M.K.J.; Investigation, N.E.M. and M.K.J.; Resources, N.E.M. and M.K.J; Data Curation, N.E.M.; Writing-Original Draft Preparation, N.E.M. and M.K.J.; Writing-Review & Editing, N.E.M. and M.K.J.; Visualization, N.E.M.; Supervision, N.E.M. and M.K.J.; Project Administration, N.E.M. and M.K.J.; Funding acquisition, M.K.J.

Funding: We Acknowledge National Research Foundation of South Africa for their funding, which enabled the success of this research.

Conflicts of Interest: We declare that there is no conflict of interest with any other organization.

Appendix A

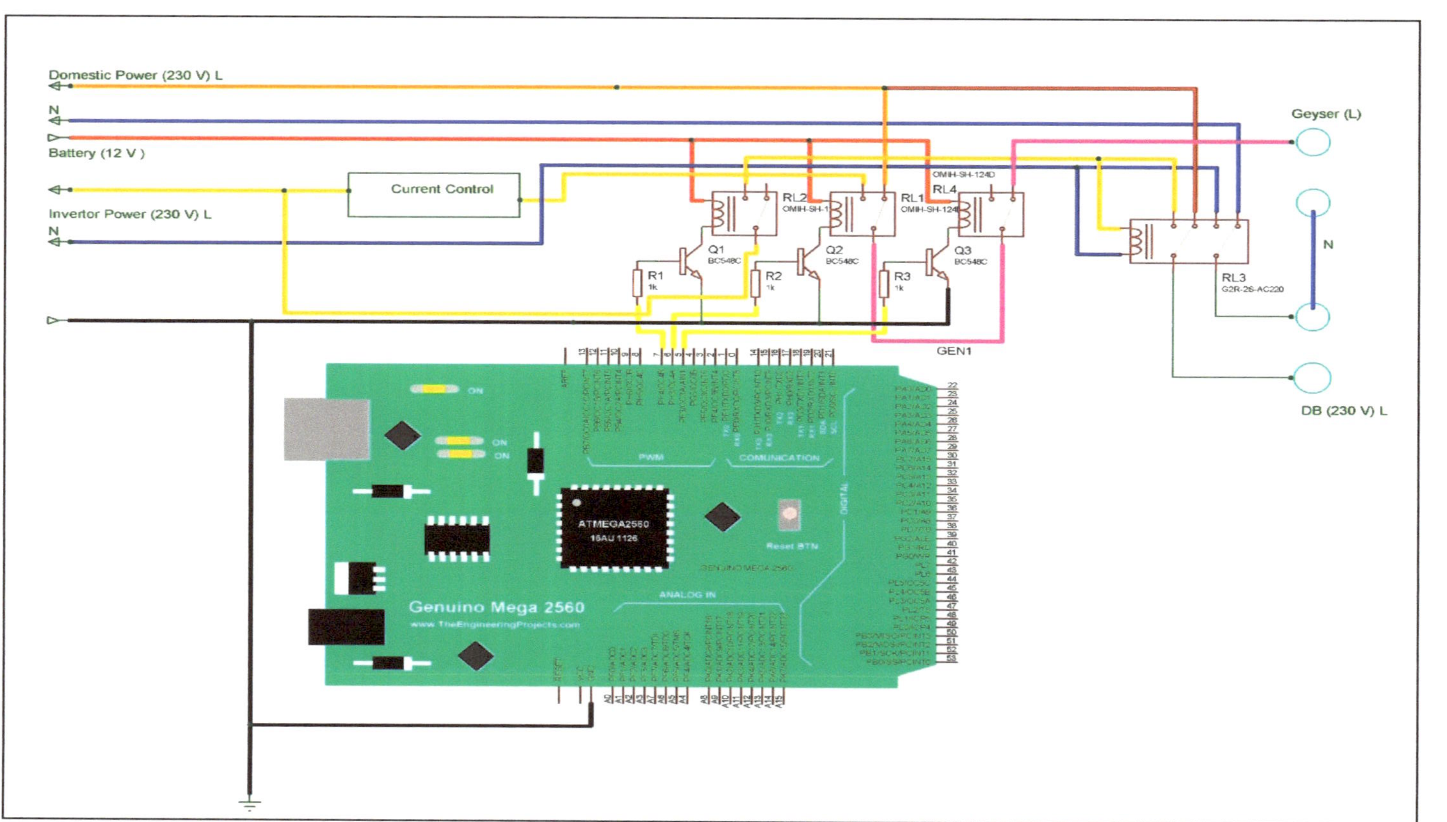

Figure A1. Schematic diagram of the complete controlling circuit.

Appendix B

```
/*
GeyserCTRL V1.00
The software can be used as a function or independently
Geyser control software code that works */

// Include the libraries we need
#include <OneWire.h>
#include <DallasTemperature.h>
#include <EmonLib.h>              // Include Emon Library
EnergyMonitor emon1;
boolean flag = FALSE;
float batt =100;
double energy = 0;
int Tempe, Temp;
int count = 100;
double Irms;
int power;

// Data wire is plugged into port 2 on the Arduino
#define ONE_WIRE_BUS 2

// Setup a oneWire instance to communicate with any
//OneWire devices(not just Maxim/Dallas temperature ICs)
OneWire oneWire(ONE_WIRE_BUS);

Pass our oneWire reference to Dallas Temperature.
DallasTemperature sensors(&oneWire);

/*The setup function. We only start the sensors here*/

void setup(void)
{
   pinMode(6,OUTPUT);
   pinMode(7,OUTPUT);
   // start serial port
   Serial.begin(9600);
   emon1.current(1, 111.1); // Current: input pin, calibration.
   // Start up the library
}

/*Main function, get and show the temperature */

void loop(void)
{
   sensors.requestTemperatures();//issue a global //temperature
      // request to all devices on the bus
      //Serial.print("Requesting temperatures...");
      // sensors.requestTemperatures(); //Send the command to
       //get temperatures
      // Serial.println("DONE");
      // After we got the temperatures, we can print them here.
      // We use the function ByIndex, and as an example
```

```
        batt = analogRead(A0) * 5.00 / 1023.00; //read battery
        batt ==(batt-2.90) * 0.016;


//get the temperature from the first sensor only.
        Tempe =sensors.getTempCByIndex(0);
        Temp = Tempe;

while (Tempe < 50)
   {
   batt = analogRead(A0) *5.00 /1023.00;//converts 5V Max
   batt =(batt-2.90) * 0.016
      sensors.requestTemperatures();
      Tempe =sensors.getTempCByIndex(0);
      Irms = emon1.calcIrms(1480);     // Calculate Irms only
      while(Irms*230 >600)        //Test for pwer > 600 W
           {
               Irms = emon1.calcIrms(1480); // Calculate Irms only
               Irms = abs((Irms - 0.0)*0.7491);
               Serial.print(energy);
               LCD();
               lcd.setCursor(12, 1);
               lcd.print("*HI*");
               Serial.println(",HIGH");
               digitalWrite(7, LOW);//switch on fast element
           }
Irms = abs((Irms -0.25)*0.7491);

energy = energy + (Irms*230)/3600000;
   digitalWrite(7, HIGH);//switch on fast element
   digitalWrite(6, LOW);// switch off slow element
   if (batt < 50) //stop and go to charging
   flag= false;
   delay (1000);
   }
while ((Temp >=50) && (flag==1)) //keep water warm
   {
      batt = analogRead(A0) * 5.00 / 1023.00;
      batt=(batt-2.90) * 0.016;
      sensors.requestTemperatures();
   Tempe =sensors.getTempCByIndex(0);
   Temp = Tempe;
   Irms = emon1.calcIrms(1480); // Calculate Irms only
   Irms = abs((Irms -0.25)*0.7491);
   energy = energy + (Irms*230)/3600000;
   delay (1000);
   digitalWrite(6, HIGH);//switch off fast element
   digitalWrite(7, LOW);// switch on slow element
   if ((Tempe > 72) || (batt <=50))
   flag= false; //cutoff at 72
   }
   while ( (Temp >= 50) && (batt < 100)&&(!flag) )
// charge battery
```

```
    {
      batt = analogRead(A0) * 5.00 / 1023.00;
      batt=(batt-2.90) * 0.016;
      sensors.requestTemperatures();
      Tempe =sensors.getTempCByIndex(0);
      Temp = Tempe;
      Irms = emon1.calcIrms(1480); // Calculate Irms only
      Irms = abs((Irms -0.25)*0.7491);
      energy = energy + (Irms*230)/3600000;
      delay (1000);
      digitalWrite(6, LOW); //switch off fast element
      digitalWrite(7, LOW);// switch off slow elements
      if (batt > 75) //stop charging
      flag= true;
    }
  }// end program
```

References

1. Espinar, B.; Aznarte, J.L.; Girard, R.; Moussa, A.M.; Kariniotakis, G. Photovoltaic Forecasting: A state of the art. In Proceedings of the 5th European PV-Hybrid and Mini-Grid Conference, Tarragona, Spain, 29–30 April 2010.
2. Badwawi, R.A.; Abusara, M.; Mallick, T. A Review of Hybrid Solar PV and Wind Energy System. *Smart Sci.* **2015**, *3*, 127–138. [CrossRef]
3. Wan, C.; Zhao, J.; Song, Y.; Xu, Z.; Lin, J.; Hu, Z. Photovoltaic and Solar Power Forecasting for Smart Grid Energy Management. *CSEE J. Power Energy Syst.* **2015**, *1*, 38–46. [CrossRef]
4. Davidson, C.; Margolis, R. *Selecting Solar: Insights into Residential Photovoltaic (PV) Quote Variation*; NREL, Denver West Parkway: Golden, CO, USA, 2015.
5. Cengiz, M.S.; Mamiş, M.S. Price-Efficiency Relationship for Photovoltaic Systems on a Global Basis. *Int. J. Photoenergy* **2015**, *2015*, 256101. [CrossRef]
6. Khatib, T.; Elmenreich, W. An Improved Method for Sizing Standalone Photovoltaic Systems Using Generalized Regression Neural Network. *Int. J. Photoenergy* **2014**, *2014*, 748142. [CrossRef]
7. Shuhrawardy, M.; Ahmmed, K.T. The feasibility study of a grid connected PV system to meet the power demand in Bangladesh—A case study. *Am. J. Energy Eng.* **2014**, *2*, 59–64. [CrossRef]
8. Šály, V.; Váry, M.; Packa, J.; Firický, E.; Perný, M.; Kubica, J. Performance and testing of a small roof photovoltaic system. *J. Electr. Eng.* **2014**, *65*, 15–19.
9. Shinde, S.; Savkare, S. Automatic Battery Charging in Solar Robotic Vehicle. *SSRG Int. J. Electron. Commun. Eng.* **2017**, *4*, 1–4.
10. Traboulsi, H. Optimization of Auxiliary Power Supply (APS) Systems with Photovoltaic Modules. *Int. J. Sci. Technol. Res.* **2012**, *1*, 71–75.
11. Sulaiman, O.; Saharuddin, A.H.; Nik, W.B.W.; Ahmad, M.F. Techno Economic Study of Potential Using Solar Energy as a Supporting Power Supply for Diesel Engine for Landing Craft. *Int. J. Bus. Soc. Sci.* **2011**, *2*, 113–119.
12. Gowtham, M.; Seenivasagan, V.; Manikandan, P.; Manikandan, P. Design and implementation of solar energy with grid interfacing. *Int. J. Sci. Eng. Technol. Res.* **2013**, *2*, 1526–1530.
13. Gonti, P.Y.; Prasad, S. Power Management Strategy in Hybrid PV-FC and Wind- Power Generation Systems by Using Multi Input Single-Control (MISC) Battery. *Int. J. Res. Comput. Commun. Technol.* **2014**, *3*, 1152–1157.
14. Rauf, S.; Khan, N. Application of DC-AC Hybrid Grid and Solar Photovoltaic Generation with Battery Storage Using Smart Grid. *Int. J. Photoenergy* **2017**, *2017*, 6736928. [CrossRef]
15. Matuska, T.; Sourek, B. Performance Analysis of Photovoltaic Water Heating System. *Int. J. Photoenergy* **2017**, *2017*, 7540250. [CrossRef]
16. Moradi, K.; Ebadian, M.A.; Cheng-Xian, L. A review of PV/T technologies: Effects of control parameters. *Int. J. Heat Mass Transf.* **2013**, *64*, 483–500. [CrossRef]

17. Huang, C.-Y.; Huang, C.-J. A study of photovoltaic thermal (PV/T) hybrid system with computer modeling. *Int. J. Smart Grid Clean Energy* **2014**, *3*, 75–79. [CrossRef]

18. Herrando, M.; Markides, C.N. Hybrid PV and solar-thermal systems for domestic heat and power provision in the UK: Techno-economic considerations. *Appl. Energy* **2015**, *161*, 512–532. [CrossRef]

19. Garrab, A.; Bouallegue, A.; Bouallegue, R. An Agent Based Fuzzy Control for Smart Home Energy Management in Smart Grid Environment. *Int. J. Renew. Energy Res.* **2017**, *7*, 559–612.

20. Al_Issa, H.A.; Thuneibat, S.; Abdesalam, M. Sensors Application Using PIC16F877A Microcontroller. *Am. J. Remote Sens.* **2016**, *4*, 13–16. [CrossRef]

21. Von-Ketelhodt, A.; Wöcke, A. The impact of electricity crises on the consumption. *JESA* **2008**, *19*, 1–12.

22. DuPlessis, W. Energy efficiency and the law: A multidisciplinary approach. *S. Afr. J. Sci.* **2015**, *111*, 1–8. [CrossRef]

23. Mzini, L.; Tshombe, L.-M. An assessment of electricity supply and demand at Emfuleni Local Municipality. *J. Energy South. Afr.* **2014**, *25*, 20–26. [CrossRef]

24. Wang, S.; Liu, J.; Chen, J.-J.; Liu, X. PowerSleep: A Smart Power-Saving Scheme with Sleep for Servers Under Response Time Constraint. *IEEE J. Emerg. Sel. Top. Circuits Syst.* **2011**, *1*, 289–298. [CrossRef]

25. Fuentes, M.; Vivar, M.; Burgos, J.M.; Aguilera, J.; Vacas, J.A. Design of an accurate, low-cost autonomous data logger for PV system monitoring using Arduino™ that complies with IEC standards. *Sol. Energy Mater. Sol. Cells* **2014**, *130*, 529–543. [CrossRef]

26. Zahurula, S.; Mariuna, N.; Grozescub, V.; Lutfia, M.; Hashima, H.; Amrana, M.; Izham. Development of a prototype for remote current measurements of PV panel using WSN. *Int. J. Smart Grid Clean Energy* **2013**, *3*, 241–246. [CrossRef]

27. Lawan, B.; Samaila, Y.A.; Tijjani, I. Automatic Load Sharing and Control System Using a Microcontroller. *Am. J. Mod. Energy* **2017**, *3*, 1–9. [CrossRef]

28. Oghenemine, D.H.; Ilogho, F.; Folorunso, O. Hybrid Power Control System. *IOSR J. Eng.* **2017**, *7*, 17. [CrossRef]

29. Zhao, C.; Dong, S.; Li, F.; Song, Y. Optimal home energy management system with mixed types of loads. *CSEE J. Power Energy Syst.* **2015**, *1*, 29–37. [CrossRef]

30. Sabry, A.H.; Zainal, W.Z.W.H.M.; Amran; Shafie, S.B. High efficiency integrated solar home automation. *ARPN J. Eng. Appl. Sci.* **2015**, *10*, 6424–6434.

31. Friansa, K.; Haq, I.N.; Santi, B.M.; Kurniadi, D.; Leksono, E.; Yuliarto, B. Development of Battery Monitoring System in Smart Microgrid Based on Internet of Things (IoT). *Procedia Eng.* **2017**, *170*, 482–487. [CrossRef]

32. Kalaiarasi, N.; Paramasivam, S.; Dash, S. Implementation of Switching Circuit between Grid and Photovoltaic system with fixed and Movable Tracking. *Int. J. ChemTech Res.* **2016**, *9*, 367–375.

33. Odigwe, I.A.; Ologun, O.O.; Olatokun, O.A.; Awelewa, A.A.; Agbetuyi, A.F.; Samuel, I.A. A microcontroller-based Active Solar Water Heating System for Domestic Applications. *Int. J. Renew. Energy Res.* **2013**, *3*, 838–845.

34. Ahmed, M.S.; Mohamed, A.; Homod, R.Z.; Shareef, H.; Khairuddin, K. Modeling of Electric Water Heater and Air Conditioner for Residential Demand Response Strategy. *Inter. J. Appl. Eng. Res.* **2016**, *11*, 9037–9046.

35. Yin, Z.; Che, Y.; Li, D.; Liu, H.; Yu, D. Optimal Scheduling Strategy for Domestic Electric Water Heaters Based on the Temperature State Priority List. *Energies* **2017**, *10*, 1425. [CrossRef]

36. Dallas Semiconductor, DS18B20-PAR Digital Thermometer Features 15 July 2017. Available online: http://datasheets.maximintegrated.com/en/ds/DS18B20-PAR.pdf (accessed on 15 July 2017).

37. Cotek, SK Series, Pure Sine Wave Inverter, User Manual. Available online: https://www.solacity.com/docs/Cotek/Cotek_SK_Series_Installation_Manual.pdf (accessed on 18 June 2017).

38. 14core, Wire Temperature Sensor. Available online: http://www.14core.com/wiring-the-ds18b20-1-wire-temperature-sensor/ (accessed on 18 June 2017).

39. Bar, D.; Mert, T.; Deniz, Y.; Canbolat, U. A Dimmer Circuit for Various Lighting Devices. In Proceedings of the 2013 8th International Conference on Electrical and Electronics Engineering (ELECO), Bursa, Turkey, 28–30 November 2013.

40. Mabunda, N.; Joseph, M. Embedded Data Acquisition Systems for tracking energy consumption from renewable sources. In Proceedings of the 2016 IEEE International Conference on Emerging Technologies and Innovative Business Practices for the Transformation of Societies (EmergiTech), Balaclava, Mauritius, 3–6 August 2016.

Article

Design of Adaptive Controller Exploiting Learning Concepts Applied to a BLDC-Based Drive System

Pierpaolo Dini and **Sergio Saponara** *

Department of Information Engineering, University of Pisa, 56122 Pisa, Italy; pierpaolo.dini@phd.unipi.it
* Correspondence: sergio.saponara@unipi.it

Received: 29 March 2020; Accepted: 13 May 2020; Published: 15 May 2020

Abstract: This work presents an innovative control architecture, which takes its ideas from the theory of adaptive control techniques and the theory of statistical learning at the same time. Taking inspiration from the architecture of a classical neural network with several hidden levels, the principle is to divide the architecture of the adaptive controller into three different levels. Each level implements an algorithm based on learning from data and therefore we can talk about learning concepts. Each level has a different task: the first to learn the required reference to the control loop; the second to learn the coefficients of the state representation of a model of the system to be controlled; and finally, the third to learn the coefficients of the state representation of the actual controller. The design of the control system is reported from both a rigorous and an operational point of view. As an application example, the proposed control technique is applied on a second-order non-linear system. We consider a servo-drive based on a brushless DC (BLDC) motor, whose dynamic model considers all the non-linear effects related to the electromechanical nature of the electric machine itself, and also an accurate model of the switching power converter. The reported example shows the capability of the control algorithm to ensure trajectory tracking while allowing for disturbance rejection with different disturbance signal amplitude. The implementation complexity analysis of the new controller is also proposed, showing its low overhead vs. basic control solutions.

Keywords: adaptive control techniques; statistical learning; electric-drive control; brushless DC (BLDC) motor

---

1. Introduction

In the field of industrial automation and vehicle electrification, which obviously includes both robotics and automotive applications, it is required that modern control systems are able to predict anomalous behavior and compensate for it as much as possible, to maintain a certain desired behavior by the process itself. Anomalous behaviors include all those behaviors due to the introduction in the control loop of variations of the plant itself, such as sensors and actuators failures (which from a mathematical point of view are equivalent to a change in the model of the dynamic system itself) or degradation of the components, which then translates into parametric variations when thinking about the dynamic model of the process to be controlled. Additionally, it can include all those uncontrolled exogenous actions that are in fact classified as external disturbances, which can affect both actuators and sensors. Think, for example, of the trajectory control of the end-effector of a robotic manipulator subject to involuntary interactions with the external environment. In fact, this translates into a non-deterministic change in the mechanical load on the actuators, which leads to an anomalous behavior of the electric motor supply currents. These effects can be modelled, with some effort on the part of the designer, in order to take them into account when planning the robot's trajectory. If the trajectory of the end-effector is subject to external actions falling within a predetermined range of external disturbances, then it could be partly compensated for. This is the robust control approach [1,2], where the system under control can be modeled in a linearized way, and limited noise inputs at certain points of the control loop are expected. If the disturbances applied to the system are within the range of permitted variance, then a certain trade-off in terms of required performance and stability can be guaranteed. The same can be done for any parametric variations, which, if confined within a certain confidence zone, can be partly absorbed by the control algorithm. The greater limitation of the robust control approach is the dependence on the process model for the description at nominal level, which in fact disappears once the parametric variations leave the confidence bands. There are many robust optimization techniques, such as the optimal control H-2, H-infinite and mu-synthesis [3–6], which, however, have the great limit of strong dependence on the deterministic model and the limited range of uncertainties that can be compensated, even if, in spite of this, these are adopted in many fields, such as aeronautics or mobile robotics in a non-anthropic environment.

The evolution of robust control in this sense is the adaptive control [7–14], which basically involves linearized process modeling at many operational points of application interest. On each sub-model of the process, a controller is designed with simple control techniques that therefore apply locally. If, then, the system to be controlled is simple enough and does not have to work in too different operating conditions, it is preferred to do the so called gain scheduling [15], in which a unique structure of the controller is provided, but in which appropriate parameters can be chosen according to the operating condition itself. When the operating condition of the process changes, then the control system switches from one controller model to another, to ensure performance and stability. The change from an operating condition can occur either because a different performance is required, which in fact provides for a change of references, or because there has been an external action that is interpreted as a different working point of the system to be controlled. The limit of the procedure is precisely that of being able to incorporate the effect of external disturbances and/or parametric variations within the sub-models. Therefore, although we can theoretically cover a wider range of uncertainties, including unforeseen external disturbances and parametric variations, there is clearly a dependence on the system models that describe it under various operating conditions.

Summing up, for the comparison with the technique we propose, we only consider adaptive feedback control techniques, which are robust to parametric variations and to any kind of model uncertainty. These techniques can be schematized in two macro categories [16,17], which will be better explained in the state of art section: Gain Scheduling and "Architecture" Scheduling.

The first category fixes a control system structure, in which some parameters are modified, based on the measurements of reference signals, control signal and process outputs.

The second category considers a set of possible control architectures, each of which is activated based on a decision taken after the measurements of the reference, control and output signals. The big disadvantage of the second category is highlighted if you think about an embedded platform implementation based on microcontroller, where you have undoubted limits of memory resources. This is because, in order to have available all the architectures necessary to control the process, when operating conditions change, all the necessary structures must be stored in memory.

Moreover, still talking about computational disadvantages, to carry out the change of controller (be it only its coefficients or its entire architecture), it is necessary for the presence of a "Decision Maker" system, which to all intents and purposes, must perform an inferential statistical analysis based on the reference measurements, control action and process output, and identify the adaptation action of the controller.

This means that it must implement a real operating conditions classifier: after a certain number of measurements collected, it must verify the hypotheses of belonging to confidence intervals and choose a class that will coincide with one of the expected operating conditions. Moreover, this approach requires a preliminary analysis to set decision thresholds, i.e., a priori knowledge. Therefore, an important objective is also to reduce the computational complexity due to the presence of a functional block in the control structure that must process the measures to make a decision.

To overcome the above limits, this work presents a methodology that takes advantage of an innovative control structure based on simple iterative tuning rules of the parameters of the control structure itself. The proposed control methodology is inspired by the classic neural network structure (in particular of a Multi-Layer Perceptron), in which there is a hierarchy among the subsystems/layers that constitute it. By exploiting simple concepts linked to the application of learning algorithms, the aim of this work is to contribute in terms of adaptive control techniques to reduce the limits linked to the dependence of a priori knowledge on the system to be controlled.

To be noted is that the proposed technique does not consider artificial intelligence techniques since, as anticipated, the objective is to limit the computational complexity, both in terms of methods for the acquisition of knowledge a priori (offline) and learning techniques in service (online).

In fact, artificial intelligence techniques based on neural networks are not very suitable for control applications, since once the network training is done in the preliminary phase (off-line training) on the basis of "enough" data previously accumulated, they are in fact open loop algebraic systems, not very robust to parametric variations. Much more advanced techniques of reinforced learning [18] instead lend themselves to control applications. However, they have the big disadvantage of requiring a high dimension in terms of the number of recursive equations necessary to learn the behaviour of the system every time it is solicited in a new way.

The article is divided as follows: in Section 2 is reported the state of the art on robust and adaptive control techniques applied in the field of industrial and vehicle automation in general; in Section 3 is described the proposed methodology through the explanation of some essential mathematical steps and the explanation of the complete new control architecture; in Section 4 is explained how to apply the proposed technique, to a system of reduced order and SISO (single input single output), in order to report results of simple interpretation by the reader. As an application case study, the control of a servo-drive based on a brushless DC (BLDC) motor is presented, whose dynamic model considers all the non-linear effects related to the electromechanical nature of the electric machine itself. Finally, conclusions are drawn in Section 5.

2. State of the Art on Robust and Adaptive Control

A summary of the state of the art in adaptive control techniques is reported to create the context for the method proposed in the following sections. The basic concepts and methods currently in use for the adaptive control law project are described below. The need to design adaptive control algorithms stems from the fact that the designer is not always able to completely model the dynamics of the system to be controlled, and in any case, a very high effort may be required, both in terms of costs of the characterization procedures of a process, and in terms of time spent [1]. Furthermore, also in the case of relatively simple models, it is difficult to prevent a certain type of disturbances and where those signals are inserted in the system architecture (affecting control signal or measured quantities, or whatever). It is convenient, therefore, to consider the possibility to modify with certain update criteria some parts or some parameters of the control architecture, in order to create a real adaptation to those not modelled or unexpected events of the physical process to be controlled. Basically, it is possible to exploit two different approaches, the robust one and the adaptive one.

With the application of the robust control theory [2], the goal is to design an optimal control law vector, that, by satisfying a certain optimal criterion, is also able to guarantee the nominal performance request within bounded parametric variation and uncertainly (including external disturbances). A typical approach is to exploit a nominal model of the plant [3], related to a specific operating condition, and apply both optimal and robust criteria, which include models of uncertain conditions. If the plant will work in different operating conditions, it is also possible to directly exploit a non-linear dynamic model [4]. It is important to note that the solution derived by the application of robust control techniques, for example as in [5,6], have often a greater dimension with respect to the controlled plant. This is a disadvantage from an engineering point of view, because it means that to control a certain dynamic system, a more complex control system is needed. Anyway, the robust control approach in general is a power methodology in the case of parametric variation of the dynamic system. Under this point of view, if the goal is designing a control system that provides robustness stability and performance, including tolerance to some type of fault/disturbance, maintaining a limited size of the control system itself, then it is suggested to use the advanced control technique.

There are many adaptive control architectures, so in the following, we briefly report the various choices and its application, in order to make the reader, in condition, to understand that our proposal in Sections 3 and 4 introduces new concepts vs. the state of art.

A type of adaptive control architecture is the MRAS (model reference adaptive system), which can be differentiate between direct and indirect [7]. Based on a reference model of the control loop, it is computed the output assumed as reference output signal for the closed loop. Based on the correction on the error between reference output and actual output, the nominal controller gains are adjusted. The difference between direct and indirect is described in the following. In the direct MRAS technique, the controller is based on the reference model, meanwhile in indirect MRAS technique, the controller is based on the identification process which is done on-line. Furthermore, the direct MRAS technique exploits the error between the references model output and the actual one, while in indirect MRAS technique, the error between actual and estimated output signal is used for the updating of the controller gains. As explained in Fereka et al. [8], the MRAS method is suggested in case of relatively slow behavior of the system itself, and furthermore, it does not guarantee from a formal point of view the asymptotic stability of the closed loop. The modern application of the MRAS control paradigm is associated with the power drive control systems, such as three-phase motor control in sensor-less conditions. Examples of this kind of applications can be found in [8–10].

It is therefore well known that, although MRAS type techniques give satisfactory results and can be applied in areas of industrial interest, such as electric power drives, these techniques are more suitable when low dynamic performance is required and therefore suitable, for example, for electric drives based on asynchronous or reluctance motors (applications with medium-high loads, but at near constant speeds).

Our proposal is more suitable when higher dynamic performance is required, because the adaptation of the controller parameters is done instant by instant, in fact, unlike MRAS adaptive algorithms, we use differential equations and not recursive equations to make the algorithm as reactive as possible.

Other types of adaptive architectures are called deterministic and stochastic adaptive control. The first one is based on neglecting the contribution of external disturbance and measuring errors on the system response. Instead, stochastic adaptive control is based on the statistical interpretation of external disturbance and errors on the system response included in the closed loop system model. As an example of the application of this paradigm, Tian et al. [11] present an interesting case study in which they consider an analysis on the quantization procedure effects.

The disadvantage of statistical control, however, is the limited speed in the adaptation time of the closed-loop system, because, based on statistics, accumulating a certain amount of information before updating the parameters of distribution models describing external disturbances is required, through statistical inference and regression techniques. In fact, it is not possible to obtain good adaptation results when signals change abruptly and the number of accumulated data used to update the controller model again is not large enough. As mentioned, and as will be explained in detail in the method description section, the proposed control technique is able to adapt quickly to sudden changes in the reference signal without having to make an inferential analysis, taking advantage of statistical learning theory and the instantaneous gradient algorithm.

Another type of adaptive paradigm is the MMAC (multi-model adaptive control) [12]. Basically, in MMAC some operating conditions of interest concerning plant operation are foreseen. In MMAC, a linearized model of the dynamic system is computed with respect to each operating condition and, based on its a controller, is designed, for example, exploiting linear control theory technique. A supervisor system takes in input the control signal and the measured output signal in order to decide which operating condition occurs, with the goal to select the opportune controller. Some examples of modern application of this control concept can be founded in Zengin et al. [13], where the authors propose an application to the control of the vehicle lateral dynamic model, and in Outeiro et al. [14], where the authors present the application of the control strategy on a quad-rotor flying trajectory tracking.

Here, the disadvantage is the one briefly mentioned in the introduction, so this type of technique pays a computational weight, both for the actual complexity of the block operations that decides which control architecture to activate depending on the operating condition, but also from a waste of resources point of view when we start talking about implementation on low-cost embedded platforms, which have limited memory resources.

With respect to this problem, the advantage of the proposed method is that complex structures are not allocated in memory, but simply the number of variables needed to define the representation in the state space for the reference model, process and controller.

Gain scheduling is an empirical solution to make an adaptive controller, used in aeronautical and then automotive applications [15]. Thanks to theoretical arguments, it is possible to synthesize adaptive control algorithms that provide greater robustness and better performance. The idea behind gain scheduling is to design the controller for different points of operation of the system to be controlled; the different configurations, being the result of an approximation, can ensure compliance with specifications only locally at the point of work. Therefore, the parameters obtained in the different configurations are interpolated, making them variable with the operating point.

Gain scheduling techniques can be a solution in case the system you want to control has to work in a pre-established operating condition, and it is expected that it may be subject to limited parametric variations or variations in the input signal waveforms due, for example, to non-ideal effects such as saturation.

The proposed technique, instead, does not present a limit in the choice of the operating condition, that is, it can change during the operation of the process itself and, at least theoretically, there are no limitations to parametric variations, if not those that would lead to the breakage of the components of the process itself.

It is possible to demonstrate stability only for LTV (linear time variant) systems and under particularly stringent conditions; for this reason, the controller is subjected to numerous experimental validation tests. Examples of application of this paradigm can be founded in Hakim et al. [19], where the authors present an interesting application of a fuzzy PID (proportional-integral derivative) gain scheduling to an inverted pendulum model. In Poksawa et al. [20], the authors propose a gain scheduled PID control system for fixed-wing UAVs where a family of PID cascade control systems is designed for several operating conditions of airspeed. Other adaptive control techniques are the auto-tuning [21,22] or self-tuning, model-free [23,24], neuro-control [25] and fuzzy logic [26,27] and the iterative learning control (ILC) [28,29].

The proposed control technique could be classified as model-free, in which a multi-target optimization problem is iteratively solved.

3. Adaptive Controller Exploiting Learning Concepts

In this Section, we present the architecture of the proposed control algorithm, which exploits both adaptive learning concept and architecture. The learning concepts, such as the gradient descent algorithm [30] in the instantaneous version of it, are used to set adaptive rules for the coefficients updating. Furthermore, we inspired the proposed control architecture to the classical NN (neural network) constitutive architecture [31], dividing into different layers each one that has a task in terms of learning, by the data elaborated from the previous layer.

The state space representation is useful to write an approximated (but in closed form) solution for what concerns the output function. Through the expression of the output signal, meaning as the output of the controlled plant, it is possible to build the operating procedure based on simple learning concepts, to derive an adaptive control algorithm. As we explain in the rest of this paper, the procedure is clearly easier to set in the case of linear model of the plant, but is not limited to this case.

Basically, the control system it is able to compute every time a new linear model to approximate the plant, finding new parameters both for plant, control and reference signal models. In this way, the controller is robust (clearly within certain limits), both to the application of external disturbance signals and to parametric uncertainties. We present an innovative architecture, based on the usage of the simple learning concept, such as the online gradient descent algorithm to adapt the various parameters of the entire control system, repeated on different levels.

Our control system architecture is composed of three different subsystems, each one with a different functionality. As we explain better through mathematical formulation, a level for the reference signals approximation is required for our solution, as well as a level for the assessment of the plant model, and a final level for the control parameters adaptation. Schematically, the starting point is represented in Figure 1, where a simple control loop is reported, valid also for MIMO and the general non-linear system.

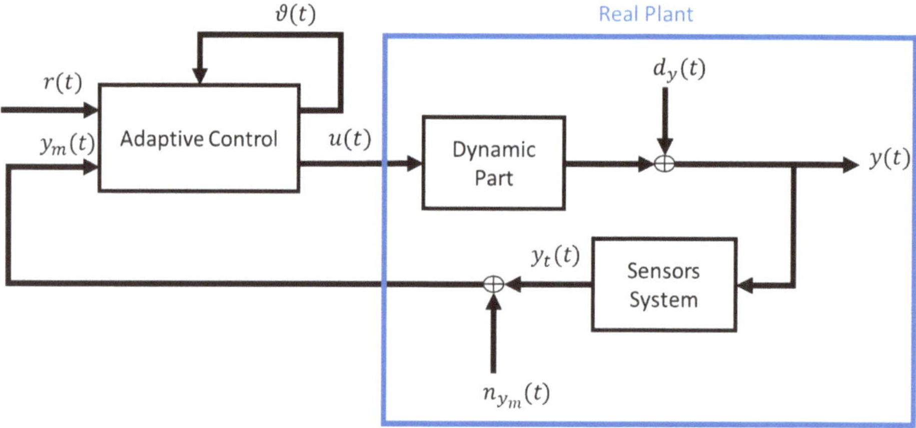

Figure 1. Schematic representation of the high-level Architecture of the proposed method.

In Figure 1, the high level architecture is reported, where we take into account the adaptive control, which takes, as input signals, the measured output vector from the physical process $y_m(t)$, the reference output vector $r(t)$ and the actual value of the internal parameters vector $[\theta_i(t)]_r$, $[\theta_i(t)]_p$ and $[\theta_i(t)]_c$. By real plant, we mean the union of the dynamic part, which is the part of the system usually modelled through a differential equations system, the dynamic of the sensors system including the effect of external disturbance $d_y(t)$ vector and measured noise $n_y(t)$.

We want to highlight below the difference with the adaptive control techniques that are used until now, through an explicit schematization of the criterion shown in the figures below. As anticipated, we can enclose the adaptive control techniques in two macro categories: the first where only some parameters are modified, and the second where a change from one control structure to another is expected; in both cases, the change occurs downstream of a decision algorithm, which will perform inferential operations based on a certain amount of collected data.

For simplicity, in Figure 2, the case of an adaptive PID controller according to the gain scheduling paradigm is represented. As can be seen, the parameters are modified according to the signal coming from the "Adaptation Algorithm" block, which has both the process output signal and the control action itself as inputs. Apart from the computational problems mentioned above, linked to the operations required for the inference part, it is noted that, for the calculation of the operating condition, it is necessary to measure the control action.

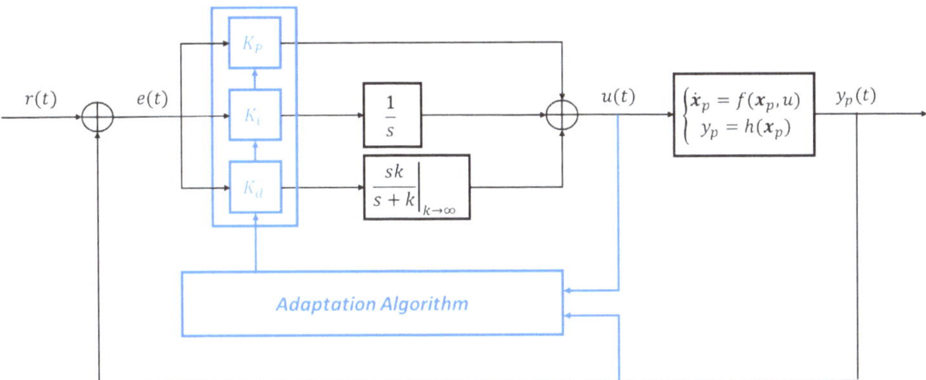

Figure 2. A single input single output (SISO) proportional-integral derivative (PID) controller with Gain Scheduling Architecture.

In the case of electrical drives in general, the presence of voltage sensors as well known is something we try to avoid in the design phase, because, in order to appreciate the effect of modulation and the presence of the inverter, the sensors must have a high bandwidth, which makes them expensive.

The same considerations are even more valid in the case of using adaptive control strategy based on architecture scheduling, as shown in Figure 3, where the computational cost is further increased, as it will be necessary to memorize all controllers' rules on which it is expected to switch according to the estimated operating condition.

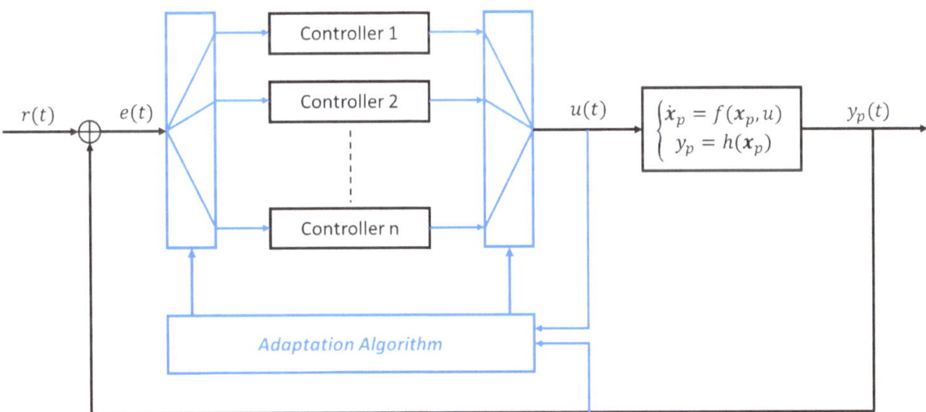

Figure 3. A SISO Architecture Scheduling controller schematization.

Another advantage of the proposed control architecture, as shown in Figure 1, is that it does not require the measurement of the input signals to the system to be controlled.

The design of the control system (adaptive or not) is based on the modelling of the dynamic part of the plant, without a care about the sensors dynamic. Clearly, this is a valid assumption if the dynamic of the sensors is higher than the plant one. This assumption is usually valid, since in the preliminary sizing phase of the global system, the designers select electronic components (sensors, controllers, actuators) that have a higher dynamic than the plant under control.

As anticipated, in this work we consider the state form representation both for what concerns the plant system and the controller system. We assume a continuous time domain representation for the design procedure, also with respect of the algorithms that regard learning concepts.:

$$
\begin{aligned}
P &: \begin{cases} \dot{x}_p(t) = A_p x_p(t) + B_p u(t) \\ y(t) = C_p x_p(t) \end{cases} \\
C &: \begin{cases} \dot{x}_c(t) = A_c x_c(t) + B_c e(t) \\ u(t) = C_c x_c(t) \end{cases}
\end{aligned}
\tag{1}
$$

In Equation (1), the state form representation of the plant is reported, in the system of equations P, as well as the controller, in the system of equations C. In Equation (1), $x_p(t)$ and $x_c(t)$ are the relative state vectors; A_p and A_c are the transition state matrix; B_p and B_c are the state-input matrix, which map the contribute of input vector signals in the state vector dynamic; and C_p and C_c are the output-state vectors, which map the contribute of the state vector on the output of the system itself.

The term in input to the controller is the trajectory error, that can be defined as $e(t) = y(t) - r(t)$, where $r(t)$ represents the reference for the output of the plant.

As explained in the following, we refer to linear dynamic state form representation, but the method can be applied also to a non-linear dynamic system. This is true because creating adaptive rules for the model parameters, basically the elements of all the matrix of the state form, the control algorithm will be able to compute a linear local approximation valid in a certain moment.

We assume that the size of the state space related to the plant must be greater than, or at least equal to, the dimension of the state space of the control model, $size(x_p) \geq size(x_c)$. This assumption is also reasonable from an engineering point of view, related to the fact that it is not acceptable to control a dynamic system with another dynamic system that is more complicated in terms of realization.

Imposing that the output of the plant is the input of the control system together with the reference signal, and that the input of the plant is the output of the controller, it is possible to write the augmented system, as in the following equations.

$$
S : \begin{cases} \dot{\tilde{x}}(t) = \tilde{A}\tilde{x}(t) + \tilde{B}r(t) = \begin{pmatrix} A_p & B_p C_c \\ -B_c C_p & A_c \end{pmatrix} \begin{pmatrix} x_p(t) \\ x_c(t) \end{pmatrix} + \begin{pmatrix} 0 \\ B_c \end{pmatrix} r(t) \\[2mm]
y(t) = \tilde{C}\tilde{x}(t) = \begin{pmatrix} C_p & 0 \end{pmatrix} \begin{pmatrix} x_p(t) \\ x_c(t) \end{pmatrix} \end{cases}
\tag{2}
$$

In Equation (2) is reported the augmented state form derived by the connection of the plant model and the control model. The cardinality of the new state space is clearly the sum of the two disjointed state spaces: $size(\tilde{x}) = size(x_p) + size(x_c)$. Calling $N_{x_p} = size(x_p)$, $N_{x_c} = size(x_c)$, $N_u = size(u)$, $N_y = size(y)$, in the above equations we have $A_p \in R^{N_{x_p} * N_{x_p}}$; $B_p \in R^{N_{x_p} * N_u}$; $C_p \in R^{N_y * N_{x_p}}$; $A_c \in R^{N_{x_c} * N_{x_c}}$; $B_c \in R^{N_{x_c} * N_y}$; and $C_c \in R^{N_y * N_{x_c}}$.

Another reasonable assumption is the presence of enough control variables to control all the output vector components, or in other words that $N_u \geq N_y$; and this is an assumption for all the following formal considerations.

For our control technique, we need a method to make it possible to write the explicit solution of the augmented state form independently to the reference signal r. In this work, we refer to a polynomial approximation both for the reference signal and the exponential matrix needed to write the explicit solution of the state form representation. In particular, we consider a second order approximation.

$$
\begin{aligned}
r(t) &\cong r_0 + r_1 t + r_2 t^2 \\
e^{At} &\cong I + At + A^2 \frac{t^2}{2}
\end{aligned}
\tag{3}
$$

In Equation (3), the two truncated series of the reference signal and exponential matrix are reported, where t represents the time variable. The proposed control technique provides a continuous time domain implementation, represented by a linear dynamic system depending on the update of the parameters with respect to the time. From an engineering point of view, this is not a big limitation, because, also in the low-cost embedded platform, there is the availability of adequate clock speed to run the algorithm. In the equation above, with r_0, r_1 and r_2, the coefficients of the polynomial approximated form, which in general can be functions of time, are indicated.

$$\widetilde{x} = e^{\widetilde{A}t}\widetilde{x}_0 + \int_0^t e^{\widetilde{A}(t-\alpha)}\widetilde{B}r(\alpha)d\alpha \cong \left(I + \widetilde{A}t + \widetilde{A}^2\tfrac{t^2}{2}\right)\widetilde{x}_0 + \int_0^t\left(I + \widetilde{A}(t-\alpha) + \widetilde{A}^2\tfrac{(t-\alpha)^2}{2}\right)\widetilde{B}\left(r_0 + r_1\alpha + r_2\alpha^2\right)d\alpha =$$

$$= \int_0^t \widetilde{B}\left(r_0 + r_1\alpha + r_2\alpha^2\right)d\alpha + \int_0^t \widetilde{AB}(t-\alpha)\left(r_0 + r_1\alpha + r_2\alpha^2\right)d\alpha + \int_0^t \widetilde{A}\tfrac{(t-\alpha)^2}{2}\widetilde{B}\left(r_0 + r_1\alpha + r_2\alpha^2\right)d\alpha = \quad (4)$$

$$= \widetilde{B}\left(r_0 t + r_1\tfrac{t^2}{2} + r_2\tfrac{t^3}{6}\right) + \widetilde{AB}\left(r_0\tfrac{t^2}{2} + r_1\tfrac{t^3}{6} + r_2\tfrac{t^4}{24}\right) - \widetilde{A}^2\widetilde{B}\left(r_0\tfrac{t^3}{12} + r_1\tfrac{t^4}{12} + r_2\tfrac{t^5}{30}\right) =$$

$$= \widetilde{B}r_0 t + \left(\widetilde{B}r_1 + \widetilde{AB}r_0\right)\tfrac{t^2}{2} + \left(2\widetilde{B}r_2 + 2\widetilde{AB}r_1 - \widetilde{A}^2\widetilde{B}r_0\right)\tfrac{t^3}{12} + \left(2\widetilde{AB}r_2 - \widetilde{A}^2\widetilde{B}r_1\right)\tfrac{t^4}{12} - \widetilde{A}^2\widetilde{B}r_2\tfrac{t^5}{20}$$

The above chain of equality in Equation (4) allows one to write a model for the output vector signals with respect to the plant, inserted in a control loop architecture. The output of the global feedback system can be founded with the algebraic relationship, with the state space vector of the augmented dynamic model in Equation (5).

$$y = \widetilde{C}\widetilde{x} = \begin{pmatrix} C_p & 0 \end{pmatrix}\begin{pmatrix} x_p(t) \\ x_c(t) \end{pmatrix} = C_p x_p(t). \quad (5)$$

Clearly, the goal of the proposed method is to find the value of the components of the dynamic matrix of the control model in state form. Furthermore, it is possible to use the $y(t)$ expression to compute online the value of the state space state representation of the plant itself if we consider the parameters of the control model and the reference signal polynomial approximation coefficients to be known. In this mode, we can rewrite the output model $y(t) = \widetilde{C}\widetilde{x}(t)$ as a function of all the needed parameters.

For convenience, in the next sections we also use the following notation:

- $[\theta_i(t)]_r$: the ith parameters of the reference vector approximation;
- $[\theta_i(t)]_c$: is the ith parameters of the state space representation related with the controller model;
- $[\theta_i(t)]_p$: is the ith parameters of the state space representation of the plant dynamic model;

In the following, we show the architecture of the control system, describing in detail the fundamental operation inside every single functional block.

As schematically represented in Figure 4, the control system is divided in three subsystems: first the subsystem dedicated to the approximation of the reference signals vector (we choose the polynomial form, but it is not mandatory); second, the subsystem that provides an estimation of the state space representation of the plant (it takes, as input, the parameters of the reference approximation and the parameters from the last subsystem); the third subsystem takes as input the result of the previous estimated parameters and provides an estimation of the control dynamic system with which build the control vector $u(t)$, with respect of the Equation (1).

In the following, we describe more in detail every single subsystem and the relative formalism needed to define the adaptive rules, exploiting simple machine learning concepts.

3.1. Learning Desired Signal

In the functional block that provides an estimation of the second order polynomial approximation, we exploit the instantaneous version of gradient descent algorithm in the following way. Fixing an objective function to minimize (at every time instant) the differential equation that makes possible the update of the coefficients is reported in Equation (6). In Equation (6) with $J_r(t)$, we indicate the objective function to be minimized at every time instant. In Equation (6), α_r is the learning rate, which for simplicity, is equal for all the parameters ($r_0(t), r_1(t)$ and $r_2(t)$ in Equation (6)), and clearly must be a negative real part value for stability condition. Equation (6) can be summarized in the "compressed" form of Equation (7). A schematic representation of the implementation reported in Equation (6) is showed in Figure 5.

$$Layer_1 : \begin{cases} J_r = (r(t) - \hat{r}(t))^2 = \left(r(t) - \left(r_0(t) + r_1(t)t + r_2(t)t^2\right)\right)^2 \\ \frac{dr_0(t)}{dt} = -\alpha_r \frac{\partial J_r}{\partial r_0}(t) = -2\alpha_r\left(r_0(t) + r_1(t)t + r_2(t)t^2\right) \\ \frac{dr_1(t)}{dt} = -\alpha_r \frac{\partial J_r}{\partial r_1}(t) = -2\alpha_r\left(r_0(t) + r_1(t)t + r_2(t)t^2\right)t \\ \frac{dr_2(t)}{dt} = -\alpha_r \frac{\partial J_r}{\partial r_2}(t) = -2\alpha_r\left(r_0(t) + r_1(t)t + r_2(t)t^2\right)t^2 \end{cases} \tag{6}$$

$$\frac{d[\theta_i(t)]_r}{dt} = -\alpha_c \frac{\partial(r(t) - \hat{r}([\theta_i(t)]_r))^2}{\partial[\theta_i(t)]_r}, \forall i \tag{7}$$

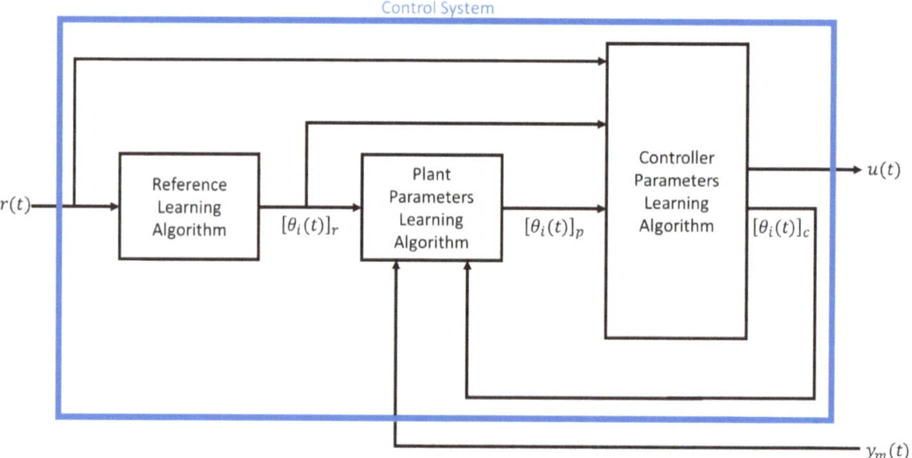

Figure 4. Schematic representation of the internal architecture of proposed control system.

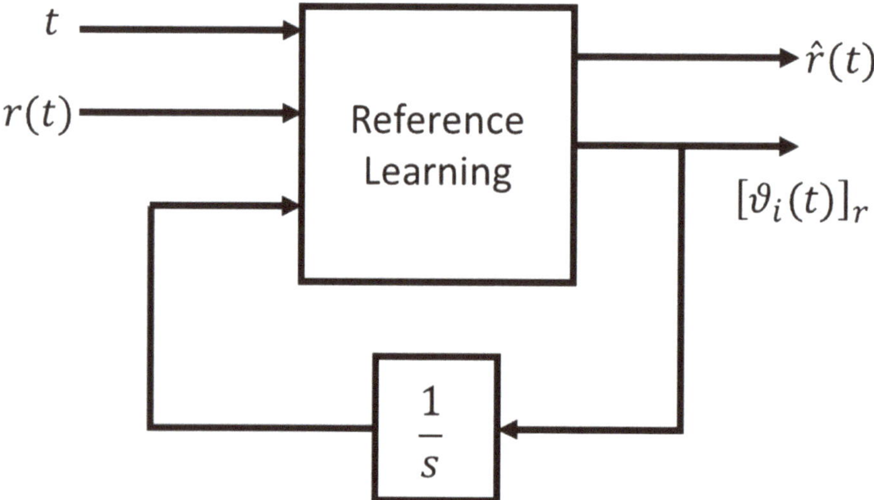

Figure 5. Schematic representation of the implementation reported in Equation (6).

3.2. Learning Plant Model

In this section, we describe the subsystem in which the instantaneous state form representation of the plant is estimated, which can also be interpreted as the local linear approximation valid in a certain instant of time for a non-linear dynamic plant system. As shown in Figure 6, the subsystem takes, as input, the result provided by the other subsystem and the measured plant output. In this part of the control system, it is assumed that the parameters learned by the other subsystem are like constant values, which are basically all the coefficients of the matrix A_c, B_c and C_c and the polynomial coefficients of the desired signal (or more in general, a vector of desired signals) estimation $[\theta_i(t)]_r$; meanwhile, the coefficients of the matrix A_p, B_p and C_p are the variable of the current block on-line learning phase. Basically, in this part of the algorithm, $\hat{y} = \hat{y}(A_p, B_p, C_p, t) = \hat{y}([\theta_i(t)]_p \ \forall i)$, where $\hat{y}(t)$ is the model of the output, which, in this block, is implemented with the following algorithm:

$$
\begin{cases}
J_p = \left(y(t) - \widetilde{C}\left(\widetilde{Br}_0 t + \left(\widetilde{Br}_1 + \widetilde{ABr}_0 \right)\frac{t^2}{2} + \left(2\widetilde{Br}_2 + 2\widetilde{ABr}_1 - \widetilde{A}^2\widetilde{Br}_0 \right)\frac{t^3}{12} + \left(2\widetilde{ABr}_2 - \widetilde{A}^2\widetilde{Br}_1 \right)\frac{t^4}{12} - \widetilde{A}^2\widetilde{Br}_2 \frac{t^5}{20} \right) \right)^2 \\[4pt]
\dfrac{d[A_p]_{ij}(t)}{dt} = -\alpha_p \dfrac{\partial J_p}{\partial [A_p]_{ij}}(t) \\[4pt]
\dfrac{d[B_p]_{ij}(t)}{dt} = -\alpha_p \dfrac{\partial J_p}{\partial [B_p]_{ij}}(t) \\[4pt]
\dfrac{d[C_p]_{ij}(t)}{dt} = -\alpha_p \dfrac{\partial J_p}{\partial [C_p]_{ij}}(t)
\end{cases}
\tag{8}
$$

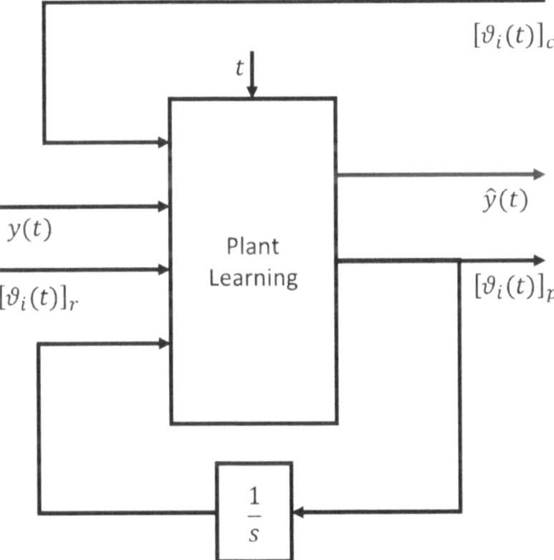

Figure 6. Schematic representation of the implementation reported in Equation (8).

As in the previous block description, we can compact the formulation as $\dfrac{d[\theta_i(t)]_p}{dt} = -\alpha_p \dfrac{\partial(y(t) - \hat{y}(t))^2}{\partial[\theta_i]_p}$; clearly, we represent all the coefficients $[A_p]_{ij}$, $[B_p]_{ij}$ and $[C_p]_{ij}$ with a more compact one $[\theta_i(t)]_p$, $i = 1, \ldots, N_x^2 + N_x N_u + N_y N_x$.

In Equation (8) are reported the update dynamic rules based on the instantaneous gradient descent algorithm. The goal for this subsystem is to fit best as possible, based on the optimal chosen criterion, the real output of the plant through the approximated model explained before. In this case, the learning rate is equal for all the updated components (but this is not mandatory; the only mandatory condition of the learning rate is on the sign of its real part, that must be negative for stability reasons).

3.3. Learning Controller Model

In this functional block the learning of the control state space representation parameters is done, in which the learnt coefficients in other subsystems are considered as constant values. In this part of the control algorithm the coefficients of the matrix A_p, B_p and C_p are considered as known; meanwhile, the coefficients of the matrix A_c, B_c and C_c are updated with the following rules. In this way, $\hat{y} = \hat{y}(A_c, B_c, C_c, t) = \hat{y}([\theta_i(t)]_c, \forall i)$. This approach is different from the previous functional block in Section 3.2, where, instead, the reference was to variables with subscript label "p" plant, instead of "c" controller.

$$\begin{cases} J_c = (r(t) - \hat{y}([\theta_i(t)]_c, \forall i))^2 \\ \frac{d[\theta_i(t)]_c}{dt} = -\alpha_p \frac{\partial J_c}{\partial[\theta_i]_c}(t), \forall i \end{cases} \tag{9}$$

In Figure 7 are reported in schematic way the update rule equations of this subsystem, where the objective function J_c is set with the aim to provide a control vector able to manipulate the behavior of the plant closest as possible to the reference signal, changing the coefficients of the state space representation of the controller itself instantaneously.

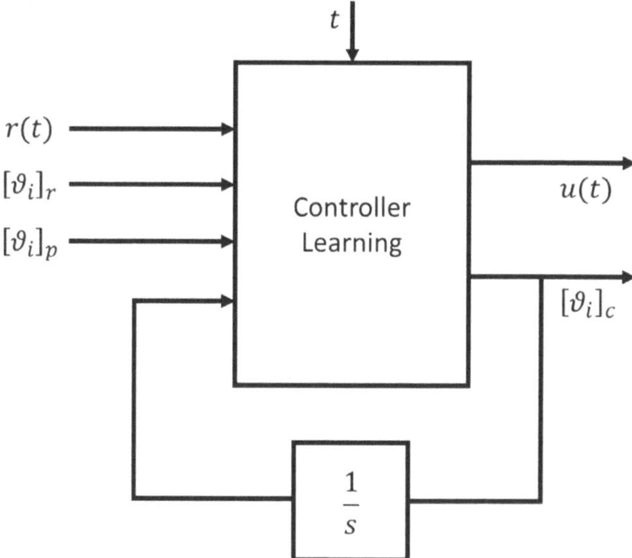

Figure 7. Schematic representation of the control learning subsystem.

4. Case Study: Nonlinear Model of a BLDC Motor Power Drive System

As an application case study of the innovative control technology described above, we present in this Section an electric drive based on a BLDC motor. To make the case study as realistic as possible, we consider the main intrinsic non-linearity effects in the dynamic BLDC motor model. In particular, we consider both the presence of the cogging torque [32–34] phenomenon and the torque due to the streabeck effect, which makes the dynamic model of the electric motor non-linear [35–38]. We also consider a model of the inverter (needed to generate the BLDC synchronous command signals from the DC power source), with an H-bridge driven with the bipolar PWM technique.

The complete dynamic model is reported in the set of equations in Equation (10): the first equation refers to the electric equilibrium model; the second and third equations refer to the mechanical equilibrium model for the rotation of the rotor axis.

$$\begin{cases} v_a(t) = R_a i_a(t) + L_a \frac{di_a(t)}{dt} + K_e \omega(t) \\ K_t i_a(t) + \sum_{k=1}^{N} C_k sin\left(k\frac{\vartheta(t)}{\vartheta_{cog}} + \varphi_k\right) - C_L = I\frac{d\omega(t)}{dt} + \left[C_c + (C_s - C_c)e^{-\left(\frac{\omega(t)}{\omega_s}\right)^2} + b\omega(t)\right]sgn(\omega(t)) \\ \frac{d\vartheta(t)}{dt} = \omega(t) \end{cases} \quad (10)$$

In particular, the third equation is a congruence equation between the angular position and the angular velocity. The supply voltage of the armature circuit (which is the control variable of the system) has been indicated with $v_a(t)$; $i_a(t)$ is the armature circuit current; $\omega(t)$ and $\vartheta(t)$ are the speed and angular position of the rotor axis respectively; R and L represent the resistance and the inductance of the impedance of the armature circuit; K_e and K_t represent the counter-electromotive force and torque constants, respectively; I is the inertia of the rotor; b is the viscous friction coefficient; C_s is the static friction torque; C_c is the coulomb friction torque; ω_s is the streabeck speed ($\omega_s \cong \omega_{max}$); C_L represents the load torque.

With the term $\sum_k C_k sin\left(k\frac{\vartheta(t)}{\vartheta_{cog}} + \varphi_k\right)$ the cogging torque contribute on the mechanical equilibrium is represented. The cogging torque model consists in a limited Fourier series, where C_k represent the amplitude of the kth harmonic and φ_k is the phase of it. ϑ_{cog} is the cogging period which is a function of the internal structure, in particular of the number of stator teeth and magnets arranged on the rotor iron.

We are referring to the features of a real DC motor reported in Table 1 [39]. The results obtained in simulation are shown hereafter, for a current control in which it is requested at the same time to maintain the current at a desired value and to reject different types of current disturbance.

Table 1. Characteristic of the considered brushless DC (BLDC).

Operating Feature	Value
Continuous Stall Torque	0.94 (Nm)
Peak Stall Torque	1.44 (Nm)
Continuous Stall Current	4.7 (A)
Maximum Pulse Current	16.7 (A)
Maximum Terminal Voltage	60 (V)
Maximum Speed	6000 (rpm)
Rotor Moment of Inertia	$5.3 \, 10^{-5}$ (Kgm^2)
Mechanical Time Constant	8 (ms)
Motor Mass	1.6 (Kg)
Thermal Resistance (armature to ambient)	4 ($^\circ$C/W)
Maximum armature Temperature	155 ($^\circ$C)
Torque Constant	0.086 (Nm/A)
Voltage Constant	9 (V/Krpm)
Armature Resistance	1.1 (Ω)
Thermal Resistance	1.5 (Ω)
Armature Inductance	2.4 (mH)
Electrical Time Constant	1.6 (ms)

In this work, a second order model was considered for both the controller state space representation and the plant model. Below are the reduced equations for the presented case which has the peculiarity of being a SISO (single-input-single-output) system.

$$\hat{P}: \begin{cases} \dot{x}_p(t) = \begin{bmatrix} a_{11} & a_{12} \\ a_{21} & a_{22} \end{bmatrix} x_p(t) + \begin{bmatrix} b_1 \\ b_2 \end{bmatrix} u(t) \\ \hat{y}(t) = \begin{bmatrix} c_1 & c_2 \end{bmatrix} x_p(t) \end{cases}$$
$$C: \begin{cases} \dot{x}_c(t) = \begin{bmatrix} m_{11} & m_{12} \\ m_{21} & m_{22} \end{bmatrix} x_c(t) + \begin{bmatrix} n_1 \\ n_2 \end{bmatrix} (r(t) - \hat{y}(t)) \\ u(t) = \begin{bmatrix} p_1 & p_2 \end{bmatrix} x_c(t) \end{cases} \tag{11}$$

In Equation (11) the status representations are reported for the model necessary for the learning of the system to be controlled and of the actual controller.

$$S: \begin{cases} \dot{\tilde{x}}(t) = \widetilde{A}\tilde{x}(t) + \widetilde{B}r(t) \\ \hat{y}(t) = \widetilde{C}\tilde{x}(t) \end{cases}$$
$$\widetilde{A} = \begin{bmatrix} A_p & B_pC_c \\ -B_cC_p & A_c \end{bmatrix} = \begin{bmatrix} \begin{bmatrix} a_{11} & a_{12} \\ a_{21} & a_{22} \end{bmatrix} & \begin{bmatrix} b_1 \\ b_2 \end{bmatrix}\begin{bmatrix} p_1 & p_2 \end{bmatrix} \\ -\begin{bmatrix} n_1 \\ n_2 \end{bmatrix}\begin{bmatrix} c_1 & c_2 \end{bmatrix} & \begin{bmatrix} m_{11} & m_{12} \\ m_{21} & m_{22} \end{bmatrix} \end{bmatrix}, \widetilde{B} = \begin{bmatrix} 0 \\ 0 \\ n_1 \\ n_2 \end{bmatrix}, \widetilde{C} = \begin{bmatrix} c_1 & c_2 & 0 & 0 \end{bmatrix} \tag{12}$$

In Equation (12), the state representation of the closed loop system model is reported, where the internal structures of the dynamic matrices are explained. The internal parameters of the matrices are in turn updated according to the criteria explained in general in the previous section. To avoid overloading the computational complexity, the learning parameters α_r, α_p and α_c, have been left fixed for each sub-system of the control architecture. This requires studying which combination of the learning parameters triad is the best among those tested in simulation. Clearly, it is also possible to add a differential learning parameter update equation for each of the control architecture sub-blocks.

Figure 8 shows the response to the current step of the closed loop system, which is simultaneously required to reject a 10% amplitude disturbance with respect to the reference signal. Figure 9 shows a particular view of the transitory phase and of the rejection phase of Figure 8. The step responses in Figure 8 are superimposed as the learning parameter combinations change; in particular, it can be noted that a valid combination is that given by $\left[\alpha_r, \alpha_p\alpha_c\right] = [1, 100, 100]$.

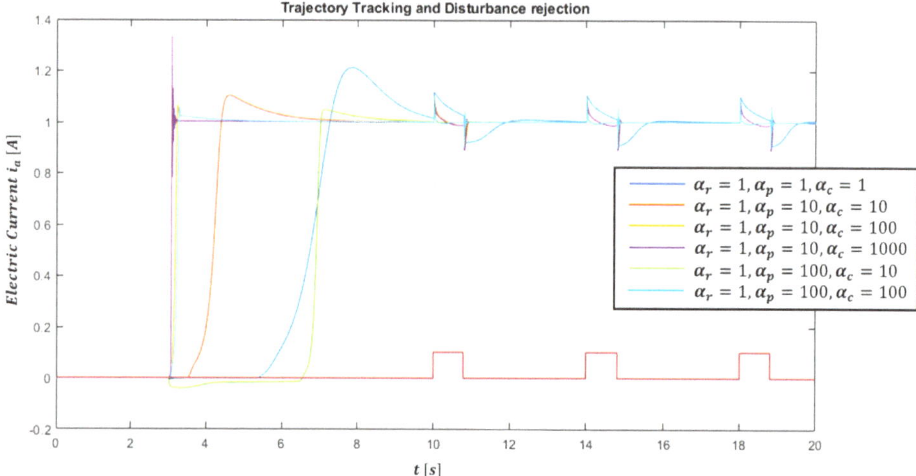

Figure 8. Step response and Disturbance rejection with respect to different learning rates combination.

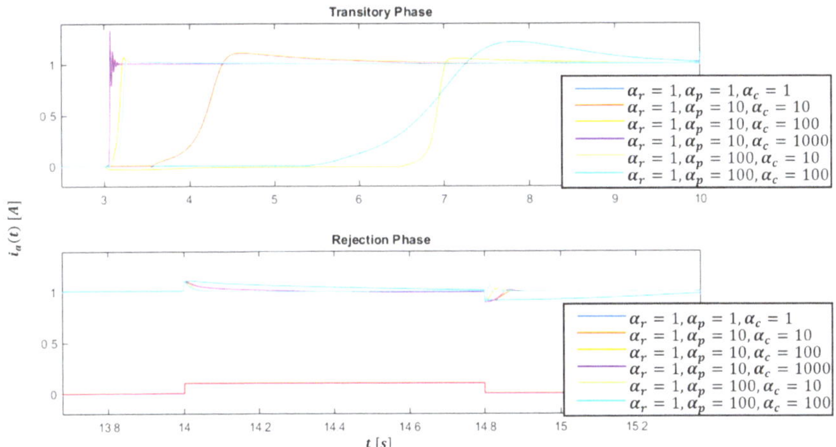

Figure 9. A particular view of the Transitory Phase (**a**) and Rejection Phase (**b**) of Figure 6.

Clearly it is possible to avoid a preliminary analysis on the combinations of the learning parameters, by inserting for each level of the proposed control architecture an additional update law.

Although true that it would increase the computational complexity of the whole algorithm, undoubtedly the procedure of estimating the reference signal and the representations in the form of state, both the process and of the controller, would become even more autonomous.

Taking inspiration from the theory of the statistical learning, there are some simple methods from an implementation point of view, which, however, have the limitation to be strongly dependent on the choice of the conditions of the learning parameters themselves. A robustness analysis to the variation of the initial conditions would be required anyway, so in this work it has been chosen to leave just these ones as free parameters.

In any case, this does not change the systematic design procedure of the controller through the structure that we propose.

It is important to point out that the result obtained does not require knowledge of any parameter of the electric motor or of the external disturbance model, which is instead necessary to apply the principle of the internal model to try to reject a disturbance through a linear controller.

Once a valid combination has been found for the learning parameters, the second test necessary to validate the control algorithm is the robustness to disturbance variation.

Figure 10 shows the result obtained by varying the amplitude of the disturbance, with a progressively higher percentage. In the legend, the percentages of the disturbance amplitude are relative to the amplitude of the current reference. Clearly, as the amplitude of the disturbance increases, the performance of the closed loop system also deteriorates. However, the rejection of the disturbance is still satisfactory, with a relative amplitude of 20%.

This is difficult to achieve through the classic cascade control structure, but also through non-linear control techniques that are based on deterministic models of both the system to be controlled and on assumptions about the disturbance, both the shape and the entry point within the control architecture. It is necessary to highlight again that the proposed control technique does not use any working hypothesis about the system to be controlled and no a priori knowledge about external disturbances.

For completeness in the validity analysis of the proposed control algorithm, the result of current stepped trajectory tracking in which it is required to reject a sinusoidal shaped disturbance is also reported. This further explains that the result obtained is not a function of the chosen waveforms, considering that the step function is the worst in terms of discontinuity of the derivative among all the reference signal choices. Figure 11 shows the rejection result of a sinusoidal current disturbance, with a relative amplitude of 10%, relative to the step reference.

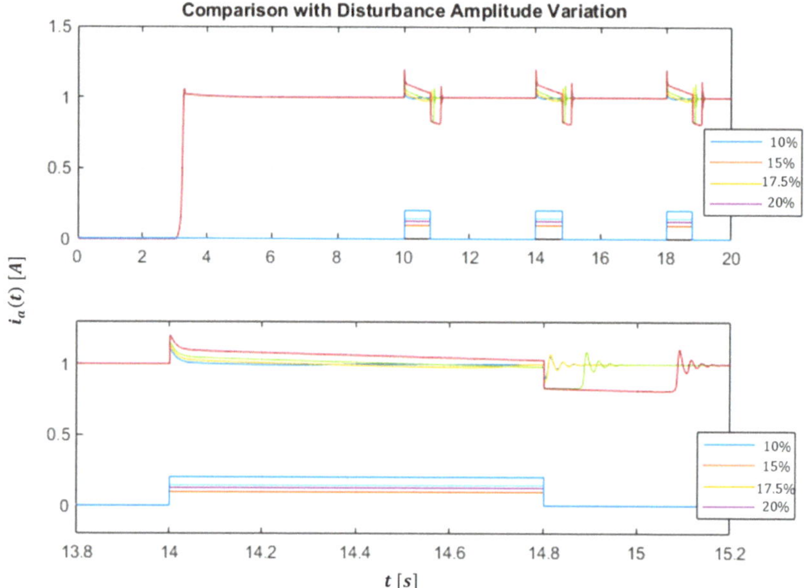

Figure 10. Disturbance Rejection with different disturbance signal amplitude.

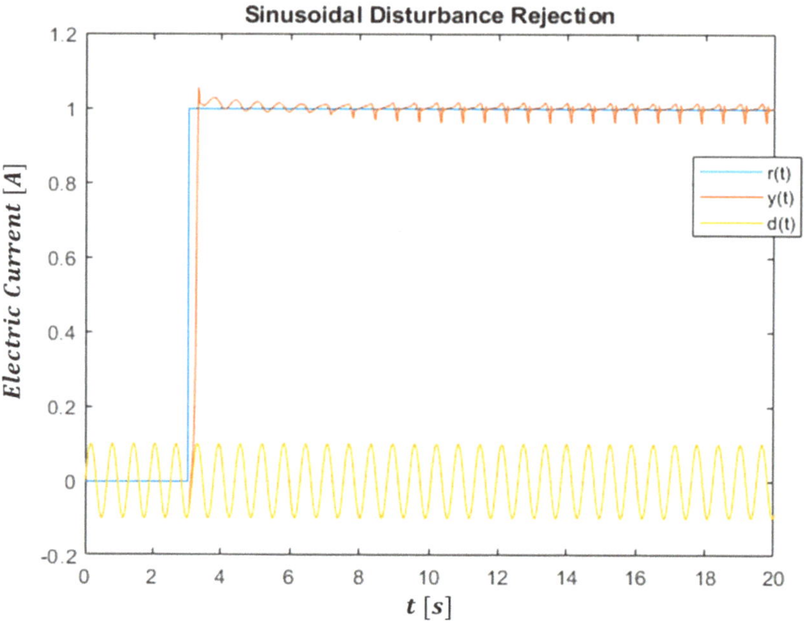

Figure 11. Comparison between current reference $r(r)$, current supplied by the motor $y(t)$ and current disturbance $d(t)$.

Below is also the verification made on the tracking of the trajectory in terms of the desired armature current, under conditions of uncertainty about the value of the cogging torque. We report the trajectory tracking graph in which the rejection of a piecewise current disturbance is required, with a 15% increase in the maximum intensity of the cogging torque model used in the BLDC dynamic model.

In Figure 12, it can be seen, both in the transient phase in terms of following the desired current trajectory and in the phases of rejection of the disturbance, that a slightly degraded result is obtained. In fact, as far as the trajectory tracking is concerned, with the same combination of the learning parameters, an over-elongation and a slightly longer settling time are obtained. As far as the rejection of the disturbance is concerned, there is the addition of an additional oscillatory behaviour, due to the cogging torque. In fact, both on the uphill and downhill front, there is a second order behaviour.

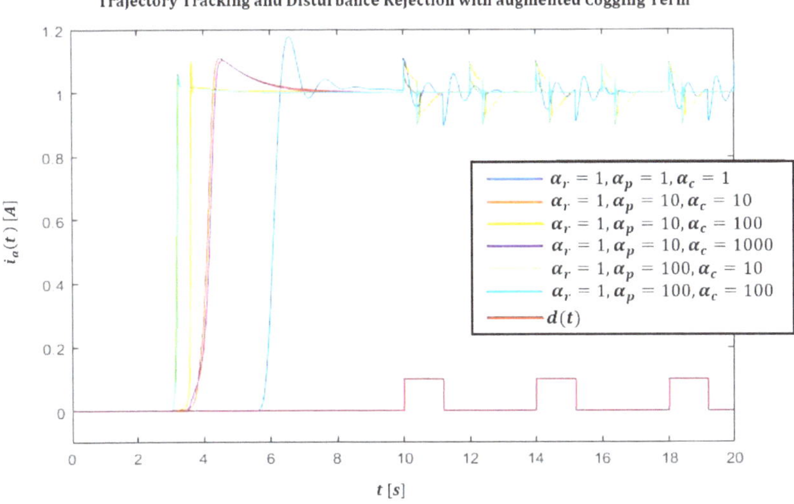

Figure 12. Trajectory Tracking and disturbance rejection with a higher intensity of Cogging Torque effect.

In any case, the disturbance continues to be partially rejected, despite the total lack of a priori knowledge about it.

The addition of an automatic update of the learning parameters could be a strong point also to solve eventual degradations of the performances, due to strong uncertainties of model in simulation phase, because the algorithm would be able to find independently the appropriate learning parameters triads, while in this case, the combinations are the same of the previous tests.

Last analysis that is reported is relative to the computation analysis of the used control architecture, using a SW tool like the Simulink profiler, that gives us some indications on the resources required by the algorithm, also in view of a future implementation on an embedded platform.

In the Table 2, Time field means the total time spent executing all invocations of the specific function as an absolute value and as a percentage of the total simulation time; Time/Calls field represents the average time required for each invocation of the specific function, including the time spent in functions invoked by the function itself; Calls field means the number of time the specific function was invoked; and Self-Time field represents the average time required to execute the specific function, excluding time spent in functions called by this function.

This preliminary complexity analysis of the controller code confirms that the proposed technique, for the BLDC example case, can be implemented in real-time on the low cost and low power NXP S32K14x family of automotive-qualified microcontrollers, which are based on the Arm-Cortex 4MF 32bit core.

Table 2. Complexity summary.

Functional Block	Time		Calls	Time/Call	Self-Time	
PID-Cascade Control	43.8672	1.60%	400,036	~0.0001184	30.1587	1.60%
Leaning-Based	265,047	9.70%	2,800,221	~0.0000947	137.578	5.00%

5. Conclusions

In this work, an innovative adaptive control structure was presented, partly inspired by the layered structure of neural networks. From a technical and formal point of view, the control structure consists of three levels of learning. Each level uses statistical learning concepts to update the parameters of the controller and of the process model state representations, and the coefficients of the polynomial representation of the reference. Each subsystem of the control architecture solves a different task, using the instantaneous gradient algorithm, learning any type of reference and adapting to any type of disturbance.

In conclusion, ours is an adaptive control technique classified as "Model Free", as justified in the article, in which, however, compared to classical techniques, the contribution of the theory of learning has been introduced, in order to keep the computational complexity limited, compared to modern methods that use architectures based on neural networks.

In fact, three optimization problems are solved at every step of the algorithm, so not only is it an adaptive control technique that exploits learning concepts with low computational impact, but it could also be considered an optimal adaptive control technique with high robustness characteristics, especially to parametric variations.

For this reason, a certainly interesting extension of this work should consider direct comparison with robust optimal control techniques in application contexts where the last one is applied, such as the control of vehicle dynamics.

Another interesting extension could be an integration of the control architecture proposed in a context of non-linear control techniques, replacing non-systematic methods such as Lyapunov's, to make adaptive all or part of the control laws that are designed through advanced criteria, such as Feedback linearization that is based on concepts of differential geometry, but still limited by the knowledge of a model of the system to be controlled.

As anticipated, in this work, no reference has been made to artificial intelligence methods, because those based on neural networks are onerous at computational level in the learning phase, and very little robust in the exercise phase. Clearly, artificial intelligence theory has developed modern techniques such as those of learning for reinforcement, with which it would certainly be interesting to compare the proposed technique, even in a proof of concept like the one proposed in this article.

Simulation results are presented in terms of current/torque control of a BLDC motor, in which the mathematical modelling of nonlinearity effects, such as the cogging torque and streback effect, are considered and considering the electric drive components, such as the effect of modulation and power supply through the single-phase inverter. The results achieved verify the robustness and quality of the response of the closed-loop system, both in terms of learning parameters and the amplitude of the applied disturbances. The implementation complexity analysis of the new controller is also addressed, showing its low overhead vs. basic control solutions. As a development of the work presented, it is under verification the implementation of the proposed control algorithm on a low-cost embedded platform using automotive qualified processors such those of the NXP S32K14x family, equipped with an Arm-Cortex 4MF microprocessor, and exploiting the NXP model design toolbox.

Author Contributions: Conceptualization, P.D. and S.S.; methodology, P.D. and S.S.; software, P.D. and S.S.; validation, P.D. and S.S.; formal analysis, P.D. and S.S.; investigation, P.D. and S.S.; resources, P.D. and S.S.; data curation, P.D. and S.S.; writing—original draft preparation, P.D. and S.S.; writing—review and editing, P.D. and S.S.; visualization, P.D. and S.S.; supervision, S.S.; project administration, S.S.; funding acquisition, S.S. All authors have read and agreed to the published version of the manuscript.

Funding: This work was partially supported by the project Crosslab- Dipartimenti di Eccellenza, MIUR.

Conflicts of Interest: The authors declare no conflict of interest.

References

1. Ayyub, B.M.; Klir, G.J. *Uncertainty Modeling and Analysis in Engineering and the Sciences*; CRC Press: Boca Raton, FL, USA, 2006.
2. Francis, B.A.; Malcolm, C.S. *Control of Uncertain Systems: Modelling, Approximation, and Design: A Workshop on the Occasion of Keith Glover's 60th Birthday*; Taylor & Francis: Abingdon, UK, 2006; Volume 329.
3. Zhang, W.; Xie, L.; Chen, B.-S. *Stochastic H2/H∞ Control: A Nash Game Approach*; CRC Press: Boca Raton, FL, USA, 2017.
4. Abu-Khalaf, M.; Jie, H.; Frank, L.L. *Nonlinear H2/H-Infinity Constrained Feedback Control: A Practical Design Approach Using Neural Networks*; Springer Science & Business Media: Berlin/Heidelberg, Germany, 2006.
5. Santoso, F.; Ming, L.; Gregory, E. Linear Quadratic Optimal Control Synthesis for a UAV. In Proceedings of the 12th Australian International Aerospace Congress, AIAC12, Melbourne, Australia, 19–22 March 2007.
6. Santoso, F.; Liu, M.; Egan, G.K. Robust μ-synthesis loop shaping for altitude flight dynamics of a flying-wing airframe. *J. Intell. Robot. Syst.* **2014**, *79*, 259–273. [CrossRef]
7. Radwan, A.G.; Taher Azar, A.; Vaidyanathan, S.; Munoz-Pacheco, J.M.; Ouannas, A. "Fractional-order and mem restive nonlinear systems: Advances and applications". *Complexity* **2017**. [CrossRef]
8. Fereka, D.; Zerikat, M.; Belaidi, A. MRAS Sensor-Less Speed Control of an Induction Motor Drive Based on Fuzzy Sliding Mode Control. In Proceedings of the 2018 7th International Conference on Systems and Control (ICSC), Valencia, Spain, 24–26 October 2018.
9. Zhang, M.; Ming, C.; Bangfu, Z. Sensor-Less Control of Linear Flux-Switching Permanent Magnet Motor Based on Improved MRAS. In Proceedings of the 2018 IEEE 9th International Symposium on Sensor-less Control for Electrical Drives (SLED), Helsinki, Finland, 13–14 September 2018.
10. Yassine, B.; Zidani, F.; Larbi, C.-A. New MRAS Approach for Sensor-less control of IM. In Proceedings of the 2019 19th International Conference on Sciences and Techniques of Automatic Control and Computer Engineering (STA), Sousse, Tunisia, 24–26 March 2019.
11. Tian, Y.; Hui, Z. On Stochastic Adaptive Control Under Quantization. In Proceedings of the 11th World Congress on Intelligent Control and Automation, Shenyang, China, 29 June–4 July 2014.
12. Kevin, D.S.; Bequette, B.W. Multiple Model Adaptive Control (MMAC). In *Nonlinear Model Based Process Control*; Springer: Berlin/Heidelberg, Germany, 1998; pp. 33–57.
13. Zengin, H.; Zengin, N.; Fidan, B.; Khajepour, A. Blending Based Multiple-Model Adaptive Control for Multivariable Systems and Application to Lateral Vehicle Dynamics. In Proceedings of the 2019 18th European Control Conference (ECC), Naples, Italy, 25–28 June 2019; pp. 2957–2962.
14. Outeiro, P.; Cardeira, C.; Oliveira, P. MMAC Height Control System of a Quadrotor for Constant Unknown Load Transportation. In Proceedings of the 2018 IEEE/RSJ International Conference on Intelligent Robots and Systems (IROS), Madrid, Spain, 1–5 October 2018; pp. 4192–4197.
15. Damiano, R. *Advances in Gain-Scheduling and Fault Tolerant Control Techniques*; Springer: Berlin/Heidelberg, Germany, 2017; ISBN 978-3-319-62902-5.
16. Ganesh, A.; Jaswinder, S. Adaptive Hybrid Control for Noise Rejection. In Proceedings of the 9th WSEAS International Conference on Neural Networks, Stevens Point, WI, USA, 2–4 May 2008.
17. Ioannou, P.A.; Jing, S. *Robust Adaptive Control*; Courier Corporation: North Chelmsford, MA, USA, 2012.
18. Kamalapurkar, R.; Walters, P.; Rosenfeld, J.; Dixon, W. *Reinforcement Learning for Optimal Feedback Control*; Springer: Cham, Switzerland, 2018.
19. Hakim, A.A.M.; Ibrahim, M.H.S. Adaptive Control for x Inverted Pendulum Utilizing Gain Scheduling Approach. In Proceedings of the 2018 International Conference on Computer, Control, Electrical, and Electronics Engineering (ICCCEEE), Khartoum, Sudan, 12–14 August 2018.
20. Poksawat, P.; Wang, L.; Mohamed, A. Gain Scheduled Attitude Control of Fixed-Wing UAV With Automatic Controller Tuning. *IEEE Trans. Control Syst. Technol.* **2018**, *26*, 1192–1203. [CrossRef]
21. Dulf, E.; Muresan, C.I.; Both, R.; Fustos, C.; Dulf, F. Auto-Tuning Fractional Order Velocity Control of a DC Motor. In Proceedings of the 2015 Intl Aegean Conference on Electrical Machines & Power Electronics (ACEMP), 2015 Intl Conference on Optimization of Electrical & Electronic Equipment (OPTIM) & 2015 Intl Symposium on Advanced Electromechanical Motion Systems (ELECTROMOTION), Side, Turkey, 2–4 September 2015; pp. 159–162.

22. Tang, W.-J.; Liu, Z.-T.; Wang, Q. DC Motor Speed Control Based on System Identification and PID Auto Tuning. In Proceedings of the 2017 36th Chinese Control Conference (CCC), Dalian, China, 26–28 July 2017; pp. 6420–6423.

23. Jin, S.-T.; Hou, Z.-S.; Chi, R.-H. A Model-Free Adaptive Switching Control Approach for a Class of Nonlinear Systems. In Proceedings of the 11th World Congress on Intelligent Control and Automation, Shenyang, China, 29 June–4 July 2014.

24. Zen, Z.; Cao, R.; Hou, Z. MIMO Model-Free Adaptive Control of Two Degree of Freedom Manipulator. In Proceedings of the 2018 IEEE 7th Data Driven Control and Learning Systems Conference (DDCLS), Enshi, China, 25–27 May 2018; pp. 693–697.

25. Harikumar, K.; Mohamad, A.H.; Mahardhika, P.; Feng, N.B. Robust Evolving Neuro-Fuzzy Control of a Novel Tilt-Rotor Vertical Takeoff and Landing Aircraft. In Proceedings of the 2019 IEEE International Conference on Fuzzy Systems (FUZZ-IEEE), New Orleans, LA, USA, 23–26 June 2019; pp. 1–6.

26. Daniel, L.; Charles, A.; Daniel, P.; Gustavo, S.; Igor, S. Nonlinear Fuzzy State-Space Modeling and LMI Fuzzy Control of Overhead Cranes. In Proceedings of the 2019 IEEE International Conference on Fuzzy Systems (FUZZ-IEEE), New Orleans, LA, USA, 23–26 June 2019; pp. 1–6.

27. Lu, Y.K. Adaptive-Fuzzy Control Compensation Design for Direct Adaptive Fuzzy Control. *IEEE Trans. Fuzzy Syst.* **2018**, *26*, 3222–3231. [CrossRef]

28. Lv, Y.K.; Chi, R.H. Data-Driven Adaptive Iterative Learning Predictive Control. In Proceedings of the 2017 6th Data Driven Control and Learning Systems (DDCLS), Chongqing, China, 26–27 May 2017; pp. 374–377.

29. Armin, N.; Charles, R.K. Robotic Manipulator Control Using Pd-Type Fuzzy Iterative Learning Control. In Proceedings of the 2019 IEEE Canadian Conference of Electrical and Computer Engineering (CCECE), Edmonton, AB, Canada, 5–8 May 2019; pp. 1–4.

30. Bousquet, O.; von Luxburg, U.; Gunnar, R. *Advanced Lectures on Machine Learning: ML Summer Schools 2003, Canberra, Australia, 2–14 February 2003, Tübingen, Germany, 4–16 August 2003, Revised Lectures*; Springer: Berlin/Heidelberg, Germany, 2011; Volume 3176.

31. James, G.; Witten, D.; Hastie, T.; Tibshirani, R. *An Introduction to Statistical Learning*; Springer: New York, NY, USA, 2013; Volume 112.

32. Dini, P.; Saponara, S. Cogging Torque Reduction in Brushless Motors by a Nonlinear Control Technique. *Energies* **2019**, *12*, 2224. [CrossRef]

33. Dini, P.; Saponara, S. Control System Design for Cogging Torque Reduction Based on Sensor-Less Architecture. In *Applications in Electronics Pervading Industry, Environment and Society, Lecture Notes in Electrical Engineering*; Springer: Berlin/Heidelberg, Germany, 2020; Volume 627.

34. Tudorache, T.; Trifu, I.; Ghiţă, C.; Bostan, V. Improved Mathematical Model of PMSM Taking Into Account Cogging Torque Oscillations. *Adv. Electr. Comput. Eng.* **2012**, *12*, 59–64. [CrossRef]

35. Zhang, S.; Gu, W.; Hu, Y.; Du, J.; Chen, H. Angular Speed Control of Brushed DC Motor Using Nonlinear Method: Design and Experiment. In Proceedings of the 2016 35th Chinese Control Conference (CCC), Chengdu, China, 27–29 July 2016.

36. Katsioula, A.G.; Yannis, L.K.; Yiannis, S.B. An Enhanced Simulation Model for DC Motor Belt Drive Conveyor System Control. In Proceedings of the 2018 7th International Conference on Modern Circuits and Systems Technologies (MOCAST), Thessaloniki, Greece, 7–9 May 2018.

37. Buechner, S.; Schreiber, V.; Amthor, A.; Ament, C.; Eichhorn, M.; Eichhorn, M. Nonlinear Modeling and Identification of a Dc-Motor with Friction and Cogging. In Proceedings of the IECON 2013 39th Annual Conference of the IEEE Industrial Electronics Society, Vienna, Austria, 10–13 November 2013.

38. Dumitriu, T.; Culea, M.; Munteanu, T.; Ceangă, E. Friction Compensation for Accurate Positioning in Dc Drive Tracking System. In Proceedings of the 2006 3rd International Conference on Electrical and Electronics Engineering, Veracruz, Mexico, 6–8 September 2006.

39. Available online: https://www.pamoco.it/catalogo-prodotti/motori-rotativi/motori-dc/motori-dc-electrocraft-e-mae (accessed on 14 May 2020).

Article

Cogging Torque Reduction in Brushless Motors by a Nonlinear Control Technique

Pierpaolo Dini and Sergio Saponara * 

Department of Information Engineering, University of Pisa, Via G. Caruso 16, 56127 Pisa, Italy;
pierpaolo.dini@phd.unipi.it
* Correspondence: sergio.saponara@iet.unipi.it

Received: 1 April 2019; Accepted: 9 June 2019; Published: 11 June 2019

Abstract: This work addresses the problem of mitigating the effects of the cogging torque in permanent magnet synchronous motors, particularly brushless motors, which is a main issue in precision electric drive applications. In this work, a method for mitigating the effects of the cogging torque is proposed, based on the use of a nonlinear automatic control technique known as feedback linearization that is ideal for underactuated dynamic systems. The aim of this work is to present an alternative to classic solutions based on the physical modification of the electrical machine to try to suppress the natural interaction between the permanent magnets and the teeth of the stator slots. Such modifications of electric machines are often expensive because they require customized procedures, while the proposed method does not require any modification of the electric drive. With respect to other algorithmic-based solutions for cogging torque reduction, the proposed control technique is scalable to different motor parameters, deterministic, and robust, and hence easy to use and verify for safety-critical applications. As an application case example, the work reports the reduction of the oscillations for the angular position control of a permanent magnet synchronous motor vs. classic PI (proportional-integrative) cascaded control. Moreover, the proposed algorithm is suitable to be implemented in low-cost embedded control units.

Keywords: cogging torque reduction; precision robotics; nonlinear control techniques; embedded control systems; permanent magnet synchronous motors

---

1. Introduction

The phenomenon of cogging torque is an intrinsic characteristic of permanent magnet synchronous motors (PMSMs), in both interior (IPM) and surface-mounted (SPM) electric machine types. This phenomenon is due to the magnetic interaction between the teeth of the stator slots and the permanent magnets arranged on the rotor surface in the SPM case, or internally embedded in the rotor iron in the IPM case [1].

This is because the nonuniformity of the air gap is characterized by an equivalent magnetic circuit, which describes the magnetic interaction between rotor and stator. The equivalent magnetic circuit depends on the mutual position between the teeth of the stator cavities and the permanent magnets. This implies that the reluctance of the magnetic circuit is a function of the rotor angle. Figure 1 shows the physical origins of the cogging couple generating the cogging torque.

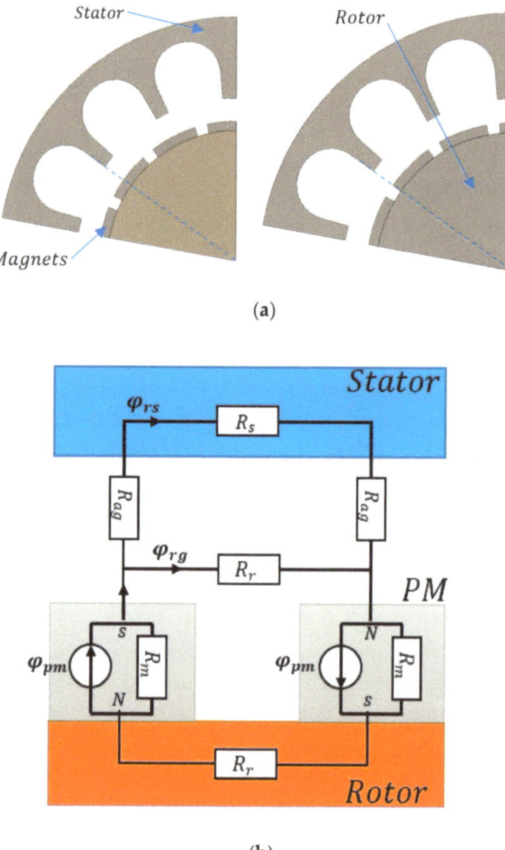

Figure 1. (**a**) By changing the angular position, the caves see different configurations. (**b**) Equivalent magnetic circuit. PM: permanent magnet.

In Figure 1, the physical quantity φ_{pm} represents the magnetic flux generated by the permanent magnets, while φ_{rs} and φ_{rg} are the portions closed in the iron of the stator and in the air gap, respectively, and R_m, R_r, R_s, and R_{ag} represent the reluctances of the permanent magnets, rotor, stator, and air gap, respectively.

Figure 1a shows that during the rotation of the rotor, the stator slot will alternately see a polar compartment (the space between two magnets) and a section of a magnet. This confirms that the magnetic circuit changes according to the angular position. Figure 1b represents the path of the magnetic flux in the ideal case of a uniform stator surface.

This phenomenon is due exclusively to the dependence of the magnetic circuit on the angular position and on the magnetic and geometric characteristics of the motor itself. Indeed, it is possible to appreciate the cogging torque even by simply rotating the motor axis manually without electrically feeding it. Starting from an angular position of the rotor, after a rotation of $\vartheta_{cog} = 2\pi/mcm(p, N_c)$, where p is the number of pole pairs and N_c is the number of caves, the same situation is encountered. Indeed, along the air gap, the same configuration in terms of interaction between teeth and magnets can be encountered several times. Consequently, the cogging couple is interpretable as a noise disturbance, with an angular frequency dependent on the number of slots and number of magnets and therefore polar pairs. The intensity of the cogging noise disturbance is a function of the geometry of the magnets and the teeth of the slots and of the magnetic characteristics of the materials involved. This disturbance is measurable by means of a test bench where, usually, a brushless motor is mechanically connected to a DC motor (not affected by the cogging effect) and passively rotated. In the mechanical connection of the two rotation axes, a torque meter is incorporated; then, from it, is possible to obtain the torque exerted in the absence of power supply. This torque is therefore equivalent to the torque for cogging effect.

In this work, a method for reducing the cogging torque is proposed, based on the use of a nonlinear automatic control technique. This technique, known as feedback linearization, is ideal for underactuated dynamic systems. State-of-the-art solutions are often based on the physical modification of the electrical machine to try to suppress the natural interaction between the permanent magnets and the teeth of the stator slots. However, such modifications are expensive because they require customized procedures, whereas the proposed method does not require any modification of the electric drive.

This paper is organized as follows. Section 2 reviews the state of the art of cogging reduction techniques that have been recently proposed in the literature and introduces the core idea of the solution proposed. In Section 3, an analytical description of the phenomenon to derive a model useful for the design of an automatic controller is carried out, as well as a model of a brushless motor including the cogging term. Section 4 presents a feedback linearization control to reduce cogging torque in brushless motors. Section 5 presents implementation results and a robustness analysis of the proposed control vs. motor parameter changes. Section 6 presents an analysis of the algorithmic complexity and the suitability for integration in a low-cost embedded system. Conclusions are drawn in Section 7.

2. Review of State of the Art for Cogging Torque Reduction

In the literature, torque ripple due to the cogging couple in PMSMs is reduced, but not eliminated, by an appropriate magnetic and layout winding design, as proposed in [2,3]. The other typical approach is the reduction of the intensity of the magnetic interaction between the stator and the rotor by properly designing the machine according to different criteria and verified with finite element method (FEM) analysis tools, as discussed in [4,5]. The problem with the above approaches is that the modifications to the electrical motor are expensive because are customized, and furthermore, they do not guarantee a complete suppression of the torque ripple caused by the cogging phenomenon. Hence, reduction techniques aiming at modeling and/or estimating the cogging phenomenon and then compensating or reducing it by proper control algorithms are more suited to achieve low-cost and scalable solutions.

The control of the machine stator currents could be used to compensate the torque harmonics that cannot be easily eliminated during the design stage. For instance, a conventional field-oriented control (FOC) scheme is proposed in [6] for torque ripple mitigation. However, because of the relatively high-frequency components present in the torque ripple, high bandwidth currents could be required. In [6,7], this is achieved using dead-beat controllers. Nevertheless, in these studies, the torque ripple is compensated by adding harmonic components in the q-axis stator current reference, and by regulating the d-axis current to zero. Consequently, with $i_d = 0$, maximum torque per ampere (MTPA) cannot be achieved [8,9] and the machine is no longer operating at the optimal torque.

In [8], a valid cogging reduction solution has been proposed through the application of the model predictive control (MPC) technique, which is based on solving optimal problems in an iterative manner, which in addition to being computationally heavy, is linked to the knowledge of the model. In recent papers, the use of resonant controllers has been also proposed for the control of PMSMs, such as in [10–12].

Resonant controllers are based on the internal model principle and they can be used to regulate with zero steady-state error signals of sinusoidal nature. For this purpose, one resonant notch is required to regulate a single-frequency component of the harmonic torque. Therefore, to eliminate most of the harmonic distortion, multiresonant controllers are required, and the design and digital implementation of such a controller is complex [13]. Moreover, if the control system is implemented in the stationary frame, a tuning mechanism is required to modify the resonant frequencies in real time when variations in the PMSM rotational speed are produced [12].

Recently, the use of finite-control set model predictive control (FCS-MPC) has been proposed for applications involving electrical drives and power converters [14]. With this methodology, a discrete model of the plant and power converter is required to predict the behaviour of the system for a particular switching state of the converter. To track references, a cost function that considers the tracking error at each sampling instant is used. The optimal switching action, which minimizes the cost function, is applied to the converter. As discussed in the literature, FCS-MPC has several advantages. For instance, it is relatively simple to include nonlinearities, constraints, and variables of different nature (electrical or mechanical) in the cost function. The application of FCS-MPC to the control of PMSMs has been discussed in the literature by using predictive current control [15,16]. Additionally, in [17,18], FCS-MPC is suggested as an alternative to a direct regulation of the machine torque. The method is called model predictive direct torque control (MP-DTC), which uses an external speed loop based on a PI controller. However, in previous papers, control methods to reduce the torque ripple produced by flux linkage harmonic distortion are not addressed; neither is mitigation of the cogging torque discussed.

In [19], a torque predictive control to mitigate the high-frequency torque due to switching effects is presented. However, this paper does not address mitigation of the low-frequency torque pulsations produced by the machine.

Control of the PMSM rotational speed based on FCS-MPC is presented in [20,21]. With this approach, the nested control loops typically used in drives are replaced by a predictive control system. The results discussed in [20,21] show that a high dynamic response can be achieved with this methodology. However, to compensate the load torque, an observer has to be implemented, and this may limit the performance of the control system to compensate relatively high-frequency torque components, particularly when the machine is operating at high rotational speed. Moreover, high-bandwidth observers are sensitive to measurement noise, parameter uncertainty, quantization error of the feedback signal, etc. A Lyapunov observer-based control technique is implemented in [22]. With respect to the other approaches based on a control algorithm, the nonlinear control proposed by us allows for an offline and deterministic projection of the control laws. This approach facilitates the integration of a cogging torque model without the need for designing a complex observer to estimate cogging-related signals.

In this work, in the following Sections, a nonlinear control approach is presented as an alternative to all the previous methods. First of all, with respect to [2–5], the proposed approach has the advantage of keeping the same mechanical structure of the electric motor without adding the payload of customized structures.

The feedback linearization allows to replace the nonlinear dynamic with a linearized one through a systematic change of variables. Furthermore, the proposed control technique is formally more robust than the above-discussed algorithmic techniques for cogging torque mitigation, due to the largest region of asymptotic stability. The proposed algorithm is suitable to be implemented on low-cost embedded controllers; for example, the Arduino-Uno platform using an ATMEGA328 microcontroller.

3. Cogging Torque and Motor Modelling

3.1. Cogging Torque Formulation

In this work, the objective is not to present a method for describing the cogging torque. The starting point is the formulation in closed form, reported in [1], that exploits the Fourier harmonic analysis against results obtained with FEM analysis. In particular, reference was made to the following expression:

$$T_{cog} \cong \sum_{k=1}^{m} T_k \sin(kZ\vartheta_m + \alpha_k) \tag{1}$$

where T_k and the α_k are the coefficients of the Fourier harmonic development, Z is the number of stator teeth, ϑ_m is the angular (mechanical) position of the rotor, and m is the number of harmonics that are used to approximate the actual shape of the cogging torque, T_{cog}. As reported in Figure 2, we have carried out a simulation analysis and a comparison in terms of variance of the number of harmonics to represent the cogging torque as a function of angular position. The results in Figure 2 show that four harmonics ($m = 4$) in Equation (1) are sufficient to represent the cogging torque (i.e., a Fourier development with four terms is a good approximation to describe the cogging phenomenon).

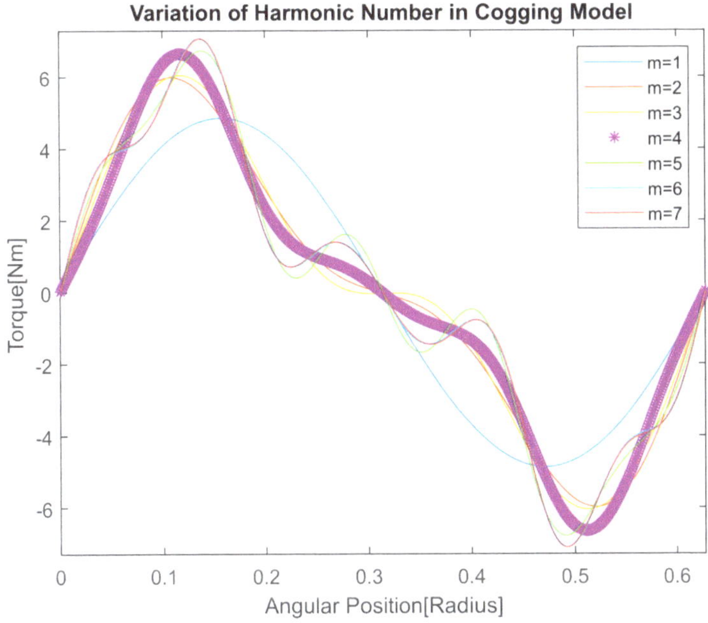

Figure 2. Comparison of the mathematical model of cogging torque with variation of the number of harmonics. The symbol * refers to the purple curve (m = 4) selected as the best trade-off between model accuracy and complexity.

The actual values of the coefficients are reported in Table 1. The values of the coefficients depend on the morphological and magnetic characteristics of the synchronous machine as well as on the magnetic exchange induced to the air gap and the magnetic permeability of the rotor and stator materials.

Table 1. Cogging torque model parameters.

Coefficient in Equation (1)	Value
T_1	4.85 Nm
T_2	2.04 Nm
T_3	0.3 Nm
T_4	0.06 Nm
α_1	0.009 rad
α_2	0.01 rad
α_3	0.017 rad
α_4	α_3

Table 2 shows the parameters of the considered electrical machine. In Table 2, R_s represents the electrical resistance of the stator's windings, L_{eq} is the equivalent inductance of the three-phase circuit that represents the dynamics of the electrical part of the machine, k_φ is the intensity of the magnetic flux generated by the permanent magnets, p represents the number of polar pairs, J_m is the motor inertia, and β is the friction coefficient.

Table 2. Motor model parameters.

Motor Model Parameter	Value
R_s	3.3 Ohm
L_{eq}	50 mH
k_ϕ	0.5 Wb
J_m	0.02 Kgm^2
p	3
β	0.01 Ns/m

The parameters shown in Tables 1 and 2 are relative to an SPM powered by 400 VAC with the features shown in Table 3. In particular, we refer to the brushless motor model Q2CA2215KVXS00M [23], using an inverter model RS1C10AL [24].

Table 3. Features of the considered electric motor.

Motor Performance Parameter	Value
Nominal Power	15 kW
Nominal Speed	1500 rpm
Maximum Speed	2000 rpm
Nominal Torque	95.5 Nm
Stall Torque	230 Nm

In particular, a motor with the characteristics in Table 3 has a torque–speed curve similar to that shown in Figure 3. In this article, we assume working in the continuous zone of the characteristic reported in Figure 3.

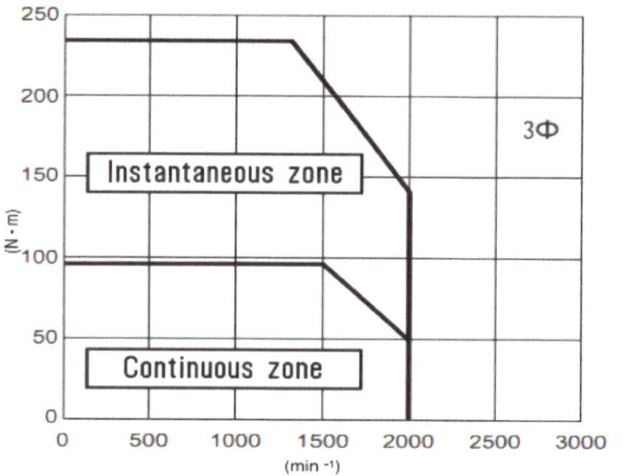

Figure 3. Torque–speed curve of the motor considered as an application example in this work. The symbol 3Φ means that it is a three-phase powered system.

It is worth noting that this work models the presence and effects of the cogging torque in an AC brushless motor with SPM architecture and uses this model in a correction algorithm, but it is not essential to know the internal geometry of the machine in terms of geometrical properties of the permanent magnets and stator caves. This because the most relevant thing is to have a mathematical model of the cogging phenomenon that highlights the dependence of the cogging torque by the rotor position. This model is then inserted into the global model of the drive system.

Instead, the dependence of the cogging torque coefficients shown in Table 1 by the internal geometry of the permanent magnets and stator caves would be needed in a work aiming at a modification of the machine structure, but this is not the case for the present study.

To make more realistic the analysis of the cogging torque reduction by the nonlinear control method proposed, a perturbated version of the cogging mathematical model of Equation (1) is inserted into the model of the mechanical equilibrium. Into the perturbation made by the noise simulation in Figure 4 is inserted the effect of the unavoidable torque ripple caused also by the nonideal commutation of the inverter switching logic. We used the model with four harmonics to project the control laws. In particular, the perturbation on the cogging model is simulated by a white Gaussian noise $w(t)$, the effects of which can be seen in Figure 4. In particular, in our case, $w(t) \sim N(\mu, \sigma)$ with $\mu = 0$, $\sigma = 1\,Nm$ is inserted.

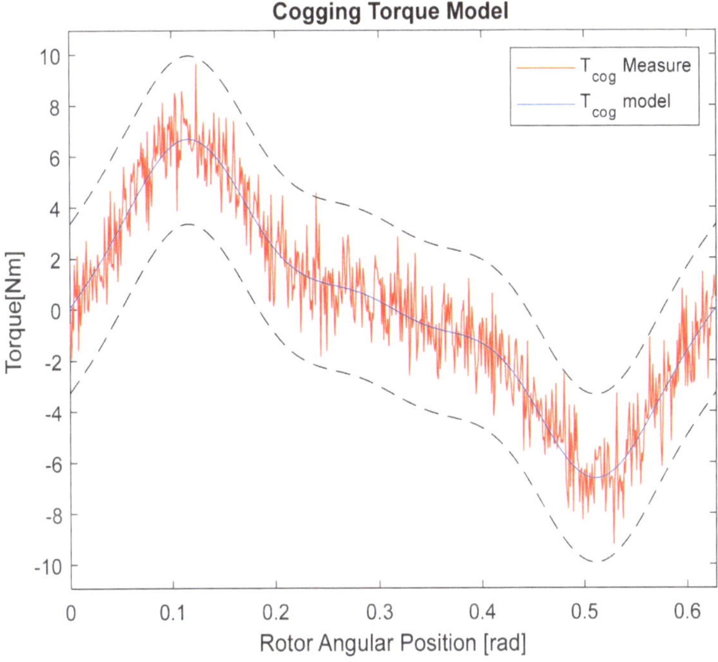

Figure 4. Comparison between the shape insert in the motor model (T_{cog} Measure), used to design the controller, and the mathematical model (T_{cog} model).

According to other works in the literature, such as [25], a reasonable value of torque ripple is around 10% of the nominal torque, i.e., $\mp 10\,Nm$ for a motor with nominal torque of about $100\,Nm$, as shown in Figure 3. In general, torque ripple clearly also includes cogging torque behavior, so in our work, we included the two terms by also adding to the model of Equation (1) the noise term $w(t)$. To respect the typical percentage of ripple entities, we set an intensity of about $7\,Nm$ for the cogging torque, adding a Gaussian noise with $3\sigma = 3\,Nm$, so the intensity of the torque ripple is in the range $\mp 10\,Nm$ (T_{cog} *by model* $\mp 3\sigma$). To better highlight the comparison of the cogging torque vs. the added torque ripple, modeled as a Gaussian noise contribution, Figure 5 shows the two separated signals—cogging torque in blue and Gaussian-like added torque ripple in red, in the time domain (top) and frequency Fourier transform domain (bottom).

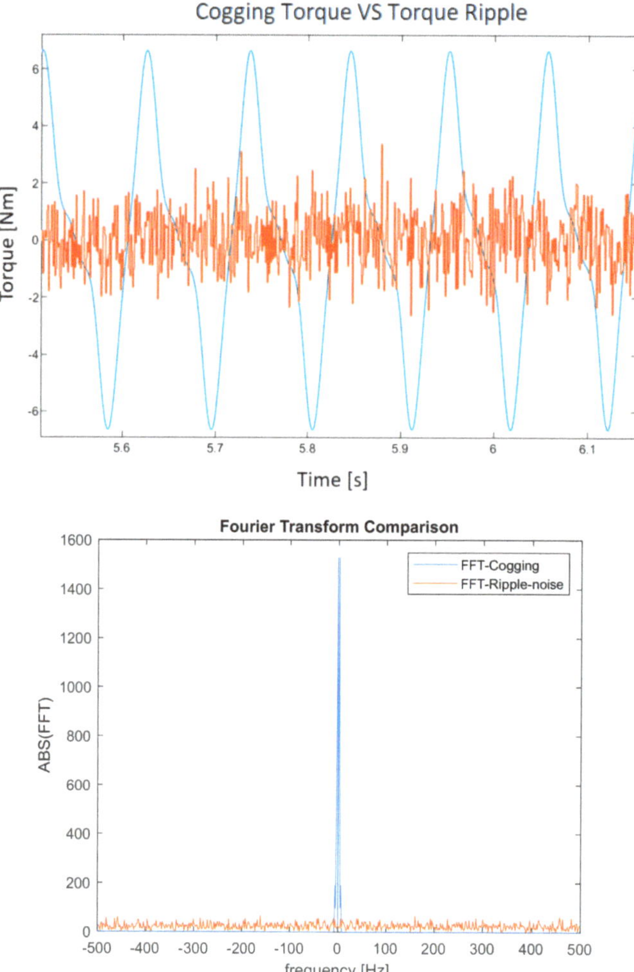

Figure 5. Cogging torque in blue vs. added torque ripple in red in the time domain (**top**) and absolute (ABS) value of the frequency Fourier transform (FFT) domain (**bottom**).

3.2. Brushless Motor Modeling with the Cogging Term

A classical model of a synchronous machine is considered. According to the unified theory of electric machines [26], and after the coordinate transformations necessary to delete the dependence on angular rotor position in electromagnetic torque, we can write the equations of electromechanical equilibrium in the following way:

$$\begin{bmatrix} u_d \\ u_q \end{bmatrix} = \begin{bmatrix} R_s & -L_q p\frac{d\vartheta_m}{dt} \\ L_d p\frac{d\vartheta_m}{dt} & R_s \end{bmatrix} \begin{bmatrix} i_d \\ i_q \end{bmatrix} + \begin{bmatrix} L_d & 0 \\ 0 & L_q \end{bmatrix} \begin{bmatrix} \frac{di_d}{dt} \\ \frac{di_q}{dt} \end{bmatrix} + \begin{bmatrix} 0 \\ p\frac{d\vartheta_m}{dt}k_\phi \end{bmatrix} \tag{2}$$

$$J_m \frac{d^2\vartheta_m}{dt^2} + \beta\frac{d\vartheta_m}{dt} = T_{em} + \sum_{k=1}^{m} T_k \sin(kZ\vartheta_m + \alpha_k) + w(t) \tag{3}$$

In the above equations, the term of cogging torque is inserted into the mechanical equilibrium as an overlapping effect in the global torque. In Equation (2) u_d *and* u_q are the voltage components in the d–q frame, i_d *and* i_q are the current components in the d–q frame, L_d *and* L_q are the inductance of the equivalent circuit transformed in the d–q frame, ϑ_m is the mechanical angular position, and T_{em} is the electromagnetic torque that in the general case, is given by $T_{em} = \frac{3}{2}p\left[\left[L_q - L_d\right]i_d i_q + k_\varphi i_q\right]$. In this work, reference is made on a SPM machine which has the characteristic of having $L_d = L_q = L_{eq}$, i.e., $\frac{L_d}{L_q} = 1$, so that the expression of the electromagnetic torque is given by $T_{em} = \frac{3}{2}pk_\varphi i_q$.

Working with $\frac{L_d}{L_q} = 1$ means that the electric machine is considered as an isotropic brushless machine, so there are not configurations of maximum or minimum reluctance and clearly, the inductance values of the equivalent circuits along the direct and quadrature axis (the frame that is obtained after the Park transformation) are equal. Such an assumption has been made for the sake of clarity, as it makes it easier to understand the expression of the electromagnetic torque. However, nothing changes if we consider an anisotropic machine with different L_d and L_q values (typically, $L_d << L_q$). Moreover, in the paper, we emulate a FOC control where the i_d current is constrained to a zero value, and this implies that the anisotropic behavior does not contribute to the final trajectory tracking result. To include the anisotropic behavior due to arrangement of the magnets in an IPM structure, then the term $3/2 \cdot p \cdot (L_q - L_d) \cdot i_d \cdot i_q$, which is zero in the case where $L_d = L_q$, should be included in the ideal electromagnetic torque, i.e., the above-reported T_{em} term in the mechanical equilibrium in the equations of the dynamic of the electrical machine.

4. Feedback Linearization Control for a Brushless Motor

For convenience in the controller design, the equations in the state form are usually assumed, as shown in Equation (4).

$$\dot{x} = \begin{bmatrix} \dot{x}_1 \\ \dot{x}_2 \\ \dot{x}_3 \\ \dot{x}_4 \end{bmatrix} = \begin{bmatrix} px_4x_2 - \frac{R_s}{L_{eq}}x_1 \\ -\frac{1}{L_{eq}}\left[R_sx_2 + px_4\left(L_{eq}x_1 + k_\phi\right)\right] \\ \frac{1}{J_m}\left[\frac{3}{2}pk_\phi x_2 - \beta x_4 + \sum_{k=1}^{4} T_k \sin(kZx_3 + \alpha_k)\right] \end{bmatrix} + \begin{bmatrix} \frac{1}{L_{eq}} \\ 0 \\ 0 \\ 0 \end{bmatrix} u_1 + \begin{bmatrix} 0 \\ \frac{1}{L_{eq}} \\ 0 \\ 0 \end{bmatrix} u_2 = f(x) + \sum_{k=1}^{m} g_k(x)u_k \quad (4)$$

It should be further noted that the state form is nonlinear, but it is "affine in the control" with the classic choice of state variables $x_1 = i_d$, $x_2 = i_q$, $x_3 = \vartheta_m$, $x_4 = \dot{\vartheta}_m$ and control variables $u_1 = u_d$, $u_2 = u_q$. The vector field $f(x)$ is called drift vector while the $g_k(x)$, with k from 1 to m, are the control vector fields.

Referring to the nonlinear theory of control systems [27], the goal is to design the control law in order to formally obtain a change of state variables that have a linear dynamic in the new base. The operating procedure consists of deriving all the expressions of the outputs for a total number of times necessary to obtain an expression where at least one of the components of the control vector appears.

Calling $y_i = h_i(x)$ the i-th control output and μ_i the i-th derivation order in which at least one element of the control vector appears, the equations useful for the systematic design of the controller and therefore the change of variables take the following form:

$$\frac{d^{\mu_i}}{dt^{\mu_i}}y_i = \frac{\partial^{\mu_i-1}h_i(x)}{\partial x^{\mu_i-1}}\left(f(x) + \sum_{k=1}^{m} g_k(x)u_k\right) = L_f^{\mu_i}h_i(x) + \sum_{k=1}^{m} L_{g_k}L_f^{\mu_i-1}h_iu_k; i = 1,2\dots,l \quad (5)$$

Inserting the linearization condition by imposing that the i-th derivative is equal to a component of the control vector $v = \begin{bmatrix} v_1 & \cdots & v_l \end{bmatrix}^T$ for the new base ($y_i^{\mu_i} = v_i \; \forall i$), thus deciding that the linear dynamics emulate a chain of multiple integrators, the following matrix form is obtained:

$$Y^\mu = \begin{bmatrix} y_1^{\mu_1} \\ \vdots \\ y_l^{\mu_l} \end{bmatrix} = \begin{bmatrix} L_f^{\mu_1} h_1 \\ \vdots \\ L_f^{\mu_l} h_l \end{bmatrix} + \begin{bmatrix} L_{g_1} L_f^{\mu_1-1} h_1 & \cdots & L_{g_m} L_f^{\mu_1-1} h_1 \\ \vdots & \ddots & \vdots \\ L_{g_m} L_f^{\mu_l-1} h_l & \cdots & L_{g_m} L_f^{\mu_l-1} h_l \end{bmatrix} \begin{bmatrix} u_1 \\ \vdots \\ u_m \end{bmatrix} = T(x) + E(x)u = \begin{bmatrix} v_1 \\ \vdots \\ v_l \end{bmatrix} = v \tag{6}$$

Inverting the equality, the expression for the controller in Equation (7) is obtained:

$$u = E^+(v - T) + \left(I - E^+E\right)\tau \tag{7}$$

A typical choice of the state vector in the new base is the following:

$$z = \begin{bmatrix} z_1 \\ z_2 \\ \vdots \\ z_\mu \end{bmatrix} = \begin{bmatrix} h_1 \\ \vdots \\ L_f^{\mu_1-1} h_1 \\ \vdots \\ h_l \\ \vdots \\ L_f^{\mu_l-1} h_l \end{bmatrix} \tag{8}$$

In Equation (8), $\mu = \sum_k \mu_k$ is the relative degree of the system, or rather the total number of derivates that is needed to have at least one component of the control laws vector in the $y_k^{\mu_k}$ expressions.

In Equations (5) and (6), with $L_f^{\mu_i} h_i(x) = \frac{\partial^{\mu_i} h_i(x)}{\partial x^{\mu_i}} f(x)$, it is indicated the μ_i-th Lie derivate of $h_i(x)$ along the vector field $f(x)$, and $L_{g_k} L_f^{\mu_i-1} h_i = \frac{\partial}{\partial x}\left(\frac{\partial^{\mu_i-1} h_i(x)}{\partial x^{\mu_i-1}} f(x)\right) g_k(x)$ is the Lie derivate along the k-th control vector $g_k(x)$.

In Equation (7), the vector τ is an arbitrary vector that is projected in the null space of the matrix $E(x)$ and is often a function to minimize. It is projected in the null space through the operator $I - E^+E$, where I indicates the identity matrix and E^+ indicates the singular value decomposition of the matrix $E(x)$. To avoid problems with the projection of the control components in the null space, we assume the case in which the $E(x)$ matrix is square and invertible. This means that starting from now, we omit the contribution of the vector τ in Equation (7).

The new controller $v(t)$ can be chosen with a linear technique that in this context, will be a static feedback of the input status made in the following way:

$$v_i = v_{ref_i} + \sum_{j=1}^{l} K_{ij}\left(z_j - z_{ref_j}\right) \tag{9}$$

In Equation (9), z_j is the j-th variable of the state vector in the new base, K_{ij} is a component of the control gain matrix, and v_{ref_i} is often used to define a reference on the derivate of the components of the state vector in the new base.

To highlight the potential of this control technique, the angular position control was taken into consideration by comparing it with a classic cascade structure FOC scheme.

For position control, therefore, control outputs $y_1 = x_1 = i_d$, $y_2 = x_3 = \vartheta_m$ are taken into account; in this way, we obtain the following expressions in terms of the control vector:

$$u_1 = L_{eq}v_1 + R_s x_1 - pL_{eq}x_1 x_2 \tag{10}$$

$$u_2 = \frac{2J_m L_{eq}}{3pk_\phi}\left\{-\frac{1}{J_m}\left\{\frac{3pk_\phi}{2L_{eq}}\left[-R_s x_2 - px_4\left[L_{eq}x_1 + k_\phi\right]\right] + Zx_4\left\{\sum_{k=1}^{4} T_k \cos\{Zkx_3 + \alpha_k\}\right\} - \frac{\beta}{J_m}\left\{\frac{3pk_\phi x_2}{2} + \left\{\sum_{k=1}^{4} T_k \sin\{Zkx_3 + \alpha_k\}\right\} - \beta x_4\right\}\right\}\right\} + \frac{2J_m L_{eq}}{3pk_\phi}v_2 \quad (11)$$

This is also the case of exact linearization feedback, because the state vector of the new base has the same dimension of the state of the model in axes d–q. This guarantees the stabilization of the starting system, because three derivations are necessary for y_2 and one for y_1, which gives us a relative degree of four. This also represents the cardinality of the initial state space.

Figure 6 schematically shows the different structure of the nonlinear control system with respect to the classic cascade architecture in Figure 7. In classic cascade architecture is considered the case of PI (proportional-integrative) control for all three concentric loops: two parallel loops for the d–q axis currents, then cascaded to angular speed loop, and angular position loop.

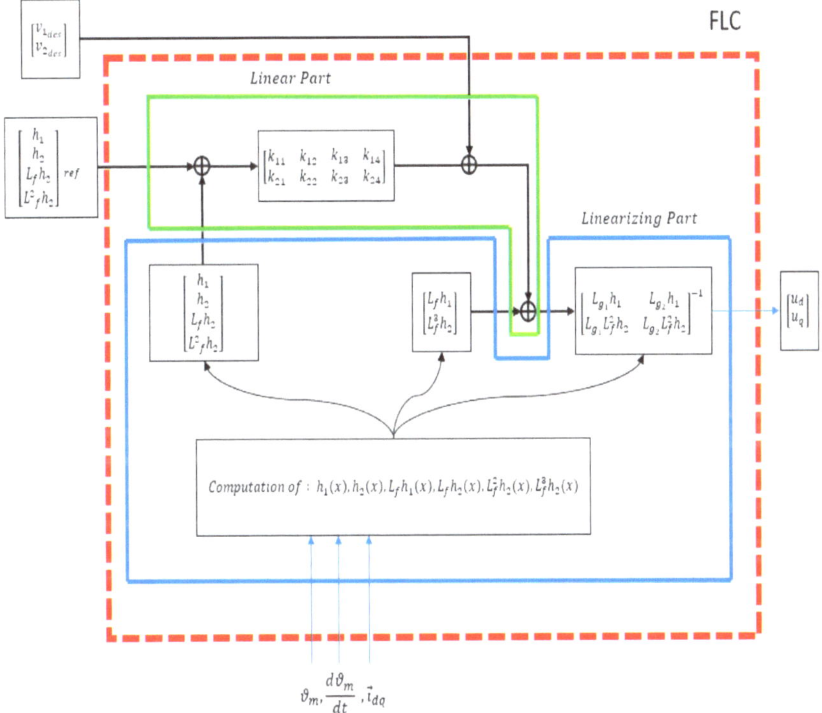

Figure 6. Schematic representation of the feedback linearization control (FLC) structure.

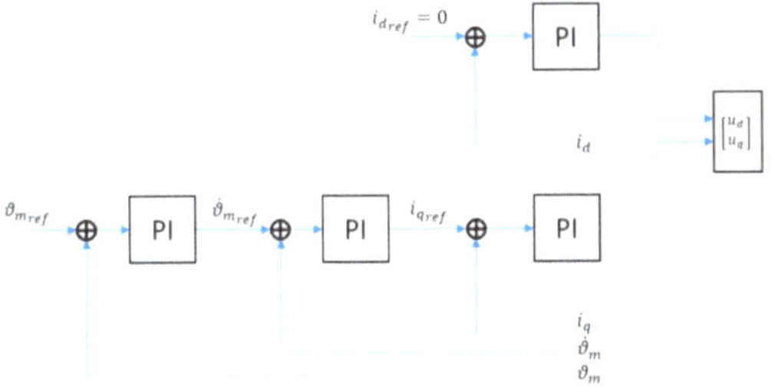

Figure 7. Classical cascade control structure. PI: proportional-integrative.

Applying u_1 and u_2, the dynamic in Equation (12) is obtained, and this representation of the new state dynamic has the property of complete controllability, which guarantees that we can change all the eigenvalue with a state feedback control.

$$\dot{z} = Az + Bv = \begin{bmatrix} 0 & 0 & 0 & 0 \\ 0 & 0 & 1 & 0 \\ 0 & 0 & 0 & 1 \\ 0 & 0 & 0 & 0 \end{bmatrix} z + \begin{bmatrix} 1 & 0 \\ 0 & 0 \\ 0 & 0 \\ 0 & 1 \end{bmatrix} v, z = \begin{bmatrix} h_1 \\ h_2 \\ L_f h_2 \\ L_f^2 h_2 \end{bmatrix}, y = Cz = [1\ 1\ 0\ 0]z \quad (12)$$

Another advantage of the linearization feedback technique with respect to all the other control techniques is that of being able to also choose non-feedback signals as control outputs; for example, the power for joule effect. The linearization feedback technique also permits choosing a zero-power reference together with a position reference, thus guaranteeing that the change of variables is formally fair for the reasons mentioned above. This allows to constrain the behavior of physical quantities that are the result of the combination of those classically used for feedback (currents and position or angular velocities).

For example, for the "joule power – position" control, we can take into account the output functions $y_1 = R_s(x_1^2 + x_2^2)$, $y_2 = x_3$, and applying the procedure described above, the following expressions are obtained for $E(x)$ and $T(x)$, from which we can obtain the control vector by applying Equation (7).

$$E(x) = \begin{bmatrix} \frac{2R_s}{L_{eq}} x_1 & \frac{2R_s}{L_{eq}} x_2 \\ 0 & \frac{2pk_\phi}{2J_m L_{eq}} \end{bmatrix} \quad (13)$$

$$T(x) = \begin{bmatrix} \frac{2R_s}{L_{eq}}[R_s x_2 + px_2(L_{eq}x_1 + k_\phi)]x_2 - 2R_s\left(px_2x_4 - \frac{R_s}{L_{eq}}x_1\right)x_1 \\ \frac{\beta}{J_m}\left(\frac{3}{2}pk_\phi x_1 + \sum_k T_k sin(kZx_3 + \alpha_k) - \beta x_4\right) - \frac{3pk_\phi}{2J_m L_{eq}}(R_s x_2 - px_4(L_{eq}x_4 + k_\phi)) - Zx_4 \sum_k T_k cos(kZx_4 + \alpha_k) \end{bmatrix} \quad (14)$$

We can observe that this is another case of exact feedback linearization, because the relative degree for the first output function is one and for the second is still three, as in the previous case.

In the case of current control or angular velocity, this is no longer guaranteed. Therefore, it will be necessary to verify that in the completion of the state vector to satisfy formal issues related to the theorem of the implicit functions, the zero-dynamics criterion is also satisfied.

5. Reducing Cogging Control and Implementation Results

The dynamic model considered for the motor is the one previously presented in Section 4. The model is parametric. For convenience, a mask has been created that encloses all the electric drive parameters. To apply the proposed technique to different electric machines is quite simple, since it is sufficient to change the parameters of the model of the engine in the mask. The parameters that have been used for the motor and for the cogging model to obtain the quantitative results presented in this work are shown in Tables 1–3. These parameters are typical of a motor for robotic manipulators for industrial applications. Reference is made to a classical AC motor with SPM structure [23,24]. As reference signals, a zero reference as justified in [26] is chosen for the direct axis curves. Meanwhile, for the position reference, a constant signal with sudden positioning changes at intervals has been selected to simulate a typical situation for a machine used in precision applications, such as, for example, a processing by a computer numerical control (CNC) machine. To make the simulation more realistic, an inverter with a pulse width modulation (PWM) modulator model has been incorporated, where the carrier is a triangular signal with amplitude of 700 V and the frequency is 10 kHz. In the following, we report the results obtained for the problem of tracking the current references in the direct axis and in the rotor angular (mechanical) position. Figure 8 shows the scheme of the implementation for the mechanical equilibrium to understand better the inclusion of the cogging term in the SPM model.

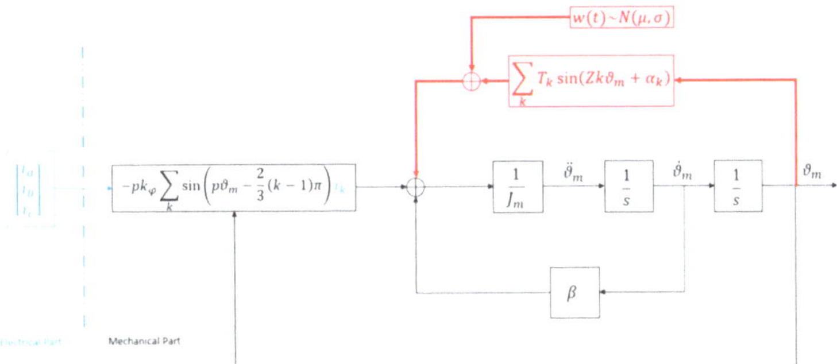

Figure 8. Model of the cogging torque (in the red box) inserted into the mechanical differential equations of the motor.

It is worth noting that the cogging torque behavior is an intrinsic nonlinearity that is already considered in this work. Other nonlinearities, such as the saturation effects on the current and on the voltage, are also taken into account with the model of the inverter and with a saturation block in the electrical dynamic equilibrium. Instead, concerning the intrinsic behavior of the magnets in terms of hysteresis curve, we assume working in the linear part of the magnetic characteristic.

In the following, we show the results in terms of trajectory tracking control for the angular mechanical position and the current for direct axis frame. The effective advantage of the proposal technique is shown by a comparison with the cascade structure for position control in a FOC structure, as explained in [26].

The goal of our work is to propose a method to mitigate the cogging torque effects on the final application (in terms of rotor axis positioning) by the projection of a control laws vector. This means that, differently from other works that propose an alternative design of internal geometry of the machine, we do not aim at deleting the presence of the cogging torque. The effect of our proposed solution can be appreciated in terms of the final result in the path-following problem, where the position, speed, or current in the quadrature axis of the machine is constrained at a precise value by the control system. This is the main difference between the proposed algorithm and classical control techniques, because we directly insert the cogging torque model inside the equations needed for the projection of the control law vector; see Figures 8 and 9.

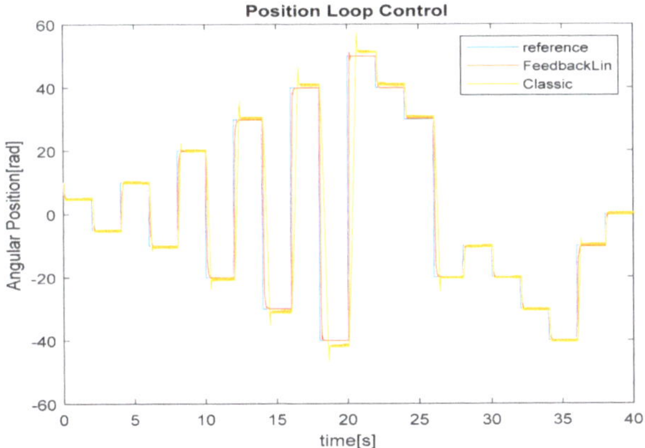

Figure 9. Comparison of the result of the control with reduction of the cogging by feedback linearization vs. a classical cascade control structure [26].

As previously said, the case of tracking trajectories of constant type has been chosen to simulate the request for rapid change of the angular position, something that often happens in industrial applications such as in automated chains or in robotized cells.

Figure 9 shows the result of the angular position reference tracking. It is shown that nonlinear control is more efficient than the classic cascade architecture in absorbing the oscillation effect due to the cogging torque. Furthermore, the proposed nonlinear control is better than the classic cascade architecture in tracking the reference, due to the fact of exploiting the intrinsic nonlinearities of the model within the control architecture. Indeed, in Figure 10, when tracking the reference from 8 s to 10 s, the classic cascade control is affected by an overshoot and by oscillations, that instead are avoided when using the proposed FLC technique. Similar results are obtained in terms of current control; see Figure 11.

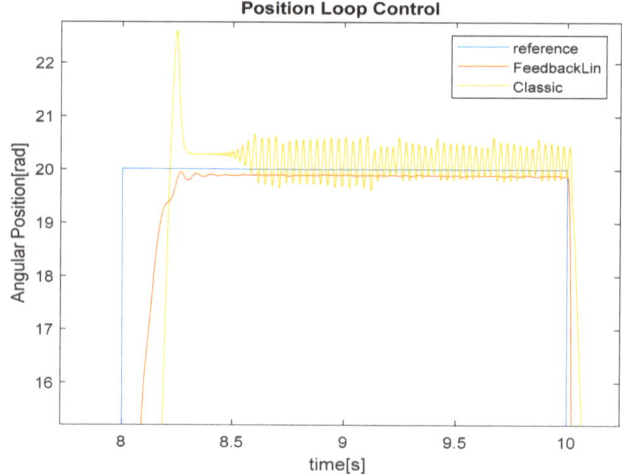

Figure 10. A detail of Figure 9 to highlight the ability of the nonlinear control to reduce the oscillation effect due to the cogging torque.

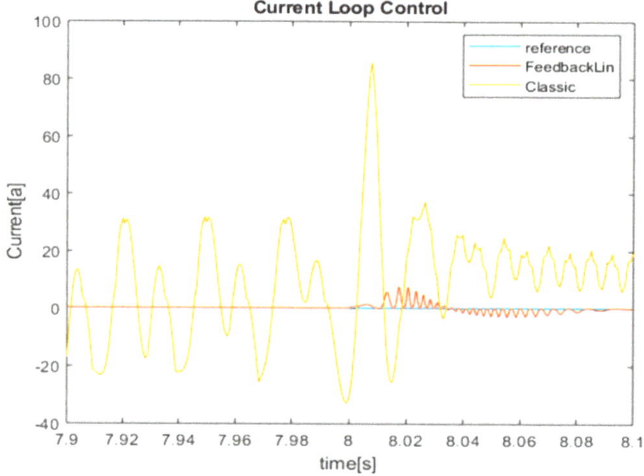

Figure 11. A detail of the current trajectory tracking to highlight the ability of the nonlinear control to reduce the oscillation effect due to the cogging torque.

Surely, the control with linearization feedback guarantees a region of asymptotic stability greater than any control based on a linear technique and is therefore more robust to parametric variations. To verify that the proposed control can be applied to a real test bench, we include a graphic analysis of robustness to the variation of the main machine parameters up to ± 30% vs. nominal values considered to design the controller; see Figures 12–14.

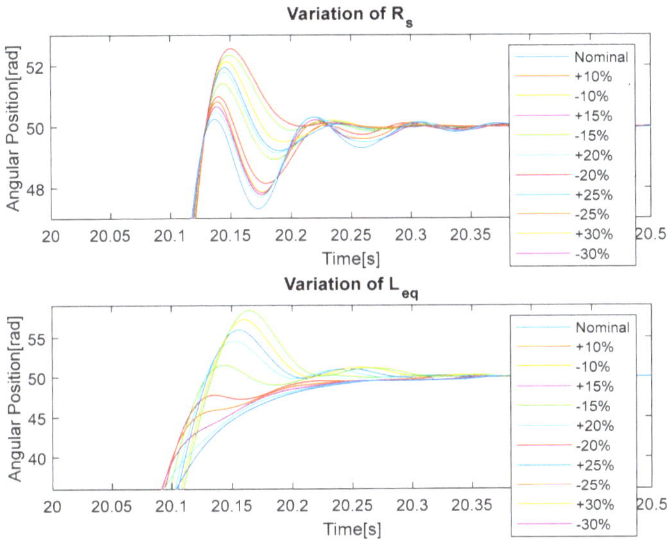

Figure 12. Robustness analysis in terms of variation of the parameters R_s and L_{eq}.

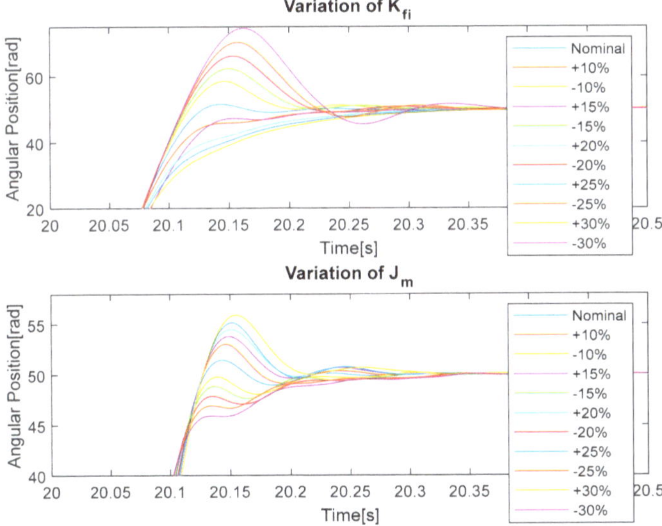

Figure 13. Robustness analysis in terms of variation of the parameters k_φ and J_m.

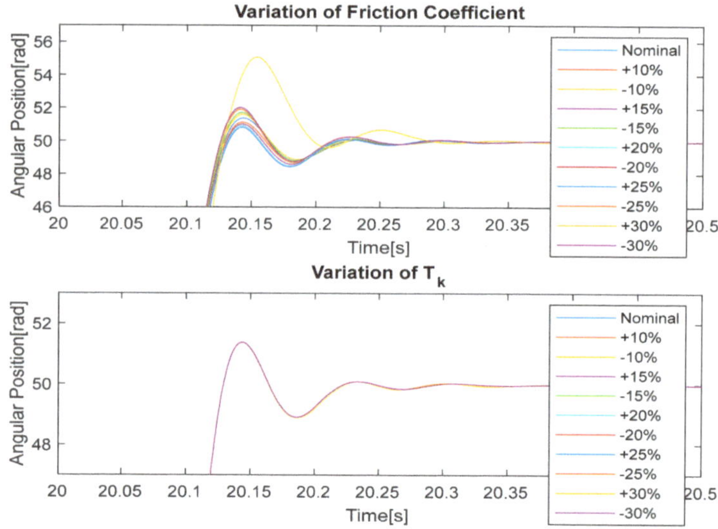

Figure 14. Robustness analysis in terms of variation of the parameters β (friction coefficient) and T_k.

It must be noted that, apart from the specific case example we have shown, the global model that we use to produce the control laws vector is a scalable model. This means that if the parameters described in Tables 1–3 are known, then the proposed algorithm can be applied to all types of synchronous motor, as well as step motors, IPMs, and BLDC (BrushLess DC) motors of all sizes and voltages.

6. Algorithm Profiling and Porting in a Low-Cost Embedded Platform

We have verified the performance of the proposed method using a Simulink profiler (on MATLAB 2018a version release) during the simulation. The profiler collects performance data while simulating the model and generates a report, called a *simulation profile*, based on the data. The simulation profile generated by the profiler shows how much time Simulink spends executing each function required to simulate the model. The profile helps to determine the parts of the model that require the most time to simulate and hence where to focus the model optimization efforts.

The summary file displays the following performance total; see Table 4.

Table 4. Description of the fields in the report.

Item	Description
Total Recorded Time	Total time required to simulate the model.
Number of Block Methods	Total number of invocations of block-level functions (e.g., Output()).
Number of Internal Methods	Total number of invocations of system-level functions (e.g., Model Execute).
Number of Nonvirtual Subsystem Methods	Total number of invocations of nonvirtual subsystem functions.
Clock Precision	Precision of the profiler's time measurement.

The summary section then shows summary profiles for each function invoked to simulate the model. For each function listed, the summary profile specifies the information in Table 5.

Table 5. Description of the fields for the single block.

Item	Description
Name	Name of the Function. This item is a hyperlink. Clicking it displays a detailed profile of this function.
Time	Total time spent executing all invocations of this function on as an absolute value and as a percentage of the total simulation time.
Calls	Number of times this function was invoked.
Time/call	Average time required for each invocation of this function, including the time spent in functions invoked by this function.
Self-Time	Average time required to execute this function, excluding time spent in functions called by this function.
Location	Specifies the block or model executed for which this function is invoked. This item is a hyperlink. Clicking it highlights the corresponding icon in the model diagram. Note that the link works only if you are viewing the profile in the MATLAB help browser.

We have compared the performance for both the control structures, with reference to cascade control and proposed FLC, to verify if there is an increase of the complexity that does not justify the adoption of the proposal algorithm. Table 6 shows the summary results of the Simulink profiler on the total scheme needed to implement the control (motor model, inverter model, control architecture, input signals generator).

Table 6. Summary of the Simulink profiler applied to the global Simulink scheme.

Profiler Parameter	Value
Total recorded time	952.69 s
Number of block methods	637
Number of internal methods	8
Number of model methods	11
Number of nonvirtual subsystem methods	32
Clock precision	30 ns
Clock speed	3800 MHz

Furthermore, to verify that the profiler is platform-independent, we tried it on two different processors, and we achieved the same results. The testing platforms were an Intel®Core™ i3-6300 CPU @ 3.80 GHz with two cores and an Intel®Core™ i7-8550U CPU @ 1.80 GHz with four cores.

Table 7 shows a comparison in terms of complexity of the FLC and of the Cascade PI controllers, measured as % (1.3% for FLC and 1.2% for PI Cascade) vs. the overall time spent of the whole model according to the Simulink profiler. The results in Table 5 prove that the algorithmic complexity of the FLC technique is similar to that of a classic cascade control. The increasing of complexity is very low compared to the obtained gain in terms of performance in the cogging torque rejection, as proved in Section 5.

Table 7. Comparison by summary of the profiler.

Algorithm	Time		Calls	Time/Call	Self-Time	
FLC	11.953125	1.3%	930736	0.00001284272	7.3593750	0.8%
PI Cascade	11.203125	1.2%	930736	0.00001203685	11.203125	1.2%

In Table 7, the self-time value is lower in the case of nonlinear control, just because is implemented by a MATLAB function that requires a certain amount of time spent to invoke it by the other functions. Meanwhile, this time frame does not exist for the PI control because it is implemented by a native block in Simulink.

Is important to remark that the usage of the Simulink profiler is only helpful to verify the computational cost of our proposed control algorithm with respect to the classical PI controller.

This just to decide if it is reasonable or not to implement the nonlinear control technique on a low-cost platform.

To implement the proposed FLC controller on a low-cost embedded system, we exploited the automatic code generation capabilities of Simulink tools. The building procedure from Simulink allows the generation of the code to be embedded in real-time in a low-cost Arduino-Uno platform equipped with the 8-bit microcontroller ATMEGA328 by Atmel, with a clock speed of 16 MHz, a flash memory of 32 KB, a SRAM volatile memory of 2 KB, and an EEPROM non-volatile memory of 1 KB.

Loading the equivalent program of the proposed control algorithm, we have verified that the sketch used 6042 bytes (18%) of the space available for the programs, where the maximum space is 32,256 bytes. The global variables used 226 bytes (11%) of dynamic memory, so 1822 bytes are available for the local variables, where the maximum space is 2048 bytes. Hence, the program and data memory capabilities of a low-cost embedded platform are enough to sustain the proposed control techniques. From a processing point of view, the computational cost of the proposed FLC algorithm when running on the low-cost Arduino-Uno platform is comparable to the computational cost of the classic PI cascade control, as already predicted by the results obtained with the profiler in Table 7.

Using a processor-in-the-loop verification approach (i.e., the motor and inverter are modeled and simulated and are interacting with the control algorithm implemented on the target real platform, Arduino-Uno using the ATMEGA328 MCU), Figure 15 presents a comparation between the PWM signal from the full simulated system and the measured signal from the Arduino in the processor-in-the-loop verification testbed. More specifically, Figure 15 shows the voltage measured with an oscilloscope on the three pins of the Arduino for the 3 PWM control phases. This can help in understanding that the low-cost platform is capable to run the control algorithm and to generate the PWM control signals to drive the inverters in the same mode that the simulation has shown. In Figure 15, it is important to notice that the signal in blue is logical (zero or one), since it is the result of the PWM operation by a state machine in the Simulink scheme. Instead, the red signal is the measurement of the physical level, so is an analog signal in the range 0–5 V.

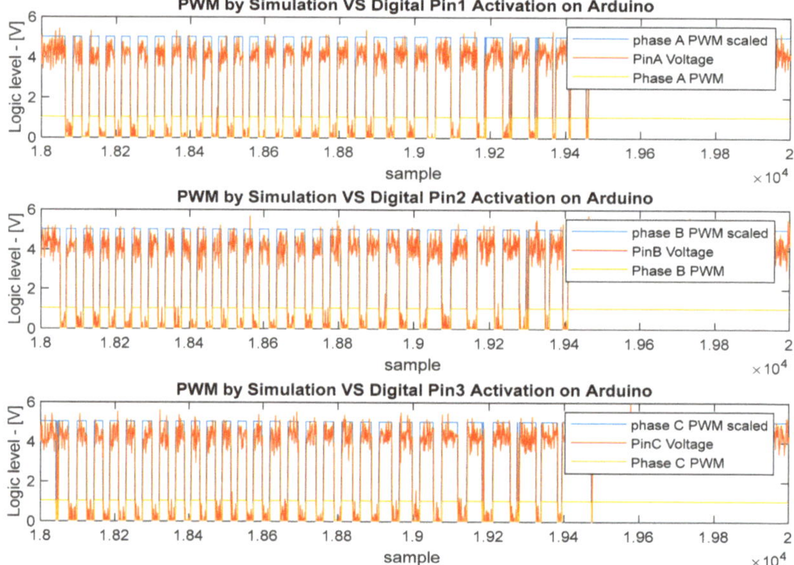

Figure 15. Comparison between the logical signal in output from the pulse width modulation (PWM) block in the simulated scheme and the physical measures of the driven Arduino digital pins (sample time of 0.0001 s) in a processor-in-the-loop validation test bed.

7. Conclusions and Future Work

In this work, a method for mitigating the effects of the cogging torque in permanent magnet synchronous motors has been presented by means of a control technique based on a feedback linearization strategy. The idea is presenting a valid alternative to the techniques of reduction of this phenomenon through physical modifications of the machine, which are expensive. The proposed nonlinear control algorithm is deterministic and easy to derive. This technique allows us to choose which are the control outputs from which to obtain the change of variables. This makes possible to emulate a FOC or DTC control technique; for example, choosing the current and speed or position as outputs. Moreover, the proposed approach allows references to be made in terms of their composition, such as choosing the power for joule effect, control output, and designing u_1 and u_2 incorporating a power feedback due to the change of base. Instead, this is not possible with a classical cascade control structure that moreover has less ability to reject the cogging effect as it does not exploit its formulation. A comparison between the proposed technique and a cascade control was proposed to show the advantages in mitigating the effect of cogging torque, with reference to a commercial motor. In addition, a parametric robustness analysis was presented to confirm the choice of the proposed control technique for the resolution of the cogging torque problem. Given the good results obtained, a future development could be the development of a sensorless-type architecture. Another future investigation in sensorless application could be to choose the best type of estimator of the state through comparative analysis, including, for example, machine learning techniques, given the increasingly frequent application of the latter.

The analysis through the Simulink profiler confirms that the complexity of the proposed control algorithm is very low. This means that it is reasonable to port the algorithm on different platforms, such as an Arduino unit with an 8-bit ATMEGA328 microcontroller.

Author Contributions: The authors contributed in equal measure to the manuscript.

Funding: This research was partially funded by University of Pisa under PRA E-TEAM project and by the Italian Government under the Crosslab program, "Dipartimenti di Eccellenza" project.

Conflicts of Interest: The authors declare no conflicts of interest.

References

1. Tudorache, T.; Trifu, I.; Ghita, C.; Bostan, V. Improved Mathematical Model of PMSM Taking Into Account Cogging Torque Oscillations. *Adv. Electr. Comput. Eng.* **2012**, *12*, 59–64. [CrossRef]
2. Xia, C.; Ji, B.; Yan, Y. Smooth speed control for low-speed high torque permanent-magnet synchronous motor using proportional-integral resonant controller. *IEEE Trans. Ind. Electron.* **2015**, *62*, 2123–2134. [CrossRef]
3. Kim, K.-C. A novel method for minimization of cogging torque and torque ripple for interior permanent magnet synchronous motor. *IEEE Trans. Magn.* **2014**, *50*, 793–796. [CrossRef]
4. Hiremath, R. Finite Element Study of Induced Emf Cogging Torque and its reductions BLDC Motor. In Proceedings of the 2017 International Conference on Intelligent Computing, Instrumentation and Control Technologies (ICICICT), Kannur, India, 6–7 July 2017.
5. Ma, G.; Qiu, X.; Yang, J.; Bu, F.; Dou, Y.; Cao, W. Structural Parameter Optimization to Reduce Cogging Torque of the Consequent Pole In-Wheel Motor. In Proceedings of the 2018 IEEE 18th International Power Electronics and Motion Control Conference (PEMC), Budapest, Hungary, 26–30 August 2018.
6. Fei, W.; Luk, P.-K. Torque ripple reduction of a direct-drive permanent-magnet synchronous machine by material-efficient axial pole pairing. *IEEE Trans. Ind. Electron.* **2012**, *59*, 2601–2611. [CrossRef]
7. Springob, L.; Holtz, J. High-bandwidth current control for torque ripple compensation in pm synchronous machines. *IEEE Trans. Ind. Electron.* **1998**, *45*, 713–721. [CrossRef]
8. Mora, A.; Orellana, A.; Juliet, J.; Cardenas, R. Model Predictive Torque Control for Torque Compensation in Variable-Speed PMSMs. *IEEE Trans. Ind. Electron.* **2016**, *63*, 4584–4592. [CrossRef]
9. Jia, H.; Cheng, M.; Hua, W.; Yang, Z.; Zhang, Y. Compensation of cogging torque for flux-switching permanent magnet motor based on current harmonics injection. In Proceedings of the 2009 IEEE International Electric Machines and Drives Conference, Miami, FL, USA, 3–6 May 2009.

10. Arias, A.; Caum, J.; Ibarra, E.; Grino, R. Reducing the Cogging Torque Effects in Hybrid stepper Machines by Means of Resonant Controllers. *IEEE Trans. Ind. Electron.* **2019**, *66*, 2603–2612. [CrossRef]

11. Ni, R.; Xu, D.; Wang, G.; Ding, L.; Zhang, G.; Qu, L. Maximum efficiency per ampere control of permanent-magnet synchronous machines. *IEEE Trans. Ind. Electron.* **2015**, *62*, 2135–2143. [CrossRef]

12. Rashid, M.H. *Power Electronics Handbook*; Academic Press: San Diego, CA, USA, 2001.

13. Cardenas, R.; Espina, E.; Clare, J.; Patrick, W. Self-tuning resonant control of a 7-leg back-to-back converter for interfacing variable speed generators to 4-wire loads. *IEEE Trans. Ind. Electron.* **2015**, *62*, 4618–4629. [CrossRef]

14. Hasanzadeh, A.; Edrington, C.; Maghsoudlou, B.; Fleming, F.; Mokhtari, H. Multi-loop linear resonant voltage source inverter controller design for distorted loads using the linear quadratic regulator method. *IET Power Electron.* **2012**, *5*, 841–851. [CrossRef]

15. Rodriguez, J.; Kazmierkowski, M.P.; Espinoza, J.R.; Zanchetta, P.; Abu-Rub, H.; Young, H.A.; Rojas, C.A. State of the art of finite control set model predictive control in power electronics. *IEEE Trans. Ind. Inf.* **2013**, *9*, 1003–1016. [CrossRef]

16. Rodriguez, J.; Cortes, P. *Predictive Control of Power Converters and Electrical Drives*; Wiley: Hoboken, NJ, USA, 2012.

17. Fuentes, E.; Rodriguez, J.; Silva, C.; Diaz, S.; Quevedo, D. Speed control of a permanent magnet synchronous motor using predictive current control. In Proceedings of the 2009 IEEE 6th International Power Electronics and Motion Control Conference, Wuhan, China, 17–20 May 2009; pp. 390–395.

18. Preindl, M.; Bolognani, S. Model predictive direct torque control with finite control set for PMSM drive systems, part 1: Maximum torque per ampere operation. *IEEE Trans. Ind. Inf.* **2013**, *9*, 1912–1921. [CrossRef]

19. Preindl, M.; Bolognani, S. Model predictive direct torque control with finite control set for PMSM drive systems, part 2: Field weakening operation. *IEEE Trans. Ind. Inf.* **2013**, *9*, 648–657. [CrossRef]

20. Cho, Y.; Lee, K.-B.; Song, J.-H.; Lee, Y. Torque-ripple minimization and fast dynamic scheme for torque predictive control of permanent-magnet synchronous motors. *IEEE Trans. Power Electron.* **2015**, *30*, 2182–2190. [CrossRef]

21. Preindl, M.; Bolognani, S. Model predictive direct speed control with finite control set of PMSM drive systems. *IEEE Trans. Power Electron.* **2013**, *28*, 1007–1015. [CrossRef]

22. Chu, H.; Gao, B.; Gu, W.; Chen, H. Low-Speed Control for Permanent-Magnet DC Torque Motor Using Observer-Based Nonlinear Triple-Step Controller. *IEEE Trans. Ind. Electron.* **2017**, *64*, 3286–3292. [CrossRef]

23. Available online: https://www.rta.it/it/product/4662-azionamento-brushless-rs1c10al-400-vac (accessed on 1 May 2019).

24. Available online: https://www.rta.it/it/product/4692-motore-brushless-q2ca2215kvxs00m (accessed on 1 May 2019).

25. Zhang, P.; Sizov, G.Y.; Demerdash, N.A.O. Comparison of Torque Ripple Minimization Control Technique in Surface-mounted Permanent Magnet Synchronous Machines. In Proceedings of the 2011 IEEE International Electric Machines & Drives Conference (IEMDC), Niagara Falls, ON, Canada, 15–18 May 2011.

26. Krause, P.; Wasynczuk, O.; Sudhoff, S.; Pekarek, S. *Analysis of Electric Machinery and Drive System*, 3rd ed.; Wiley: Hoboken, NJ, USA, 2013.

27. Isidori, A. *Non-Linear Control System*; Springer: London, UK, 1995.

Article

Spatio-Temporal Model for Evaluating Demand Response Potential of Electric Vehicles in Power-Traffic Network

Lidan Chen [1,*] **, Yao Zhang** [2] **and Antonio Figueiredo** [3]

[1] School of Electrical Engineering, Guangzhou College of South China University of Technology, Guangzhou 510800, China
[2] School of Electric Power, South China University of Technology, Guangzhou 510640, China; epyzhang@scut.edu.cn
[3] Department of Electronic Engineering, University of York, York YO10 5DD, UK; ajff_08@outlook.com
* Correspondence: chenld@gcu.edu.cn

Received: 23 April 2019; Accepted: 18 May 2019; Published: 23 May 2019

Abstract: Electric vehicles (EVs) can be regarded as a kind of demand response (DR) resource. Nevertheless, the EVs travel behavior is flexible and random, in addition, their willingness to participate in the DR event is uncertain, they are expected to be managed and utilized by the EV aggregator (EVA). In this perspective, this paper presents a composite methodology that take into account the dynamic road network (DRN) information and fuzzy user participation (FUP) for obtaining spatio-temporal projections of demand response potential from electric vehicles and the electric vehicle aggregator. A dynamic traffic network model taking over the traffic time-varying information is developed by graph theory. The trip chain based on housing travel survey is set up, where Dijkstra algorithm is employed to plan the optimal route of EVs, in order to find the travel distance and travel time of each trip of EVs. To demonstrate the uncertainties of the EVs travel pattern, simulation analysis is conducted using Monte Carlo method. Subsequently, we suggest a fuzzy logic-based approach to uncertainty analysis that starts with investigating EV users' subjective ability to participate in DR event, and we develop the FUP response mechanism which is constructed by three factors including the remaining dwell time, remaining SOC, and incentive electricity pricing. The FUP is used to calculate the real-time participation level of a single EV. Finally, we take advantage of a simulation example with a coupled 25-node road network and 54-node power distribution system to demonstrate the effectiveness of the proposed method.

Keywords: electric vehicle (EV); trip chains; demand response; user participation; dynamic road network; fuzzy algorithm

---

1. Introduction

1.1. Motivation and Background

Persistent growth of the global economy is causing issues relating to energy supply, environmental pollution, and dependence on fossil fuels, all of which need to be addressed with a sense of urgency [1]. To better tackle these problems, many countries have been committing to support the development of electric vehicle technologies as well as provide incentives to encourage the use of EVs (e.g., battery electric vehicles (BEV) and plug-in hybrid vehicles (PHEV)) [2]. It means that more and more EVs will be connected to the grid and interact with the utility in the future. On the one hand, EVs are a flexible power load [3], they will require charging from the grid at different times and at different locations [4,5]. As a consequence, the coordination of EVs' charging has been widely studied in the recent period [6,7]. On the other hand, it is also a kind of distributed energy storage resource. Since EVs have a lot of time to dwell in the parking lots during the day, they have great potential to participate in the power system DR service [8,9]. However, due to the flexibility and randomness of electric vehicle behavior and the uncertainty of participation in demand response [10], it is difficult to assess the potential of participation in demand response events, especially under the power-traffic hybrid network, which is important for the planning and operation of the power grid and transportation, therefore, an assessment of spatio-temporal uncertainties and user participation uncertainties in EV-DR is inevitable. It is worth noting that because different types of EVs have a different charging time, charging power, and battery capacity, we focus on BEVs in this paper.

1.2. Literature Overview

Motivated by the above reasoning, at present, many researchers have carried out EV and grid interaction related research, in particular, user-side management, EVs' demand response. Reference [11] analyzes the users' power transfer, reduction behavior, and the response to the demand of dispatching in the context of time-of-use electricity price and pricing strategy. In [12], an algorithm for distributed EVs' DR to shape the daily demand profile in a day-ahead market is presented. The authors in [13] provided a collaborative evaluation of dynamic-pricing and peak power limiting-based DR strategies for home energy management (HEM). The authors in [14] make full use of the EVs' DR capability and propose a corresponding frequency control strategy. Also, in [15], the author investigated a charging and discharging strategy for EVs that can utilize the DR capability of V2G in residential distribution networks. However, EVs' DR capability is assumed to be activated only after the vehicles arrive home. The authors in [16] presented an intelligent energy management framework with DR capability for industrial facilities, yet the user's willingness to participate DR program was ignored. In a recent work [17], an EV parking lot energy management system is present in consideration of the uncertainties of the arrival and departure time, and the remaining state-of-energy of EVs just before charging operation.

Valuable insights of EVs demand response works were provided in previous studies. However, it is worth mentioning that in the previously cited approaches, only the EVs' time-varying charging/discharging characteristics were taken into account, they consider EVs as a type of fixed load or response resource, while the location of EVs is commonly disregarded. In addition, the previous studies' take on EVs is that they can participate in DR events when EVs have objective controlled ability, ignoring the uncertainty of user participation willingness, the subjective participation degree of users is not considered in detail.

1.3. Contributions

Thus, our focus in this paper is the EV demand response potential evaluation from the perspective of spatio-temporal distribution and vehicle owner participation capability. The main contributions of this work are summarized as follows.

(1)　Aspects beyond the characteristics of spatial distribution of EVs and travel pattern analysis have been neglected in the existing literature, we model a dynamic traffic network considering the traffic time-varying information with randomness in travel behavior based on trip chains.

(2)　A method to analyze EVs' objective and subjective participation in a DR event is developed.

(3)　Differently from the fixed demand response mode in the related research, we proposed a fuzzy logic-based mechanism, we modeled uncertainties that affect the estimation of demand response potential of a single EV and EVA. Three key factors—the remaining parking time, the remaining SOC, and incentive electricity pricing—are considered.

(4)　The real-time participation level of a single EV and EVA from a spatio-temporal scope in the power-traffic network are evaluated.

The remainder of this paper is organized as follows: In Section 2, we formulate the spatio-temporal model of electric vehicle travel patterns based on trip chains under dynamic road network. In Section 3, the objective participation ability as well as the subjective participation ability of EV users are considered, and the EVs DR mechanism is obtained by fuzzy algorithm. In Section 4, the case study and the results are presented, analyzed, and discussed. Conclusions are drawn in Section 5.

2. EVs Travel Model in Dynamic Traffic Network

The proposed electric vehicles aggregator demand response evaluation (EVA-DRE) method is illustrated in Figure 1. The first part of the method is EVs' travel modeling to get the spatial and temporal distribution, which will be introduced in this section. Another part is the EV user participation modeling and response mechanism to obtain the EV-DR power and capacity.

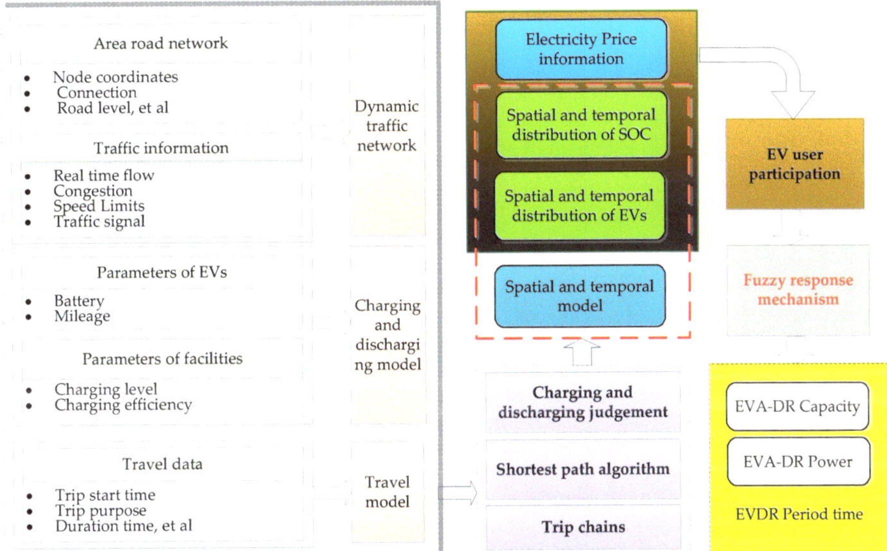

Figure 1. Scheme of proposed EVA-DRE method.

In this section, we will provide a general method for simulating the daily travel pattern with dynamic traffic network and trip chains. First, the time-dependent dynamic road network model is established by using graph theory. Travel characteristics of EVs are then analyzed. Furthermore, the process of the travel pattern simulation is presented.

2.1. Time-Dependent Dynamic Road Network Model

During the day's travel of a private electric vehicle, it will depart from the starting point which we assume the house, and it will pass through one or more trip destinations, including multiple trip, and the choice of each travel route will be affected by the road network and traffic conditions. Yet, the traffic conditions of the road network change over time as shown in Table 1. In addition, graph theory is usually adopted to describe the complicated actual road network [18]. In this work, the traffic time-varying information is considered in the road network, as in Equation (1).

$$\begin{cases} G = (V, E, W, D, T) \\ V = \{1, 2, \dots, n\} \\ E = \{e_{ij} | i, j \in V\} \\ W = \{t_r(k) | r \in E, k \in T\} \\ D = \{t_d(k) | d \in V, k \in T\} \\ T = \{k | k = 1, 2, \dots, K\} \end{cases} \qquad (1)$$

where, the vertex V of the graph G represents the intersection of the road, and the edge E of the graph represents the section between the two adjacent intersections, and the set of road weights W is used for describing various road lengths, travel times, and other attributes, D is the set of the delayed time of all intersections. $t_r(k)$ is the travel time function at time slot k of link r, $t_d(k)$ is the delayed time in the intersection d at time slot k; T represents the time set, and K is the total number of time intervals in a day.

Table 1. Dynamic travel time of each road section.

Road Sections/links	Time Intervals 1	2	$\cdots$	k	$\cdots$
1	$t_1(1)$	$t_1(2)$	$\cdots$	$t_1(k)$	$\cdots$
2	$t_2(1)$	$t_2(2)$	$\cdots$	$t_2(k)$	$\cdots$
$\cdots$	$\cdots$	$\cdots$	$\cdots$	$\cdots$	$\cdots$
r	$t_r(1)$	$t_r(2)$	$\cdots$	$t_r(k)$	$\cdots$
$\cdots$	$\cdots$	$\cdots$	$\cdots$	$\cdots$	$\cdots$

2.2. Spatio-Temporal Travel Characteristics of EVs

Suppose that EV users will go to one or more destinations during a day's travel, and the EV charging and discharging may occur in these trip destinations.

2.2.1. Trip Chains and Travel Route Planning

We use daily trip chains [19–21] which are created to show the whole travel routes in one day with spatial and temporal information, shown in Figure 2 and Equation (2).

$$Q = \{q_0(x_0, y_0), q_1(x_1, y_1), \cdots, q_s(x_s, y_s), \cdots\} \qquad (2)$$

where, Q is the set for the duration trip destinations of the trip chain, s is the number of the duration trip destinations, $q_0, q_1 \dots q_s$ indicate all of the trip destinations, (x_s, y_s) is the corresponding coordinates, q_0 is the first place of the trip chain which is considered to be the house in this paper.

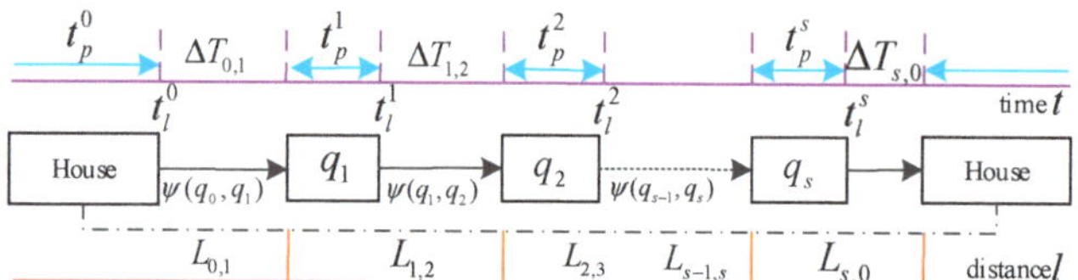

Figure 2. The daily trip chains.

A path among consecutive trip destinations in the trip chain is represented by $\psi(q_s, q_{s+1})$. The path set that characterizes the EV's spatial travel process is expressed as Π in Equation (3).

$$\Pi = \{\psi(q_0, q_1), \psi(q_1, q_2), \cdots, \psi(q_s, q_{s+1}), \cdots\} \qquad (3)$$

Let $t_l^i, t_p^i (i = 0, 1, 2, \ldots, s, \ldots)$ be the departure time and the parking duration time in the ith trip destination, respectively, and t_p^0 is the dwelling time in the house. In Figure 2, $L_{i,i+1} (i = 0, 1, 2, \ldots, s - 1, \ldots)$ and $\Delta T_{i,i+1}$ are the travel distance and travel time between two trip destinations, respectively.

2.2.2. Departure Time of the First Trip

Here, we consider the departure time t_l^0 of the EVs' first trip in a daily horizon to be randomly distributed according to probability distribution function (pdf) as

$$t_l^0 \sim f(t_l^0) \tag{4}$$

2.2.3. Traveling and Traveled Time

In a completed trip, EVs will path several links and several intersections. Hence, it is necessary to draw the required time to pass each link at a certain time when calculating the travel time between two destinations. Some existing link travel time functions are discussed in [22], and the traffic time-consuming coefficient is used to calculate the travel time of the road segment under the corresponding traffic index, which is more than the time-consuming multiple in the unblocked state. The logit-based volume delay function [23] as in Equation (5) is used for depicting travel time.

$$t(s, s + 1) = \sum_{r=1}^{n_r} t_r(k) + \sum_{d=1}^{m_d} t_d(k) \tag{5}$$

where,

$$\begin{cases} t_r(k) = t_0 \cdot c_1 \cdot \left[1 - \dfrac{c_2}{1 + \exp(c_3 - c_4 \cdot \theta_r(k))} \right]^{-1} \\ t_d(k) = t_0 \cdot p_1 \cdot \left[1 + \dfrac{p_2}{1 + \exp(p_3 - p_4 \cdot \lambda_r(k))} \right] \\ \theta_r(k) = \dfrac{q_r(k)}{C_r}, \lambda_r(k) = \dfrac{q_r(k)}{X_r} \\ t_0 = \dfrac{l_r}{v_0}, r \in E \end{cases} \tag{6}$$

where, $t(s, s + 1)$ represents the traveling time from s to $s + 1$; n_r, m_d are the total links and the total intersections between s and $s + 1$, respectively; t_0, v_0 are the free-flow traveling time and free-flow driving speed which are related to the road grades, respectively; l_r is the length of link r, in km; $q_r(k)$ is the real-time traffic of link r at time slot k; C_r and X_r represent road capacity and intersection capacity of link r, respectively; saturation of traffic volume $\theta_r(k)$ and $\lambda_r(k)$ are used to characterize the congestion factor (traffic index), the greater the values, the more congested roads and junctions; $c_i (i = 1, 2, 3, 4)$ are the adaptive coefficients that related to road grades, $p_i (i = 1, 2, 3, 4)$ are the adaptive coefficients of the intersection related to whether there is a traffic light.

2.2.4. Arrival and Departure Time at the Destination

The arrival and departure time of every trip's destination can be obtained by Equation (7).

$$\begin{cases} t_a^s = t_l^0 + \sum_{i=0}^{s-1} t(i, i + 1) + \sum_{i=1}^{s-1} t_p^i \\ t_l^s = t_a^s + t_p^s \end{cases} \tag{7}$$

where, t_a^s and t_l^s are the arrival and departure time at the trip destination, t_p^s is the parking time.

2.2.5. Parking Times

In this paper, it is also assumed that parking duration t_p^s of the EVs in non-residential areas—i.e., office, shopping mall—follows a probability distribution $t_p^s \sim f(t_p^s)$. In addition, the parking duration in the residential area can be obtained by Equation (8).

$$
t_p^h = \begin{cases} t_l^h - t_a^h, 0 < t_a^h < 0.5K \\ t_l^h - t_a^h + K, 0.5K \leq t_a^h \leq K \end{cases} \tag{8}
$$

where, t_p^s and t_p^h represent the parking time at trip destination s and in the house, respectively; t_a^h, t_l^h are the arrival and departure times at the house, respectively.

2.2.6. Route Planning

When the vehicle travels from the current location (source point) to a destination (destination point), the vehicle users tend to select the route in advance, and the users will choose different road resistances according to their different preferences, such as driving distance, travel time, road quality, congestion situation, travel expenses, etc. We assume that the user considers 'travel time' as the important basis for route selection. Therefore, the minimum travel time, which includes road travel time and traffic light delay time, is set as the target for the shortest path planning, i.e., Dijkstra's algorithm [24].

2.3. EV Battery SOC Estimate

When EVs arrive at a destination, when there is no demand response event, the user will decide whether to replenish the energy for their EV according to the current battery SOC and the next trip, is defined as,

$$
\begin{cases} S(t_a^s) \cdot E_m - \sum_{r=1}^{n_{s,s+1}} \omega_r \cdot l_r \leq \zeta_0 \cdot E_m \\ S(t_a^s) = S(t_a^{s-1}) - \left(\sum_{r=1}^{n_{s-1,s}} \omega_r \cdot l_r \right) / E_m \end{cases} \tag{9}
$$

where, $S(t_a^s)$ is the SOC at arrival time of destination s; E_m is the battery capacity of EV m, in kWh; ω_r is the energy consumption per kilometer, in kWh/km; ζ_0 is preset by EV user; $n_{s-1,s}$ and $n_{s,s+1}$ represent the number of links between two trip destinations. Likewise, the battery state of leaving the trip destination is obtained.

$$
\begin{cases} S(t_l^s) = \begin{cases} S(t_a^s), \gamma = 0 \\ \min\{S(t_a^s) + \Delta S(t_p^s), S_{set}, S_{up}\}, \gamma = 1 \\ \max\{S(t_a^s) + \Delta S(t_p^s), S_{set}, S_{low}\}, \gamma = -1 \end{cases} \\ \Delta S(t_p^s) = \gamma \cdot \delta \cdot \dfrac{P_c^s \cdot t_p^s}{E_m} \end{cases} \tag{10}
$$

where, $S(t_l^s)$ is the SOC at the departure time; P_c^s is the rated charging power; γ is a flag sign, 0, 1, and -1 are no charging, charging and discharging, respectively; δ is the charging/discharging efficiency; S_{set} is the SOC of the departure time set by EV user; S_{up} and S_{low} represent the upper limit considering the battery life and the minimum limit to support the next trip, respectively.

2.4. Travel Pattern Simulation

The temporal and geographical information of EVs in a travel day can be obtained by performing the following six key steps:

Step 1. Obtain the survey results of residents from the transportation department and analyze the structure type of the vehicles' trip chains.

Step 2. The space movement state of the vehicle is determined according to the trip chain structure that the travel destinations of the vehicle are clear.

Step 3. The first departure time of the vehicle is extracted by Equation (4) according to the type of travel chain Equation (2).

Step 4. The travel path space and time between two consecutive trip destinations are determined by the path planning algorithm and Equations (3), (5), and (6).

Step 5. Extract the dwell time of the different destinations according to Equation (8). The arrival and departure time are calculated by Equation (7).

Step 6. The state of charge of the battery in each destination is judged and calculated by Equations (9) and (10).

Through Step 1 to Step 6, we can obtain the spatio-temporal characteristics of each EV in a travel day, including the remaining SOC, the travel destination, and its dwell time. The simulation flowchart is shown in Figure 3. Monte Carlo simulation will be carried out with EVA-DRE process which will be presented in Section 3.4.

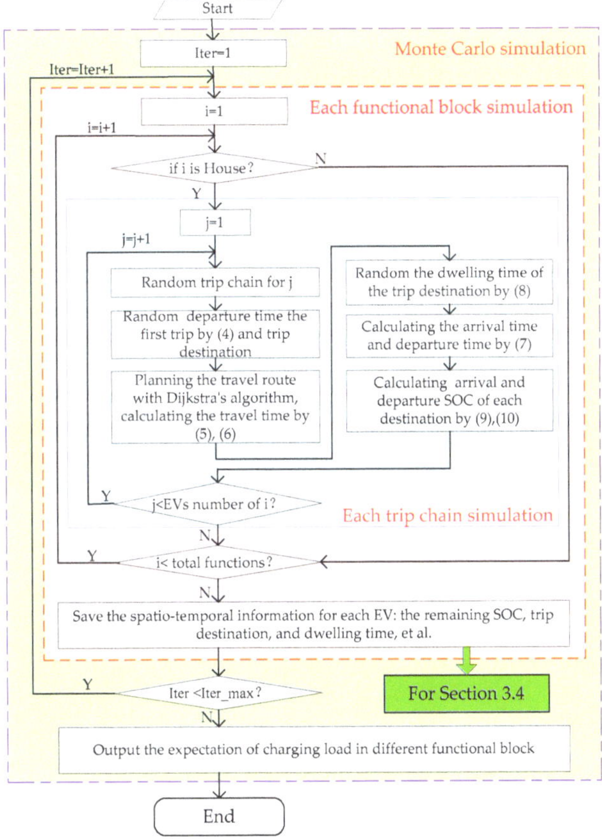

Figure 3. Flowchart of EV travel model.

3. Methodology for EVA-DRE

Having obtained the modeling of EVs travel pattern in the dynamic road network, including spatial and temporal information and SOC distribution, analyzing the user participation ability is of importance to evaluate the demand response potential upon each destination. A fuzzy rule response mechanism with three key factors are then considered in this section. The temporal and geographical distribution of single EV and EVA demand response capacity is obtained by Monte Carlo simulation.

3.1. User Participation of EVs DR

EV users usually exhibit complete rationality, limited rationality and satisfactory decision-making in the process of charging and discharging power consumption. When EVs arrive at a destination, it is only possible to participate in the actual DR event when they objectively have the DR capability. Otherwise, even if the EV user has a strong willingness to participate, it is unable to participate in the DR event. Here, the DR participation is divided into three categories: (1) participate in adjusting the charging time (delayed charging), (2) participate in the discharge case, (3) have no DR capability.

3.1.1. Objective Participation Ability

We introduce the objective participation ability $K_m^s(t)$ here to show the actual participation of EVs DR, which is presented as Equation (11) and illustrated in detail as Table 2.

$$\begin{cases} K_m^s(t) = \begin{cases} 1, & A\ or\ B \\ 0, & C \end{cases} \\ A : \Delta S \geq S_{lim} \\ B : \Delta S < S_{lim}\ \&\ t_{ch}^s < \Delta T_{sur}^s \\ C : \Delta S < S_{lim}\ \&\ t_{ch}^s \geq \Delta T_{sur}^s \\ \Delta S = S(t_a^s) - S(t(s, s+1)) \\ t_{ch}^s = \frac{\Delta S \times E_m}{\delta \times P_c^s} \end{cases} \tag{11}$$

where, ΔS is the current available SOC; $S(t(s, s+1))$ is the SOC consumed by the vehicle from s to $s + 1$; S_{lim} is the minimum residual capacity level to prolong battery life; ΔT_{sur}^s indicates the remaining time of the vehicle to the next trip; t_{ch}^s is the required charging time. EVs in cases of A and B have the capability of objective participation, but for C, regardless of parking time or SOC, it mismatches for its next trip driving requirement, thus, EVs in case C should charge the battery immediately. The charging power can be calculated by Equation (12).

$$\begin{cases} P_C^s(t) = P_c^s \sum_{m=1}^{N^C(t)} \zeta(t) \\ \zeta(t) = \begin{cases} 1, t_a^s < t < t_a^s + t_{ch}^s \\ 0, else \end{cases} \end{cases} \tag{12}$$

where, $P_C^s(t)$ is the total charging power of EVs in case C of time t at the destination of s; $N^C(t)$ is the number of EVs in case C at time t.

Table 2. Objective participation ability of EVs under different situation.

Cases	Situation	Remaining SOC	Whether to Meet the Next Trip Demand	Dwelling Time	Enough Time to Replenish	Objective Participation Ability
A		Sufficient	Yes	-	-	1
B		Insufficient	No	Long	Yes	1
C		Insufficient	No	Short	No	0

3.1.2. Subjective Participation Willingness

For different EV owners, they will make different decisions whether to participation in a DR event always based on the current state of charge, electricity price, and remaining travel time. Even the same electric vehicle owner may have different decision-making results due to random factors such as mood at the time.

We define $\rho_m(t)$ as subjective participation, to characterize the subjective willingness of EV users to participate in a DR event. Then, the subjective participation degree of A, B, and C in Section 3.1.1 can be described as Equation (13).

$$\begin{cases} A: 0 \le \rho_m^{del}(t) \le 1, 0 \le \rho_m^{v2g}(t) \le 1 \\ B: 0 \le \rho_m^{del}(t) \le 1, 0 \le \rho_m^{v2g}(t) \le 1 \\ C: \rho_m^{del}(t) = 0, \rho_m^{v2g}(t) = 0 \end{cases} \tag{13}$$

where $\rho_m^{del}(t)$ and $\rho_m^{v2g}(t)$ are the delayed and V2G participation degree of EV users, respectively.

Meanwhile, the subjective participation is limited by objective responsiveness, and it is to be satisfied as

$$\rho_m(t) = f(\alpha_m(t), \beta_m(t), \gamma_m(t) \| K_m^s(t)) \tag{14}$$

where, three essential factors $\alpha_m(t)$, $\beta_m(t)$, and $\gamma_m(t)$ are considered, which represent the remaining parking time, the remaining SOC, and the incentive price at the current time, respectively. It should be mentioned that, Equation (14) is an uncertainty function, thus, to focus on the uncertainty of EVs DR participation, fuzzy algorithm is used to calculate EV user demand responsiveness.

3.2. Responsive Mechanism Based on Fuzzy Rules

Firstly, based on the known remaining parking time and the remaining SOC from Section 2, and the incentive price are extracted to build the inputs for the fuzzy evaluator at each sampling period, these three factors should be normalized by Equation (15). Secondly, the fuzzification of the inputs is implemented based on the input membership functions and the output membership functions, which are shown in Figure 4a–c, respectively. Thirdly, the Mamdani fuzzy reasoning is carried out. The rule base is shown in Table 3, the fuzzy rules can be tuned with real tested results under different scenarios.

$$x^* = \frac{x - x_{min}}{x_{max} - x_{min}} \tag{15}$$

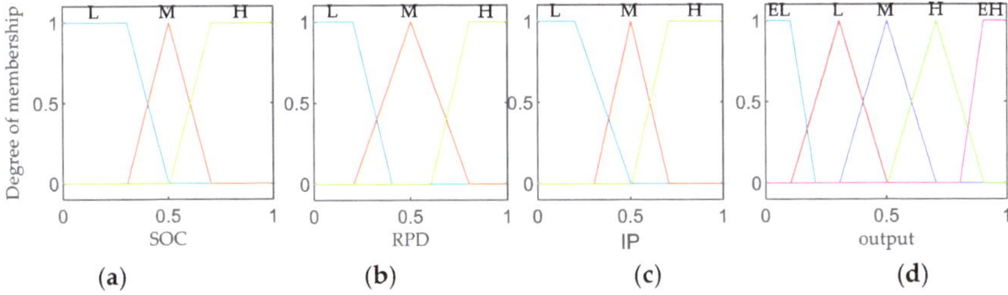

Figure 4. The membership functions. (**a**) State-of-charge (SOC); (**b**) Rest parking duration (RPD); (**c**) Incentive price (IP); (**d**) Output membership functions (EV-DR participation).

Table 3. Rule base and defuzzification.

If SOC Is	AND RPD Is	AND IP Is	Then the Participation of EVs DR Is	If SOC Is	AND RPD Is	AND IP Is	Then the Participation of EVs DR Is
L	L	L	EL	H	M	L	M
L	L	M	L	H	M	M	H
L	L	H	M	H	M	H	EH
M	L	L	L	L	H	L	L
M	L	M	M	L	H	M	H
M	L	H	M	L	H	H	H
H	L	L	L	M	H	L	M
H	L	M	M	M	H	M	H
H	L	H	H	M	H	H	EH
L	M	L	L	H	H	L	M
L	M	M	M	H	H	M	H
L	M	H	H	H	H	H	EH
M	M	L	L				
M	M	M	M				
M	M	H	H				

3.3. EVA-DR Energy and Capacity

After obtaining the demand response potential of a single EV, then we construct an aggregation model of EVs' DR. From a spatial perspective, if multiple functional blocks are powered by a certain grid node, all vehicles supplied electricity by the node are referred to herein as electric vehicle clusters which are managed by EVA. The response capability of delayed charging power and the participating discharge power at the sampling period is given by Equation (16).

$$
\begin{cases}
EVA^{del}(i,t) = \sum_{m=1}^{EV_{num,i}} \rho_m^{del}(t) \cdot P_c^s \\
EVA^{v2g}(i,t) = \sum_{m=1}^{EV_{num,i}} \rho_m^{v2g}(t) \cdot P_{dis}^s
\end{cases}
\tag{16}
$$

where, $EV_{num,i}$ is the number of EVs in the ith EVA cluster, P_c^s and P_{dis}^s are the rated charge and discharge power, respectively.

The DR capacity of an EVA and total EVAs are estimated by using Equations (17) and (18).

$$
C(i,t) = \sum_{m=1}^{EV_{num,i}} (S(m,t) - \xi_m) \cdot \rho_m(t) \cdot E_m
\tag{17}
$$

$$
C_{tot}(t) = \sum_{i=1}^{N_a} C(i,t)
\tag{18}
$$

where, $C(i,t)$ shows the DR capacity of the ith EVA at time t, $S(m,t)$ is the SOC of EV m at time t, $\rho_m(t)$ is the EVs participation in (13) and (14), ξ_m is the limited SOC set by the EVs user and N_a is the number of EVAs.

3.4. EVA-DRE Simulation Process

The steps for the proposed EVA-DRE method are provided as follows, and the simulation flowchart is described in Figure 5.

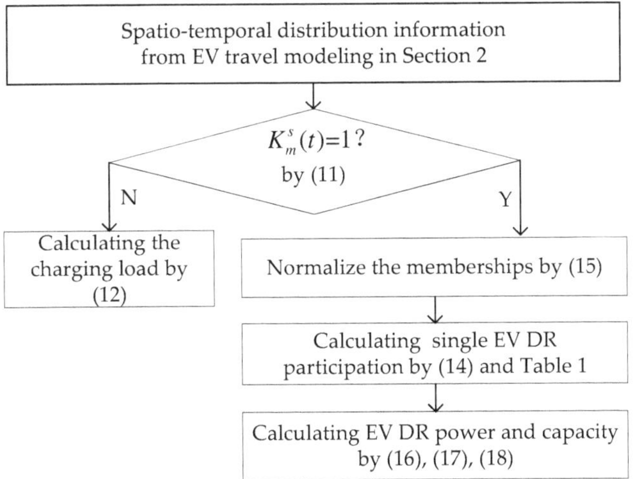

Figure 5. Flowchart of EVA-DRE method.

Step 1. The temporal and spatial distribution of the mth EV and related parameters are obtained from Section 2.

Step 2. For the participating EVs, calling the fuzzy algorithm to calculate the responsiveness of the mth EV.

Step 3. Calculate the delayed charging power and V2G power and capacity of EVA according to the location of the current time of the vehicle to the corresponding EVA.

Step 4. Accumulate the total power and capacity of EVA in the entire area.

4. Simulation Results and Analysis

In this section, we present some simulation results and the performance of the proposed method. The simulation is implemented and tested in the MATLAB software. All the results are obtained by MATLAB R2018b on a PC with Intel Core i5–4278U CPU @ 2.60 GHz, 8GB RAM memory, and 64-bit Windows 7 OS. The simulation in the case study would take 9.305 s for evaluation DR potential in each minute. In a similar fashion, to deal with large-scale dimensionality of a large scale fleet of EVs problem in [25–27], decentralized/distributed framework for evaluation can process.

4.1. Data Gathering and Parameter Settings

The parameters include road network information, traffic information, grid parameters, EV parameters, survey data of resident users, etc. A coupled network with 25-node road network [28] and 54-node distribution system [29] as shown in Figure 6 is used in the simulation. The road network in the region has 25 road nodes and 46 roads, 22 functional blocks, including 8 residential areas (H), 8 working places (W), 5 other functional areas (E), and 2 non-functional area (marked by Z1, Z2). Each functional area is powered by the distribution network node which is indicated by an arrow. For example, the gridlines in Figure 6 is the H7 block, which is powered by node 11.

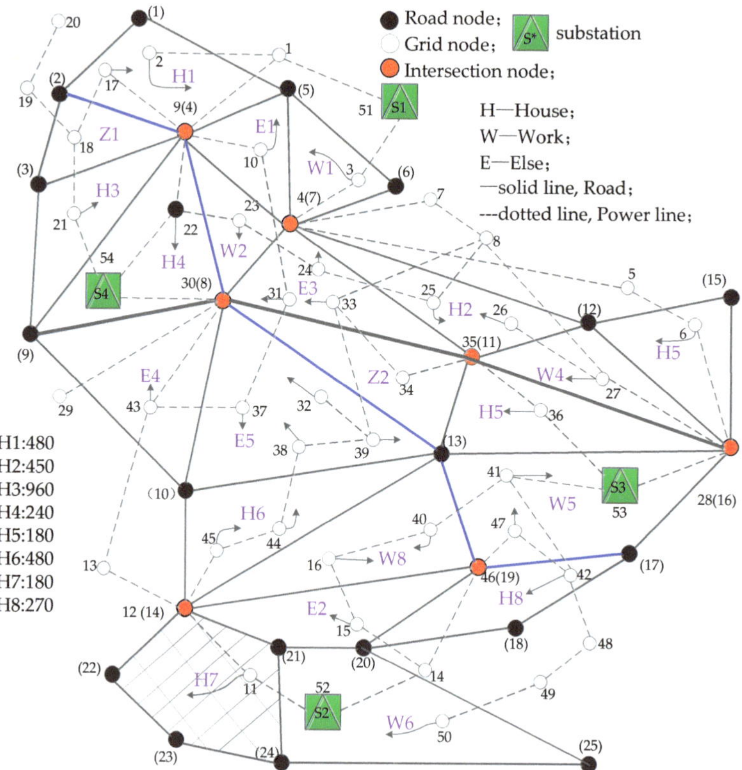

Figure 6. Topology coupled road network and distribution network.

4.1.1. Detailed Road Network Information

Detailed information including links and their corresponding area, traffic light, road grade, and the length of each link are shown in Table 4. In the column of traffic light, '1' denotes that there are traffic lights in the link, otherwise, there are no traffic lights in the link which is the first road grade with high free flow speed. In the column of area, '1' and '0' indicate the central area and other area of the city, respectively.

4.1.2. Traffic Information

In this paper, all the road sections are divided into four grades. In Figure 6, the thick black solid line is the fast track (FT), the blue sub-solid line is the main road (MR), and the rest are the secondary roads (SR). The branch roads (BR) are not reflected in the topological map which located in each functional area. Traffic lights are provided at the intersections of MR and SR. Different road grades have different free flow speeds, as shown in Table 5.

Table 4. Detailed information for the road network.

No. of Link	Original Node	Destination Node	Traffic Light	Road Grade	Length of Link	Area
1	1	5	1	3	5	0
2	1	2	1	3	4	0
3	2	3	1	3	3	0
4	2	4	1	2	4	1
5	3	4	1	3	4	0
6	3	9	1	3	4	0
7	4	5	1	3	3	0
8	4	7	1	3	5	1
9	4	8	1	2	5	1
10	4	9	1	3	7	1
11	5	6	1	3	5	0
12	5	7	1	2	5	0
13	6	7	1	3	3	0
14	7	8	1	2	3	1
15	7	11	1	3	8	1
16	7	12	1	3	9	0
17	8	9	0	1	6	1
18	8	10	1	2	6	1
19	8	11	0	1	7	1
20	8	13	1	2	7	1
21	9	10	1	3	6	0
22	10	13	1	3	6	0
23	10	14	1	2	3	0
24	11	12	1	3	2	0
25	11	13	1	3	3	1
26	11	16	0	1	7	1
27	12	15	1	3	4	0
28	12	16	1	3	4	0
29	13	14	1	3	7	0
30	13	16	1	3	7	0
31	13	19	1	2	4	0
32	14	19	1	3	7	0
33	14	21	1	3	2	0
34	14	22	1	3	4	0
35	15	16	1	3	4	0
36	16	17	1	3	4	0
37	17	18	1	3	3	0
38	17	19	1	2	3	0
39	18	20	1	3	3	0
40	19	20	1	3	3	0
41	20	21	1	3	2	0
42	20	25	1	3	4	0
43	21	24	1	3	5	0
44	22	23	1	3	3	0
45	23	24	1	3	3	0
46	24	25	1	3	8	0

Table 5. Free-Flow Speed in Different Urban Road Hierarchies.

Road Grade	FT	MR	SR	BR
Free-flow speed (km/h)	80	60	40	30

The road traffic status is divided into five levels including smooth, basically smooth, slow, medium congested, and congested. The traffic index is shown in Table 6.

Table 6. Urban Traffic Index in Different Traffic Conditions.

Status	Congested	Medium Congested	Slow	Basically Smooth	Smooth
Index	0.8–1.0	0.6–0.8	0.4–0.6	0.2–0.4	0.0–0.2

Dynamic road network parameters are updated in real time, and the weekday and weekend traffic index of Shenzhen City in southern China are used as shown in Figure 7.

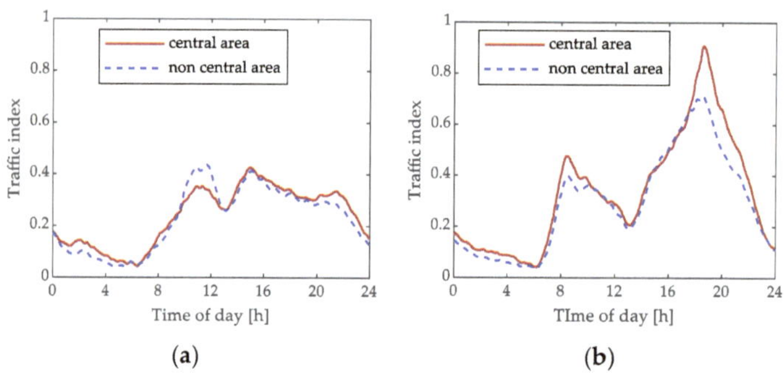

Figure 7. Traffic index of (**a**) weekend; (**b**) weekday.

4.1.3. EVs Parameters

In our simulations, the initial and final locations of the EVs are considered to be parked in residential areas in a day. The number of EVs in each residential area (H) is shown as H1~H8 in Figure 6. The BEV Nissan Leaf, with lithium-ion battery capacity of 24 kWh, is chosen as the typical private BEV used in the simulation. The initial and limited SOC is set to 0.9 and 0.3, respectively.

4.1.4. Resident Travel Parameters

The dataset for analyzing vehicle travel behavior is mainly derived from the National Household Travel Survey (NHTS) [30]. Wednesday and Sunday data are used for weekdays and weekends, respectively. A Gaussian distribution is considered for the first trip departure time with the mean and variance presented in Table 7, four types of trip chains are used for simulation as shown in Table 7.

Table 7. Parameters of start time and dwell time of each trip purpose for different trip chains.

Type of Trip Chains	Trip Chains Penetration Workday	Trip Chains Penetration Weekend	First Departure Time Workday	First Departure Time Weekend	Parking (Dwelling) Time Workday	Parking (Dwelling) Time Weekend
H-W-H	40%	10%	$(457, 142^2)$	$(550, 184^2)$	$(544, 122^2)$	$(504, 152^2)$
H-E-H	20%	70%	$(635, 220^2)$	$(744, 225^2)$	$(222, 208^2)$	$(144, 158^2)$
H-W-E-H	20%	10%	$(432, 74^2)$	$(544, 132^2)$	$(450, 179^2)$ $(57, 84^2)$	$(393, 227^2)$ $(82, 114^2)$
H-E-W-H	20%	10%	$(601, 198^2)$	$(712, 210^2)$	$(550, 184^2)$ $(179, 216^2)$	$(94, 104^2)$ $(102, 128^2)$

4.1.5. Incentive Price Information

The incentive price parameter is assumed as Table 8.

Table 8. Peak-Valley Time-of-Use Incentive Price (yuan/kWh).

Type	Time Slot	Incentive Price
Flat period	7:00–10:00 & 15:00–18:00 & 21:00–23:00	0.6832
Peak period	10:00–15:00 & 18:00–21:00	1.0558
Valley period	23:00-Next day 7:00	0.3105

4.2. Simulation Result of a Single EV DR Potential

With the time of use incentive price in Table 8, we report the simulation result of a single vehicle in a workday as shown in Figure 8a. Considering the fuzzy participation response mechanism, its response curve is shown in Figure 8b.

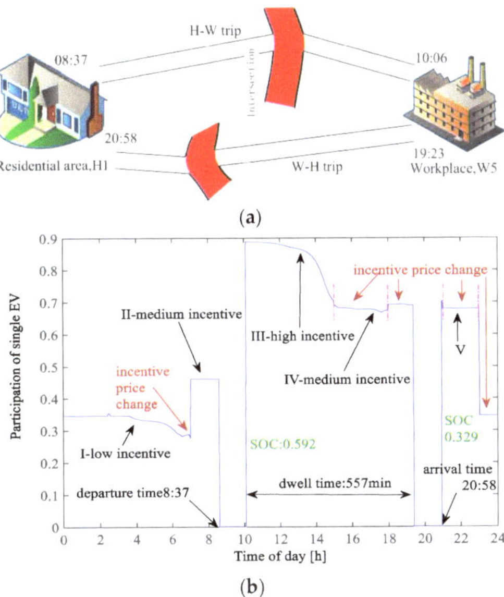

Figure 8. Probability of EV DR. (**a**) 'H-W-H' trip chain. (**b**) the demand response of a single EV.

It can be seen from Figure 8a that the EV arrived at the working place after leaving the house for 89 min and returned home after 557 min of parking. The return journey took 95 min. The path planned by the minimum travel time algorithm for the two trips is as shown in (19).

$$path \begin{cases} (H_1 \rightarrow W_6) : (1, 2, 4, 8, 11, 13, 19, 20, 25) \\ (W_6 \rightarrow H_1) : (25, 20, 19, 13, 8, 4, 2, 1) \end{cases} \tag{19}$$

where, the numbers in the brackets represent the road nodes. We can see that the round-trip routes between the two trip destinations are different. The lengths of the road segments are 34 km and 31 km, respectively, but the travel distance for calculating the power consumption is 37.01 km and 31.47 km. This is due to the large area of the functional block used in this simulation, a random length $5 \times abs(2 \times rand(1,1) - 1)km$ is added in the calculation to reflect the mileage in each functional area. Additionally, although the 'W-H' trip's driving distance was short, the traveling time was longer, which was caused by the time-consuming increase in travel time.

Figure 8b shows that the EV did not have response ability during the two-way travel period, and the lower response in the I-zone due to the low compensation of the incentive price and the closer to the departure time of the next trip. However, notwithstanding its departure time is much closer, the responsiveness in the second zone increases. This is because the increase in the incentive price has stimulated user participation.

The responsiveness of the two initial parking periods in the workplace and the residential area is relatively high, as shown in the zone III and zone V. Zone III is in a state where the incentive price is much higher, and the battery charge rate is also high. Its responsiveness is the highest throughout the day, but it gradually decreases with the declining of the remaining travel time. In zone IV, the responsiveness is further reduced due to the drop of the compensation price.

4.3. Validations

To validate the proposed method, four cases are simulated for sensitivity analysis.

4.3.1. Workdays VS Weekends

Firstly, the simulated result for the delay coefficients of the central urban and non-central areas on weekdays and weekends are shown in Figure 9. Compared with Figure 7, the trend of the two curves is basically the same which indicates that the traffic congestion caused delays, especially during peak hours, the travel time of the central urban area is nearly 1.7 times that of the free-flow speed.

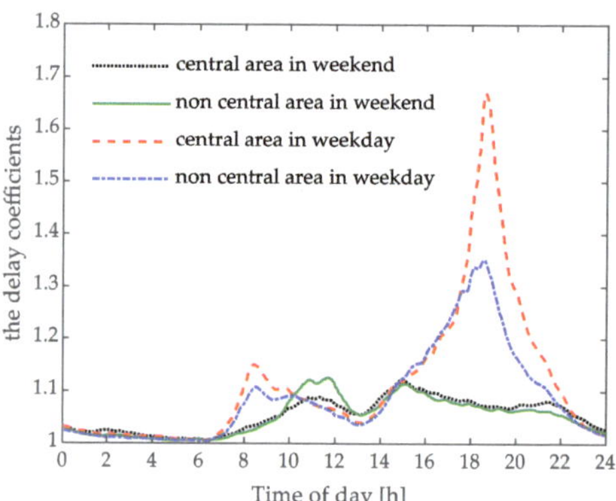

Figure 9. Delay coefficients under different traffic index.

Secondly, Table 7 shows that there is a big difference in user travel pattern between weekdays and weekends. The traffic status is also significantly different as shown in Figure 7. With the incentive price of Table 8, the corresponding V2G powers of each region are shown in Figure 10.

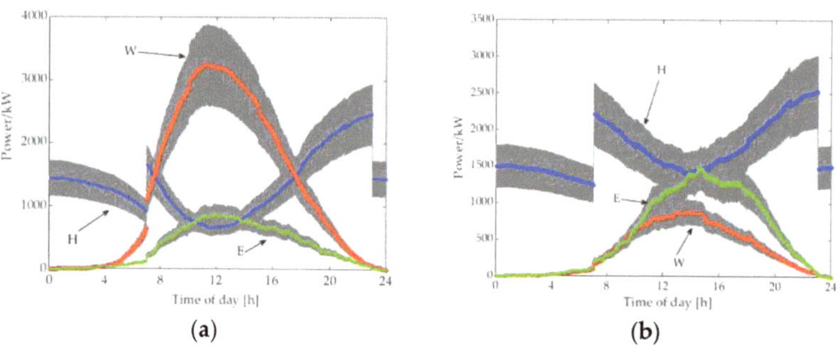

Figure 10. V2G Power demand response distribution of EVA-DR in weekday and weekend (**a**) weekday; (**b**) weekend.

4.3.2. DRN VS Static Road Network (SRN)

Table 9 displays the participation results in the dynamic road network and static road network. It is to see that the travel route, arrival time, arrival SOC, and the DR participation are different.

$$
\left\{
\begin{array}{l}
\text{Path\_S} = (25, 20, 21, 14, 10, 9, 3, 2, 1) \\
\text{Path\_19} = (25, 20, 21, 14, 10, 8, 4, 2, 1) \\
\text{Path\_23} = (25, 20, 19, 13, 8, 4, 2, 1)
\end{array}
\right.
$$

Table 9. Participation Simulation Results in Different Road Networks.

Road Network	Travel Route	Arrival Time	Arrival SOC	Participation
SRN	19:00 Path_S	20:28	0.3667	0.8806
	23:00 Path_S	23:48	0.3667	0.4143
DRN	19:00 Path_19	21:23	0.35	0.8437
	23:00 Path_23	00:31	0.3167	0.3661

4.3.3. Different Response Mechanism

Figure 11 shows the EVA-DR capacity under the proposed EVA-DRE method and the fixed response mode that EVs will participation in DR when SOC is greater than 0.3 during the parking period. From Figure 11, the capacity of fixed response mode is much larger than the proposed method.

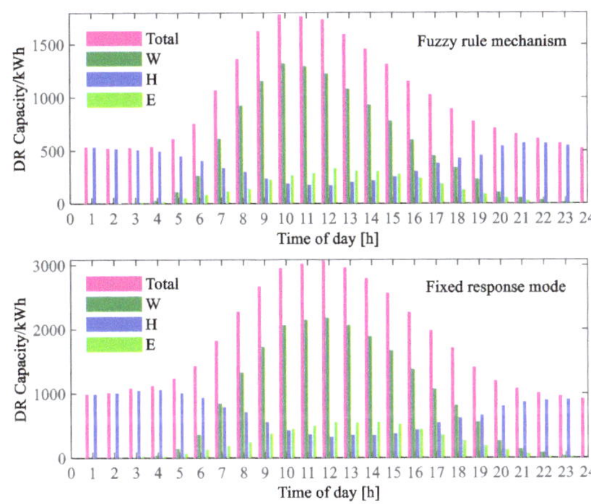

Figure 11. EVA-DR capacity expectation with different mechanism.

4.3.4. Different Incentive Price

We conducted the sensitivity analysis on the different incentive signals for the EVA-DR potential in our case study, the result is shown in Figure 12.

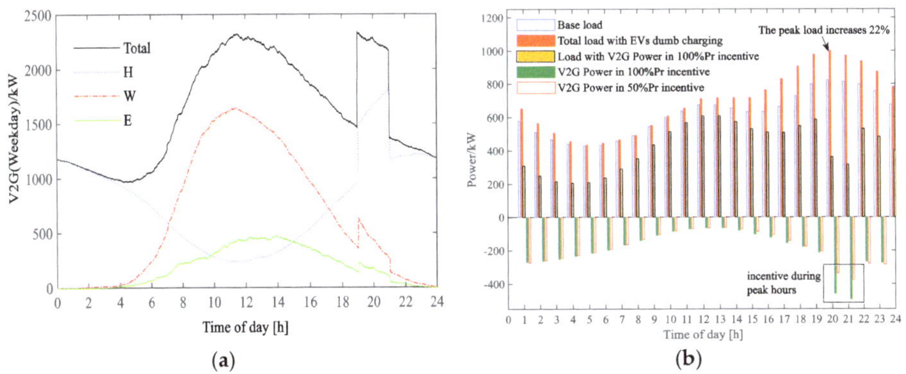

Figure 12. EVA-DR under different incentives. (**a**) V2G power expectation under high incentive compensation of peak hours (with 33% penetration); (**b**) EV charge randomly and different peak price incentives (H4: 240 EVs).

Taking EVs' demand response participation in the peak load hours as an example, the V2G incentive compensation for the peak load period (19:00–21:00) is set as 50%Pr, and V2G response power result is shown in Figure 12a. It is apparent that in the case of non-peak time uncompensated electricity price incentives, the regional V2G response capacity is significantly reduced, while during the peak hours, especially in residential areas, user participation is high due to compensation incentives, and EVs usually have returned home during this period, thus, the V2G response capacity increases dramatically.

Figure 12b provides the original load curve of the residential area H4 (powered by node 22, peak load 825 kW), the load curve with EVs charging randomly, and then we simulate the response curve of EVA-DR in the H4 functional area under different incentive signals. Figure 12b shows that the 'peak-to-peak' effect is formed with the disordered charging during the peak load period. After the demand response project is implemented, EVA-DR effectively reduces the peak load.

5. Conclusions

We have proposed a novel quantitative evaluation method for obtaining spatio-temporal projections of demand response potential from electric vehicles. The dynamic traffic network model taking over the traffic time-varying information, trip chains, the shortest path planning algorithm, and Monte Carlo simulation are employed to derive the spatio-temporal distribution of EVs dumb charging load and battery state of charge. Investigating EV users' willingness to participate in the DR event, a fuzzy logic-based user participation response mechanism is developed that takes into account various realistic factors such as the remaining dwell time, the remaining SOC and the incentive electricity pricing. Compared to the related literature, numerical results obtained in different cases of analysis demonstrate that the approach can achieve a reasonable spatio-temporal distribution of EVs dumb charging load, delayed charging, V2G power, and capacity. It can provide a reference for both the utilities and EVAs through the prediction of charging load and potential of electric vehicles participating in a DR event. At the same time, the proposed evaluation method can be used in regions with different sizes.

Our future work will enhance the EV user's decision-making process considering different battery degradation functions, investigate the pricing strategy of incentive pricing, extend the simulation analysis by presenting more realistic scenarios and comparisons with other similar approaches, and finally, large-scale dimensionality of a large scale fleet of EVs will be investigated.

Author Contributions: Conceptualization, L.C. and Y.Z.; Methodology, L.C.; Validation, L.C., A.F., and Y.Z.; Formal analysis, L.C.; Writing—original draft preparation, L.C.; Writing—review and editing, A.F.; Supervision, Y.Z.; Funding acquisition, L.C.

Funding: This research was funded by the CSC Scholarship Foundation under grant no. 201708440511.

Acknowledgments: The authors would like to thank Transportation Bureau of Shenzhen Municipality for the provision of traffic data used in this study. Its website: http://tocc.jtys.sz.gov.cn/#/.

Conflicts of Interest: The authors declare no conflict of interest.

References

1. Boulanger, A.G.; Chu, A.C.; Maxx, S.; Waltz, D.L. Vehicle electrification: Status and issues. *Proc. IEEE* **2011**, *99*, 1116–1138. [CrossRef]
2. Islam, N.; Bloemink, J. Bangladesh's energy crisis: A summary of challenges and smart grid-based solutions. In Proceedings of the 2018 2nd International Conference on Smart Grid and Smart Cities (ICSGSC), Kuala Lumpur, Malaysia, 12–14 August 2018; pp. 111–116.
3. Gerossier, A.; Girard, R.; Kariniotakis, G. Modeling and forecasting electric vehicle consumption profiles. *Energies* **2019**, *12*, 1341. [CrossRef]
4. Canizes, B.; Soares, J.; Costa, A.; Pinto, T.; Lezama, F.; Novais, P.; Vale, Z. Electric vehicles' user charging behaviour simulator for a smart city. *Energies* **2019**, *12*, 1470. [CrossRef]
5. Ammous, M.; Belakaria, S.; Sorour, S.; Abdel-Rahim, A. Optimal cloud-based routing with in-route charging of mobility-on-demand electric vehicles. *IEEE Trans. Intell. Transp. Syst.* **2018**, 1–13. [CrossRef]

6. Stüdli, S.; Crisostomi, E.; Middleton, R.; Shorten, R. A flexible distributed framework for realising electric and plug-in hybrid vehicle charging policies. *Int. J. Control* **2012**, *85*, 1130–1145. [CrossRef]
7. Carli, R.; Dotoli, M. A decentralized control strategy for optimal charging of electric vehicle fleets with congestion management. In Proceedings of the 2017 IEEE International Conference on Service Operations and Logistics, and Informatics (SOLI), Bari, Italy, 18–20 September 2017; pp. 63–67.
8. Battistelli, C. Demand response program for electric vehicle service with physical aggregators. In Proceedings of the IEEE PES ISGT Europe 2013, Copenhagen, Denmark, 6–9 October 2013; pp. 1–5.
9. Develder, C.; Sadeghianpourhamami, N.; Strobbe, M.; Refa, N. Quantifying flexibility in EV charging as DR potential: Analysis of two real-world data sets. In Proceedings of the 2016 IEEE International Conference on Smart Grid Communications (SmartGridComm), Sydney, Australia, 6–9 November 2016; pp. 600–605.
10. Acquah, M.A.; Han, S. Online building load management control with plugged-in electric vehicles considering uncertainties. *Energies* **2019**, *12*, 1436. [CrossRef]
11. Aghaei, J.; Alizadeh, M. Critical peak pricing with load control demand response program in unit commitment problem. *IET Gener. Transm. Distrib.* **2013**, *7*, 681–690. [CrossRef]
12. Rassaei, F.; Soh, W.; Chua, K. Demand response for residential electric vehicles with random usage patterns in smart grids. *IEEE Trans. Sustain. Energy* **2015**, *6*, 1367–1376. [CrossRef]
13. Erdinc, O.; Paterakis, N.G.; Mendes, T.D.P.; Bakirtzis, A.G.; Catalão, J.P.S. Smart household operation considering bi-directional EV and ESS utilization by real-time pricing-based DR. *IEEE Trans. Smart Grid* **2015**, *6*, 1281–1291. [CrossRef]
14. Izadkhast, S.; Garcia-Gonzalez, P.; Frìas, P.; Ramìrez-Elizondo, L.; Bauer, P. An aggregate model of plug-in electric vehicles including distribution network characteristics for primary frequency control. *IEEE Trans. Power Syst.* **2016**, *31*, 2987–2998. [CrossRef]
15. Liu, Y.; Gao, S.; Zhao, X.; Han, S.; Wang, H.; Zhang, Q. Demand response capability of V2G based electric vehicles in distribution networks. In Proceedings of the 2017 IEEE PES Innovative Smart Grid Technologies Conference Europe (ISGT-Europe), Torino, Italy, 26–29 September 2017; pp. 1–6.
16. Wang, J.; Shi, Y.; Zhou, Y. Intelligent demand response for industrial energy management considering thermostatically controlled loads and EVs. *IEEE Trans. Ind. Inform.* **2018**, 1. [CrossRef]
17. Şengör, I.; Erdinç, O.; Yener, B.; Taşcikaraoglu, A.; Catalão, J.P.S. Optimal energy management of EV parking lots under peak load reduction based dr programs considering uncertainty. *IEEE Trans. Sustain. Energy* **2018**, 1. [CrossRef]
18. Peng, W.; Dong, G.; Yang, K.; Su, J. A random road network model and its effects on topological characteristics of mobile delay-tolerant networks. *IEEE Trans. Mob. Comput.* **2014**, *13*, 2706–2718. [CrossRef]
19. Chen, L.; Nie, Y.; Zhong, Q. A model for electric vehicle charging load forecasting based on trip chains. *Trans. China Electrotech. Soc.* **2015**, *30*, 216–225.
20. Gong, L.; Cao, W.; Liu, K.; Zhao, J.; Li, X. Spatial and temporal optimization strategy for plug-in electric vehicle charging to mitigate impacts on distribution network. *Energies* **2018**, *11*, 1373. [CrossRef]
21. Li, M.; Lenzen, M.; Keck, F.; McBain, B.; Rey-Lescure, O.; Li, B.; Jiang, C. GIS-based probabilistic modeling of BEV charging load for Australia. *IEEE Trans. Smart Grid* **2018**, 1. [CrossRef]
22. Castillo, E.; Calviño, A.; Sánchez-Cambronero, S.; Lo, H.K. A multiclass user equilibrium model considering overtaking across classes. *IEEE Trans. Intell. Transp. Syst.* **2013**, *14*, 928–942. [CrossRef]
23. Lu, C.; Zhao, F.; Hadi, M. A travel time estimation method for planning models considering signalized intersections. In Proceedings of the ICCTP 2010: Integrated Transportation Systems: Green, Intelligent, Reliable, Beijing, China, 4–8 August 2010; pp. 1993–2000.
24. Amini, M.H.; Karabasoglu, O. Optimal operation of interdependent power systems and electrified transportation networks. *arXiv*, 2018; arXiv:1701.03487.
25. Ma, Z.; Callaway, D.S.; Hiskens, I.A. Decentralized charging control of large populations of plug-in electric vehicles. *IEEE Trans. Control Syst. Technol.* **2013**, *21*, 67–78. [CrossRef]
26. Parise, F.; Colombino, M.; Grammatico, S.; Lygeros, J. Mean field constrained charging policy for large populations of Plug-in Electric Vehicles. In Proceedings of the 53rd IEEE Conference on Decision and Control, Los Angeles, CA, USA, 15–17 December 2014; pp. 5101–5106.
27. Carli, R.; Dotoli, M. A distributed control algorithm for waterfilling of networked control systems via consensus. *IEEE Control Syst. Lett.* **2017**, *1*, 334–339. [CrossRef]
28. Hodgson, M.J. A flow-capturing location-allocation model. *Geogr. Anal.* **1990**, *22*, 270–279. [CrossRef]

29. Miranda, V.; Ranito, J.V.; Proenca, L.M. Genetic algorithms in optimal multistage distribution network planning. *IEEE Trans. Power Syst.* **1994**, *9*, 1927–1933. [CrossRef]
30. U.S. Department of Transportation Federal Highway Administration. *2017 National Household Travel Survey*; U.S. Department of Transportation Federal Highway Administration: Washington, DC, USA, 2017.

Article

Thermal Analysis of Power Rectifiers in Steady-State Conditions

Adrian Plesca [1,*] and **Lucian Mihet-Popa** [2]

[1] Faculty of Electrical Engineering, Gheorghe Asachi Technical University of Iasi, Blvd, Dimitrie Mangeron, 21-23, Iasi 700050, Romania

[2] Faculty of Engineering, Østfold University College, BRA Veien 4, 1671 Fredrikstad, Norway; lucian.mihet@hiof.no

* Correspondence: aplesca@tuiasi.ro; Tel.: +40-232-278683

Received: 26 March 2020; Accepted: 12 April 2020; Published: 15 April 2020

Abstract: Power rectifiers from electrical traction systems, but not only, can be irreversibly damaged if the temperature of the semiconductor junction reaches high values to determine thermal runaway and melting. The paper proposes a mathematical model to calculate the junction and the case temperature in power diodes used in bridge rectifiers, which supplies an inductive-resistive load. The new thermal model may be used to investigate the thermal behavior of the power diodes in steady-state regime for various values of the tightening torque, direct current through the diode, airflow speed and load parameters (resistance and inductance). The obtained computed values were compared with 3D thermal simulation results and experimental tests. The calculated values are aligned with the simulation results and experimental data.

Keywords: power rectifier; temperature distribution; mathematical model; thermal modeling and simulation

1. Introduction

A great challenge for power electronic systems is to remove the heat from the power devices in an efficient and cost-effective manner. Due to the needs of high power density and to the miniaturization of power converters, the components have to operate next to their thermal limits [1]. One of the main concerns in many power converters design procedure is to controlling the heat as the producer's moves toward all-surface-mount implementations. Thus, in the early design phase of the power converters it is very important to understand the thermal aspects of the components in order to improve the devices thermal performances [2]. The available thermal models have their boundaries to correctly estimate the thermal behavior in the IGBTs: The usually used thermal models based on one-dimensional RC node have limits to estimate the temperature variation inside devices. A new lumped 3D thermal model is suggested in [3], which can be simply described from FEM simulations and may achieve the critical thermal distribution in long-term evaluations. The boundary constraints for the thermal investigations are considered, which may be adjusted to various real-field applications of power electronic converters. In the existing losses and thermal models, just the electrical loadings for design variables are considered. In [4], a full losses and thermal model taking into consideration the device rating as input variables is also presented. A mathematical relation between the power loss, thermal impedance and the silicon area for the IGBT is estimated. Thus, all the elements with impact over the loss and thermal aspects are considered with results in an optimal design of the power converter.

For the IGBT modules three usual thermal modeling methods regarding their influence on lifetime estimation are compared in [5]. It is shown that these models have important differences in estimating the cross-coupling terms between the chips in the module.

In [6], for a 1200 A, 3.3 kV IGBT module the thermal analysis is considered by the three-dimensional transmission line matrix method. The results show a three-dimensional visualization of self-heating in the module. The temperature evolution is determined during the PWM load cycles, with a good result regarding the thermal analysis and design. A mathematical modeling for time-variant cooling systems is considered in [7], switching between such networks matched to different edge conditions. The accuracy, in terms of temperature and predicted IGBT lifetime, for Foster and Cauer networks is analyzed and compared to errors in the thermal-interface-material resistance. A power electronic converter structure based on power losses and thermal models are considered in [8] for a three level active neutral point clamped voltage source converter using press-pack IGBT-diode pairs. The thermal model is based on water-cooled press-pack switches and is simulated for a 6 MW wind turbine grid interface.

The thermal management is an important aspect for the power converters used for vehicles. For the IGBT package modules employed in hybrid electric vehicle applications it is presented in [9] a method to obtain precise models for rapid electrothermal simulations with advantages in reducing the design cycle. For the traction motors, the volume of the motors can be decreased significantly by improving the thermal flows across the inductor. The silicone resin potted high power inductor for DC/DC switch-mode boost converters can be used [10] in comparison with classical air-cooled inductors, simulations and test being presented for a 40 kW DC–DC converter. In [11] is presented an experimental analysis of multichip IGBT package modules, combining rapid transient short-circuit electrical measurements with infrared thermal mapping considering realistic operating conditions. An electrothermal concise model of the multichip system is described, considering all the important operating and particular electrothermal effects, with application in the railway industry. An electrothermal mathematical model for a real-time thermal simulation of an IGBT in an inverter power module used on hybrid vehicle is considered with estimation of the temperature and power losses [12]. Thermal simulations were realized with high precision for a long real-time, which means more than 10 min, which is important for the vehicles. The peak torque for a traction motor is limited by switching device temperatures. For an 11 kW PMSM motor, it is presented in [13] a drive control strategy that merge the active thermal management model with dynamic DC-link voltage adaptation. Thus, the switching losses can be decreased at low speed by decreasing the bus voltage resulting a considerable inverter losses decrease at low speed.

Power converters for wind turbines are highly used due to the green energy evolution. Wind turbines are using power electronic converters with multichip paralleled IGBT modules. In [14], junction temperatures of chips for various positions with a better thermal coupling impedance model are estimated, and the outcomes are compared with the results of other thermal models.

A two-level power electronic converter for wind turbine is analyzed considering the losses and the junction temperature for the power devices in the case of a large wind speed variation, resulting in the high effect of the junction temperature over the operating point of the induction generator of the turbine [15,16].

Thermal characteristics of various power switching devices are considered in [17] for their impact on the thermal cycling of a 10 MW wind power converter in different working conditions considering IGBT modules and press-pack, the thermal characteristics of the power electronic converter being significantly changed according to the power semiconductor devices technology and their configurations.

In [18] is presented also an electrothermal model of an inverter but implemented in PLECS being used to estimate the IGBT junction temperature with a mission profile for a wind power application. The thermal network is determined from the heating curves of IGBT junction and case temperature that are measured on a power-cycling rig. Junction temperature is estimated with the power dissipation and thermal network. For rapid and precise thermal simulations of power semiconductor modules it is developed a Fourier-based solution, which can estimate the temperature versus time to resolve the heat equation in two dimensions [19], resulting a fast simulation compared to the finite-element (FEM).

Microchannel coolers for thermomechanical performance of power electronic modules with IGBT are analyzed in [20] by using finite element analysis with results in increasing the lifetime modules. Power loss model is used [21] to design a 3 kVA DC/AC high output power density converter considering a connection between the circuit stray parameters and the power losses into the converter. A thermal model of a heatsink as a RC thermal equivalent network that can be embedded in any circuit simulator, it is presented in [22]. The model considers thermal time constants of the heat sink, the convection cooling, thermal hotspots on the heatsink base plate, and thermal coupling between power semiconductor modules mounted onto the heatsink.

Design procedure considerations based on various temperature distribution study cases are realized on a prototype, a 2 kW integrated power electronic semiconductor modules using Cool MOS and SiC diode [23]. The thermal model is elaborated and the power dissipation of every power semiconductors in the active IPEM is calculated through the measurement-based power loss characteristics in datasheet. With the requirements of equal temperature distribution and light thermal interaction between power dice, the modification of temperature gradient distribution with the heat transfer coefficient of heatsink and die position is analyzed. In the [24] literature review for the design and study of thermal via in PCBs for thermal control of power electronic semiconductors lead to a selection for four different models for single power devices. The experimental results match the theoretical anticipation to identify the most effective thermal model via pattern.

For the estimation of the power module temperature field contours at various temperatures, a simplified 3D model of the power module was taken into consideration [25], and steady-state thermal analysis was done. The layout of the power electronic converter is optimized to decrease the heat distribution corresponding to the analysis outcome. The simulations show that an optimal distribution of the power supply module can increase the reliability of the switching mode DC/DC converters. The three-level active-neutral-point-clamped voltage source converter is used to defeat the uneven loss distribution [26]. The junction temperature and the load current of a power electronic converter system are obtained to calculate the power losses. The thermal resistance and the thermal capacitance of different components are used to express the thermal network, which is used to estimate the power device junction temperature, and to analyze transient thermal distribution.

In [27], it proposes the certainty design to the power electronic converter's conventional compensation controller design with a new concept of a universal dual-loop controller, which uses temperature control loop as well as electric power control loop. The idea is based on a digital implementation of a variable load of power inverter system with real-time measurement approach of the chip's surface temperature. The novelty is to get a better thermal control method of carrier frequency adjustment through experimental implementation during the full life cycle of the power electronic converter. A back-to-back 2-level/3-level inverter has been designed and developed such that the power will flow in both directions while the converter will perform better under arbitrary load conditions analyzing the thermal dissipation of the semiconductors [28]. The thermal investigation has been performed in the case of an IGBT module with a specifically integrated real time current controller. In [29] it is proposed to use the active thermal management to reduce the switching losses during transient regime in order to assure a high current without overcoming the temperature limits, reducing the overdesign of power converters.

An electrothermal design methodology is proposing and a reliability study is realized [30] for converters used for photovoltaic application, which is the distributed maximum power point tracking converter. In [31] is presented an electrothermal analysis based on the reduced order modeling technique in which the zero-dimensional thermal network is obtained from the three-dimensional IGBT semiconductor module packaging structure and incorporated with the electrothermal model of the chips.

Considering the previous works of different research teams, it can be outlined that the existing thermal models have their limits to predict the thermal behavior of different types of power semiconductor devices, especially in the case of IGBTs. It has been proposed simplified 3D thermal models based on FEM and also the well-known Foster and Cauer networks to analyze inverters' behavior from thermal point of view. More, the researchers developed different methods to monitor the junction temperature of the IGBTs with the aim to check their reliability.

The aim of this paper is to investigate the thermal behavior of the power diodes from a rectifier bridge used in electrical traction during steady-state conditions.

The paper is arranged as follows, in Section 2 the mathematical model to calculate the junction and case temperature will be developed. Then, in Section 3, the calculated values will be compared with the simulated results of a 3D thermal model of a power assembly, which includes two power diodes mounted on the heatsink. Finally, the proposed mathematical model will be validated through a series of experimental tests, in Section 4 followed by the conclusion section, where the most important outcomes of the paper are highlighted.

2. Mathematical Model

The goal of the thermal analysis is to establish a mathematical model that will be used to compute the junction and the case temperature for a power diode, which is the main component from a power traction rectifier. The power loss of the diode during steady-state conditions is a sum of the following terms:

$$P = P_F + P_R + P_c \tag{1}$$

where:

P_F means the power loss during direct conduction;

P_R is the power loss in the case of locked conditions;

P_c is the power loss during commutations.

In the situation of direct conduction, the power losses can be calculated using the following expression,

$$P_F = V_T I_{TAV} + r_T I_{RMS}^2 \tag{2}$$

and it is the observed that depends both on diode intrinsic characteristics through the parameters V_T and r_T, and current waveform which flows through the power diode. In the case of a resistive inductive load, the electric current can be described by the next formula,

$$i(t) = I_m \sin(\omega t - \varphi) + I_m \sin \varphi\, e^{-\frac{\omega t}{\omega L/R}} \tag{3}$$

Thus, it results in the RMS current:

$$I_{RMS} = \sqrt{\frac{1}{2\pi} \int_0^\pi i^2(t)d(\omega t)} = \frac{I_m}{\sqrt{2\pi}} \left[\frac{\pi}{2} - \frac{2\sin\varphi\frac{\omega L}{R}}{1+\left(\frac{\omega L}{R}\right)^2}\left(\sin\varphi - \frac{\omega L}{R}\cos\varphi\right)\left(e^{-\frac{\pi}{\omega L/R}} + 1\right) - \frac{\omega L}{2R}\sin^2\varphi\left(e^{-2\frac{\pi}{\omega L/R}} - 1\right) \right]^{1/2} \tag{4}$$

Additionally, the average current has the expression,

$$I_{FAV} = \frac{1}{2\pi} \int_0^\pi i(t)d(\omega t) = I_m \frac{1}{\pi\cos\varphi} \tag{5}$$

Therefore, considering the expressions of the RMS Equation (4) and average Equation (5), the relation to calculate the power losses during direct conduction, becomes,

$$P_F = V_T I_m \frac{1}{\pi\cos\varphi} + r_T \frac{I_m^2}{2\pi}\left[\frac{\pi}{2} - \frac{2\sin\varphi\frac{\omega L}{R}}{1+\left(\frac{\omega L}{R}\right)^2}\left(\sin\varphi - \frac{\omega L}{R}\cos\varphi\right)\left(e^{-\frac{\pi}{\omega L/R}} + 1\right) - \frac{\omega L}{2R}\sin^2\varphi\left(e^{-2\frac{\pi}{\omega L/R}} - 1\right) \right] \tag{6}$$

The power losses in the case of locked conditions P_R can be calculated by the product between the maximum reverse current I_{RM} and the maximum repetitive reverse voltage V_{RRM},

$$P_R = I_{RM} V_{RRM} \tag{7}$$

Both parameters I_{RM} and V_{RRM} depend on the type of the power diode and their values are depicted in the datasheet of the diode. The commutation power loss P_c can be calculated using the following relation,

$$P_c = V_{Rmax} f Q_s = \sqrt{6} U_2 f Q_s \tag{8}$$

where the maximum reverse voltage V_{Rmax} depends on the type of the power rectifier topology. It has been considered a B6 bridge rectifier. In this case, the maximum reverse voltage is equal with the secondary voltage of the supply transformer, multiplied by the coefficient $\sqrt{6}$.

In the case of traction power rectifiers, the diodes are mounted on aluminum heatsinks with the aim to ensure an efficient cooling of the semiconductor junction. Thus, the main components of the current path within the power assembly of the bridge rectifier, includes power diode, heatsink and busbar. Therefore, in addition to the power loss of the diode P, there are also the power losses of the contact resistance between the case diode and the heatsink as main component of the current path. The power loss because of the contact resistance can be computed with the known formula,

$$P_{contact} = R_c I_{RMS}^2 \tag{9}$$

Further, it has been examined the case in which the rectifier bridge is composed from power diodes type SKN 300 threaded stud M16 $\times$ 1.5 mm. There is an inverse proportional relationship between the contact resistance R_c and the contact force F, which will be calculated knowing the tightening torque M, the screw diameter d, the thread pitch p and the friction coefficient μ, with the following formula [32],

$$F = \frac{2}{d} M \frac{\cos\left(arctg\frac{p}{\pi d}\right) - \mu \sin\left(arctg\frac{p}{\pi d}\right)}{\sin\left(arctg\frac{p}{\pi d}\right) + \mu \cos\left(arctg\frac{p}{\pi d}\right)} \tag{10}$$

The nonlinear variation between contact resistance R_c and contact force F can be represented by a mathematical function. Starting from an experimental dataset, the curve of the contact resistance has been fitted to the following expression,

$$R_c = \frac{1}{a_1 + b_1 F + c_1 F^2} \tag{11}$$

The coefficients of the fitting curve have the next values: $a_1 = 397.53$, $b_1 = 0.98625$ and $c_1 = 0.00011$. The comparison between experimental and fitting curve of the contact resistance variation vs. contact force is presented in the Figure 1.

Thus, using the expressions (4), (10) and (11), the contact power losses can be computed as follows,

$$P_{contact} = \frac{\frac{I_m^2}{2\pi}\left[\frac{\pi}{2} - \frac{2\sin\varphi\frac{\omega L}{R}}{1+\left(\frac{\omega L}{R}\right)^2}\left(\sin\varphi - \frac{\omega L}{R}\cos\varphi\right)\left(e^{-\frac{\pi}{\omega L/R}} + 1\right) - \frac{\omega L}{2R}\sin^2\varphi\left(e^{-2\frac{\pi}{\omega L/R}} - 1\right)\right]}{a_1 + b_1\frac{2}{d}M\frac{\cos\left(arctg\frac{p}{\pi d}\right) - \mu\sin\left(arctg\frac{p}{\pi d}\right)}{\sin\left(arctg\frac{p}{\pi d}\right) + \mu\cos\left(arctg\frac{p}{\pi d}\right)} + c_1\left[\frac{2}{d}M\frac{\cos\left(arctg\frac{p}{\pi d}\right) - \mu\sin\left(arctg\frac{p}{\pi d}\right)}{\sin\left(arctg\frac{p}{\pi d}\right) + \mu\cos\left(arctg\frac{p}{\pi d}\right)}\right]^2} \tag{12}$$

Finally, the total power loss considering both diode power loss and contact power loss has the expression:

$$P_{tot} = P + P_{contact} = P_F + P_R + P_c + P_{contact} \tag{13}$$

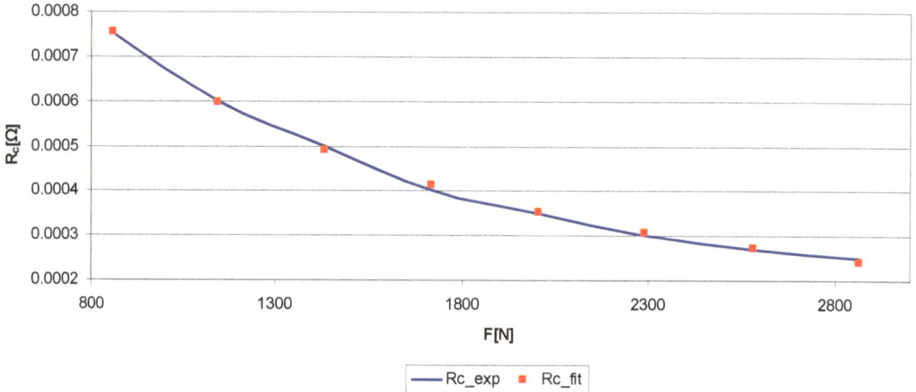

Figure 1. Contact resistance R_c vs. contact force F. Comparison between experimental (Rc_exp) and fitting curve (Rc_fit).

After the previous expressions (6), (7), (8) and (12) have been replaced in the above relation, we obtained the diode total power dissipated as:

$$P_{tot} = V_T I_m \frac{1}{\pi\cos\varphi} + r_T \frac{I_m^2}{2\pi}\left[\frac{\pi}{2} - \frac{2\sin\varphi\frac{\omega L}{R}}{1+\left(\frac{\omega L}{R}\right)^2}\left(\sin\varphi - \frac{\omega L}{R}\cos\varphi\right)\left(e^{-\frac{\pi}{\omega L/R}}+1\right) - \frac{\omega L}{2R}\sin^2\varphi\left(e^{-2\frac{\pi}{\omega L/R}}-1\right)\right] +$$

$$I_{RM}V_{RRM} + \sqrt{6}U_2 f Q_s + \frac{\frac{I_m^2}{2\pi}\left[\frac{\pi}{2} - \frac{2\sin\varphi\frac{\omega L}{R}}{1+\left(\frac{\omega L}{R}\right)^2}\left(\sin\varphi - \frac{\omega L}{R}\cos\varphi\right)\left(e^{-\frac{\pi}{\omega L/R}}+1\right) - \frac{\omega L}{2R}\sin^2\varphi\left(e^{-2\frac{\pi}{\omega L/R}}-1\right)\right]}{a_1 + b_1\frac{2}{d}M\frac{\cos\left(arctg\frac{P}{\pi d}\right)-\mu\sin\left(arctg\frac{P}{\pi d}\right)}{\sin\left(arctg\frac{P}{\pi d}\right)+\mu\cos\left(arctg\frac{P}{\pi d}\right)}+c_1\left[\frac{2}{d}M\frac{\cos\left(arctg\frac{P}{\pi d}\right)-\mu\sin\left(arctg\frac{P}{\pi d}\right)}{\sin\left(arctg\frac{P}{\pi d}\right)+\mu\cos\left(arctg\frac{P}{\pi d}\right)}\right]^2} \tag{14}$$

The above expression regarding the calculation of the total power dissipated at the power diode will be used to determine the junction and case temperature of the analyzed power diode. The following calculation formulas can be used to get the values for the junction temperature,

$$\theta_j = \theta_a + \left(R_{thjc} + R_{thck} + R_{thka}\right)P_{tot} \tag{15}$$

and for the case temperature of the power diode,

$$\theta_c = \theta_a + (R_{thck} + R_{thka})P_{tot} \tag{16}$$

The values of the junction-case and case-heatsink thermal resistance can be obtained from the power diode datasheets. Thus, in the case of SKN 300 power diode, the following values were identified: $R_{thjc} = 0.15\ ^{\circ}$C/W and $R_{thck} = 0.02\ ^{\circ}$C/W. Generally, the heatsink-environment thermal junction was not a constant value and it depended on the cooling method (forced or natural) and the heatsink type. The heatsink datasheets present variation curves of the heatsink-environment thermal junction vs. air speed. The cooling of the SKN 300 power diode can be provided by aluminum heatsink type R150-E50 manufactured by IPRS Baneasa. From the datasheet, it can be obtained the nonlinear variation of heatsink-environment thermal resistance vs. air speed. This curve can be approximated with the following mathematical expression:

$$R_{thka} = a_2 + b_2 v_{air}^2 + c_2 v_{air}^2 \ln v_{air} + d_2\frac{\ln v_{air}}{v_{air}^2} \tag{17}$$

where the parameters of the fitting curve, are: $a_2 = 0.25035$; $b_2 = -0.015359$, $c_2 = 0.006483$ and $d_2 = -0.189932$. The normal variation of decreased heatsink-environment thermal resistance for increasing air speed can be observed in Figure 2.

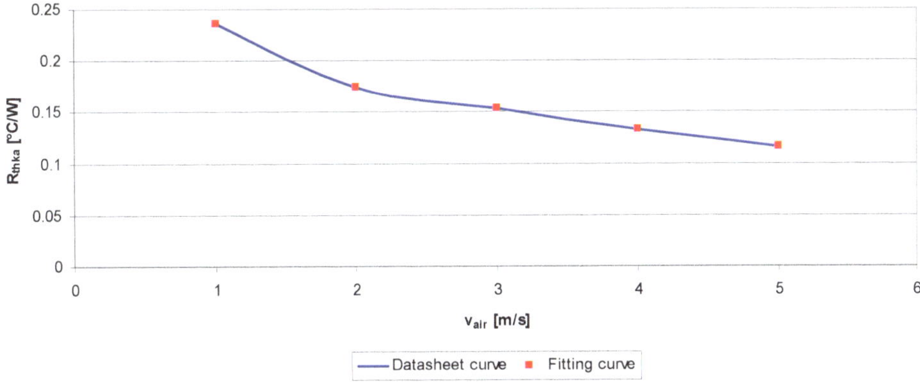

Figure 2. Heatsink-environment thermal resistance vs. air speed. Comparison between datasheet and fitting curve.

So, the junction and case temperature of the power diode will be calculated using the next expressions:

$$\theta_j = \theta_a + \left(R_{thjc} + R_{thck} + a_2 + b_2 v_{air}^2 + c_2 v_{air}^2 \ln v_{air} + d_2 \frac{\ln v_{air}}{v_{air}^2} \right) P_{tot} \tag{18}$$

and,

$$\theta_c = \theta_a + \left(R_{thck} + a_2 + b_2 v_{air}^2 + c_2 v_{air}^2 \ln v_{air} + d_2 \frac{\ln v_{air}}{v_{air}^2} \right) P_{tot} \tag{19}$$

By replacing in the above formulas, the expression of the total power loss, the calculation relations for the junction and case temperature are finally obtained:

$$
\begin{aligned}
\theta_j = \theta_a &+ \left(R_{thjc} + R_{thck} + a_2 + b_2 v_{air}^2 + c_2 v_{air}^2 \ln v_{air} + d_2 \frac{\ln v_{air}}{v_{air}^2} \right) V_T I_m \frac{1}{\pi \cos \varphi} + \\
&+ r_T \frac{I_m^2}{2\pi} \left(R_{thjc} + R_{thck} + a_2 + b_2 v_{air}^2 + c_2 v_{air}^2 \ln v_{air} + d_2 \frac{\ln v_{air}}{v_{air}^2} \right) \left[\begin{array}{c} \frac{\pi}{2} - \frac{2 \sin \varphi \, \frac{\omega L}{R}}{1 + \left(\frac{\omega L}{R} \right)^2} \left(\sin \varphi - \frac{\omega L}{R} \cos \varphi \right) \left(e^{-\frac{\pi}{\omega L/R}} + 1 \right) - \\ - \frac{\omega L}{2R} \sin^2 \varphi \left(e^{-2\frac{\pi}{\omega L/R}} - 1 \right) \end{array} \right] + \\
&+ \left(R_{thjc} + R_{thck} + a_2 + b_2 v_{air}^2 + c_2 v_{air}^2 \ln v_{air} + d_2 \frac{\ln v_{air}}{v_{air}^2} \right) \left(I_{RM} V_{RRM} + \sqrt{6} U_2 f Q_s \right) + \\
&+ \frac{\frac{I_m^2}{2\pi} \left(R_{thjc} + R_{thck} + a_2 + b_2 v_{air}^2 + c_2 v_{air}^2 \ln v_{air} + d_2 \frac{\ln v_{air}}{v_{air}^2} \right) \left[\begin{array}{c} \frac{\pi}{2} - \frac{2 \sin \varphi \, \frac{\omega L}{R}}{1 + \left(\frac{\omega L}{R} \right)^2} \left(\sin \varphi - \frac{\omega L}{R} \cos \varphi \right) \left(e^{-\frac{\pi}{\omega L/R}} + 1 \right) - \\ - \frac{\omega L}{2R} \sin^2 \varphi \left(e^{-2\frac{\pi}{\omega L/R}} - 1 \right) \end{array} \right]}{a_1 + b_1 \frac{2}{d} M \frac{\cos\left(arctg \frac{P}{\pi d} \right) - \mu \sin\left(arctg \frac{P}{\pi d} \right)}{\sin\left(arctg \frac{P}{\pi d} \right) + \mu \cos\left(arctg \frac{P}{\pi d} \right)} + c_1 \left[\frac{2}{d} M \frac{\cos\left(arctg \frac{P}{\pi d} \right) - \mu \sin\left(arctg \frac{P}{\pi d} \right)}{\sin\left(arctg \frac{P}{\pi d} \right) + \mu \cos\left(arctg \frac{P}{\pi d} \right)} \right]^2}
\end{aligned} \tag{20}
$$

and,

$$
\begin{aligned}
\theta_c = \theta_a &+ \left(R_{thck} + a_2 + b_2 v_{air}^2 + c_2 v_{air}^2 \ln v_{air} + d_2 \frac{\ln v_{air}}{v_{air}^2} \right) V_T I_m \frac{1}{\pi \cos \varphi} + \\
&+ r_T \frac{I_m^2}{2\pi} \left(R_{thck} + a_2 + b_2 v_{air}^2 + c_2 v_{air}^2 \ln v_{air} + d_2 \frac{\ln v_{air}}{v_{air}^2} \right) \left[\begin{array}{c} \frac{\pi}{2} - \frac{2 \sin \varphi \, \frac{\omega L}{R}}{1 + \left(\frac{\omega L}{R} \right)^2} \left(\sin \varphi - \frac{\omega L}{R} \cos \varphi \right) \left(e^{-\frac{\pi}{\omega L/R}} + 1 \right) - \\ - \frac{\omega L}{2R} \sin^2 \varphi \left(e^{-2\frac{\pi}{\omega L/R}} - 1 \right) \end{array} \right] + \\
&+ \left(R_{thck} + a_2 + b_2 v_{air}^2 + c_2 v_{air}^2 \ln v_{air} + d_2 \frac{\ln v_{air}}{v_{air}^2} \right) \left(I_{RM} V_{RRM} + \sqrt{6} U_2 f Q_s \right) + \\
&+ \frac{\frac{I_m^2}{2\pi} \left(R_{thck} + a_2 + b_2 v_{air}^2 + c_2 v_{air}^2 \ln v_{air} + d_2 \frac{\ln v_{air}}{v_{air}^2} \right) \left[\begin{array}{c} \frac{\pi}{2} - \frac{2 \sin \varphi \, \frac{\omega L}{R}}{1 + \left(\frac{\omega L}{R} \right)^2} \left(\sin \varphi - \frac{\omega L}{R} \cos \varphi \right) \left(e^{-\frac{\pi}{\omega L/R}} + 1 \right) - \\ - \frac{\omega L}{2R} \sin^2 \varphi \left(e^{-2\frac{\pi}{\omega L/R}} - 1 \right) \end{array} \right]}{a_1 + b_1 \frac{2}{d} M \frac{\cos\left(arctg \frac{P}{\pi d} \right) - \mu \sin\left(arctg \frac{P}{\pi d} \right)}{\sin\left(arctg \frac{P}{\pi d} \right) + \mu \cos\left(arctg \frac{P}{\pi d} \right)} + c_1 \left[\frac{2}{d} M \frac{\cos\left(arctg \frac{P}{\pi d} \right) - \mu \sin\left(arctg \frac{P}{\pi d} \right)}{\sin\left(arctg \frac{P}{\pi d} \right) + \mu \cos\left(arctg \frac{P}{\pi d} \right)} \right]^2}
\end{aligned} \tag{21}
$$

The above expressions outline that both case and junction temperature depends on conduction current, tightening torque, load resistance, load inductance and air speed.

3. Three-Dimensional Thermal Model

The thermal analysis has been focused on a power assembly made from two stud rectifier diodes type SKT 340 mounted on the same aluminum heatsink. Actually, the power assembly can be considered as one of the three legs from the three-phase full- bridge rectifier, which provides a DC current on a resistive-inductive load. The datasheet of the rectifier diode outlines the following values for the main parameters of the diode: $V_T = 0.8$ V, $r_T = 0.6$ mΩ, $I_{RM} = 60$ mA, $V_{RRM} = 1200$ V and $Q_s = 200$ µC. Considering the direct current through the power diode with the RMS value of 250 A, it can be calculated the power loss within the diode, which had the value of 119.29 W. In the case of the recommended tightening torque of 30 Nm, the contact resistance had the value of 0.45 mΩ. Thus, for the same value of the direct current flowing through the diode, the power loss because of the contact resistance between heatsink and the diode was about 28.125 W.

The structure of the power diode was highlighted by a solid copper hexagon with a threaded stud, which is used to bolt the component onto a heatsink and discharge the heat produced by the chip. A glass or ceramic cap with a bushing for the cathode terminal provides hermetical case sealing. Case parts are attached by welding or brazing. The chip is straightly soldered onto the disk made of molybdenum to prevent high mechanical pressure in the semiconductor as a result of important differences in the expansion coefficients for silicon and copper. In order to achieve the three-dimensional geometry of the analyzed power assembly, the Pro/Engineering software package has been used. The thermal model included only the components that actively participate in the heat transfer from the power diode to the environment through the heatsink. These components are (Figure 3) the copper braid, molybdenum discs, Silicom chip, solid copper hexagon with threated stud and aluminum heatsink. Due to the fact that the ceramic housing participates in a very low weight to the heat transfer from the silicon chip to the environment, it was not included in the thermal model.

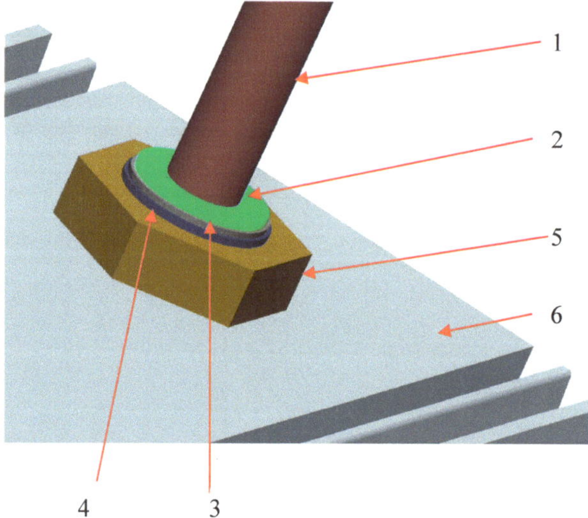

Figure 3. 3D geometry of the power diode thermal model: 1—copper braid; 2—molybdenum disc; 3—silicon chip; 4—molybdenum disc; 5—solid copper hexagon with threaded stud and 6—heatsink.

The material characteristics of each component part of the power diode and the heatsink are synthesized in the Table 1 and the complete thermal model of the power assembly is shown in Figure 4.

Table 1. Material data of power diode and heatsink.

Parameter	Material			
	Copper	Silicon	Molybdenum	Aluminum
γ (kg/m^3)	8900	2330	10220	2700
c (J/kg$^\circ$C)	387	702	255	900
λ (W/m$^\circ$C)	385	124	138	200

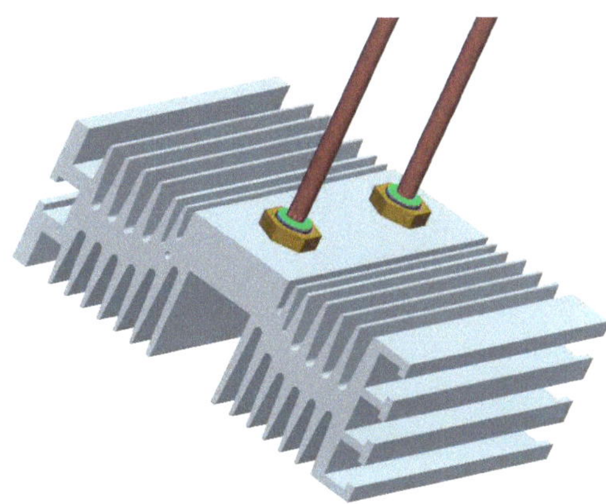

Figure 4. 3D geometry thermal model of the power assembly power diodes heatsink.

On the basis of Pro-MECHANICA software package, some thermal simulations were performed during steady-state conditions. The temperature profile of the power diode, in the case of tightening torque of 30 Nm, air speed value of 1 m/s, direct current of 200 A and the load with a resistance of 20 Ω and inductance about 60 mH, is presented in the next pictures, Figures 5 and 6. It can be seen a maximum temperature of the power assembly about 84.94 $^\circ$C on the silicon chip of the diode and a minimum temperature of 48.34 $^\circ$C on the heatsink outer surfaces.

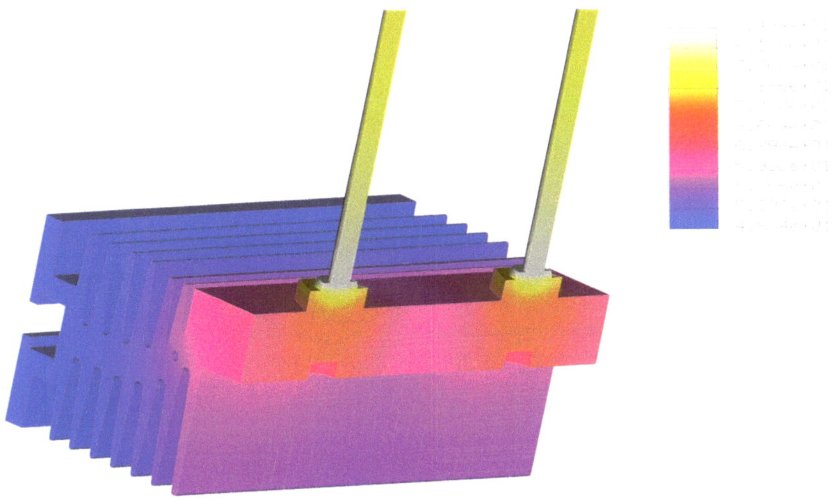

Figure 5. Temperature distribution of the power assembly. Cross-section through both power diodes.

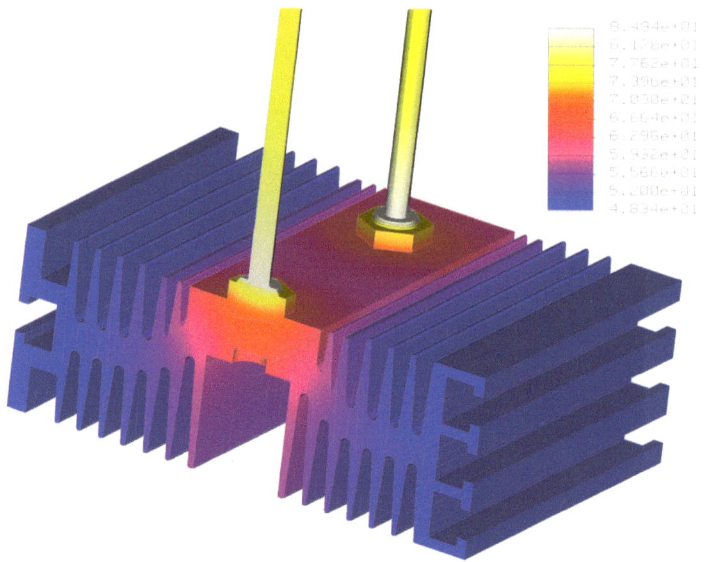

Figure 6. Temperature distribution of the power assembly. Cross-section through only one power diode.

The heat load because of the diode power loss has been applied within the volume of the silicon chip considering a uniform spatial distribution. In addition, it was considered the power loss of the contact resistance between stud copper base of the diode and the heatsink. This additional heat load was applied on the outer surface of the copper screw. It was assumed an ambient temperature about 23 °C. The value of the convection coefficient, 12.35 W/m²°C, was obtained from experimental tests. The outer surfaces of the heatsink were considered as boundary conditions, so that heat transfer by convection could be applied. It was considered a uniform spatial distribution of the convection coefficient on all outer surfaces of the heatsink and a bulk temperature of 23 °C was applied.

4. Discussion of the Results

Further on, considering the obtained solutions for the calculation of junction and case temperature of the power diode from the Equation (20), respectively (21), the influence of the direct current, airflow speed and load parameters on thermal behavior of the power diode would be investigated.

The first thermal analysis refers to the variation of junction and case temperature of the power diode against tightening torque at different current values of 200, 250 and 300 A, in the case of airflow speed of 1 m/s, load resistance about 20 Ω and load inductance of 60 mH. It can be seen a decrease in both junction and case temperatures, Figures 7 and 8, when the tightening torques increases. This can be explained because as torque increases, then the contact force between case diode and heatsink increases, which leads to lower values of the contact resistance and contact power loss, and finally, the junction and case temperature values will decrease. For instance, when tightening torque had the value of 15 Nm, the junction temperature was 109.4 °C, Figure 7, and for a torque value of 50 Nm, the junction temperature reached the value of 104.24 °C, so, a difference about 5 °C. For the same variation of the tightening torque, from 15 to 50 Nm, the case temperature decreases from 76.28 to 73.03 °C, Figure 8. At both junction and case temperature variation, the simulated values are placed on the same graph. It can be noticed that thermal simulation values were lower than the calculated ones. This can be explained because the mathematical model of junction and case temperature took into account a concentrated power loss both for power diode and contact resistance. Actually, during thermal simulations, the power loss was evenly distributed on silicon chip and contact surface between stud diode and heatsink. This led to an improved heat transfer from the power assembly to the environment, and finally, lower junction and case temperatures compared to those calculated. For instance, in the case of tightening torque of 30 Nm and 300 A, Figure 7, the simulated junction temperature was 105.1 °C, a lower value than 105.95 °C for the calculated temperature and the simulated case temperature was 73.2 °C lower than 74.11 °C, the calculated one, Figure 8.

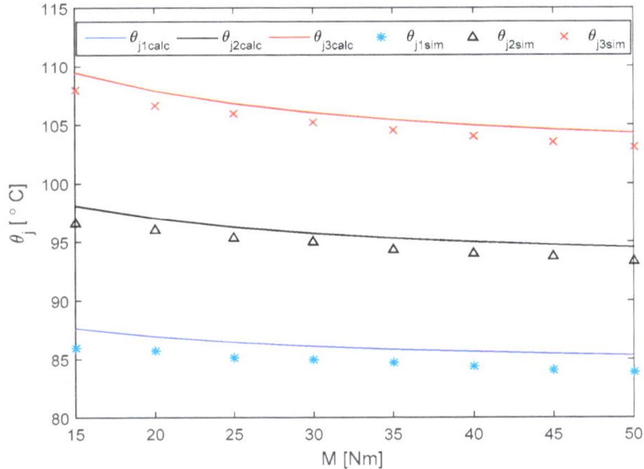

Figure 7. Junction temperature variation for different current values: 200, 250 and 300 A (v = 1 m/s, R = 20 Ω and L = 60 mH).

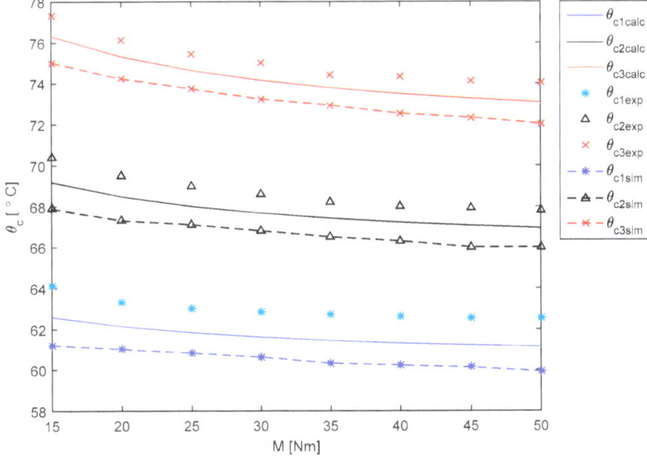

Figure 8. Case temperature variation for different current values: 200, 250 and 300 A (v = 1 m/s, R = 20 Ω and L = 60 mH).

A series of experimental tests were performed in order to validate the proposed mathematical model to calculate the junction and case temperature of power diode. A three-phase bridge rectifier with forced cooling supplied through the power transformer Tr it was considered, as can be seen in Figure 9. The main goal was to acquire the case temperatures of the power diode mounted on the aluminum heatsink. The temperature values were acquired with suitable thermocouple Th type K. It used an electronic board AS type AT2 F-16 in order to adapt the small signals provided by thermocouples to the data acquisition board type PC-LPM-16. This can be programmed by LabVIEW software. The current value can be recorded using the ammeter A.

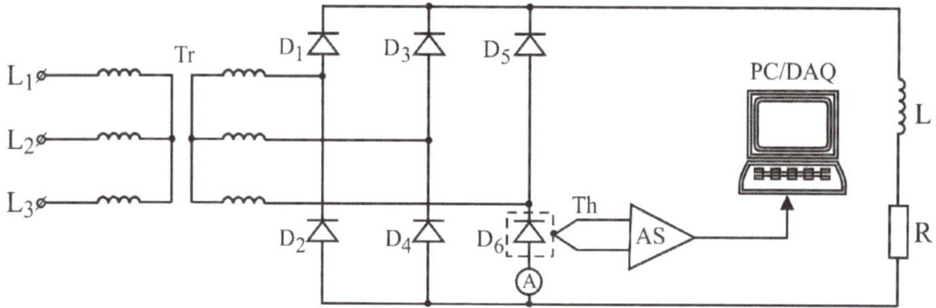

Figure 9. Experimental setup.

From the graph of case temperature variation, Figure 8, it can be noticed that experimental values were higher than the calculated ones. This is explained because the mathematical model considers a uniform cooling of the power diode case. Actually, the airflow did not uniformly cool the surface of the diode case due to the mounting technology of the cooling system for the bridge rectifier. Hence, there was not a good heat transfer from the case to the environment. As an example, in the case of tightening torque of 30 Nm and direct current of 300 A, the experimental case temperature of 75 °C was higher than 74.11 °C, the calculated value.

The second thermal analysis takes into account the junction and case temperature variation against tightening torque for different load resistance values (10, 20 and 100 Ω) when the direct current was 300 A, airflow speed of 1 m/s and load inductance of 10 mH. It can be noticed the decreasing of both junction and case temperature when the torque increased, Figures 10 and 11. The explanation was the same as in previous analyzed case.

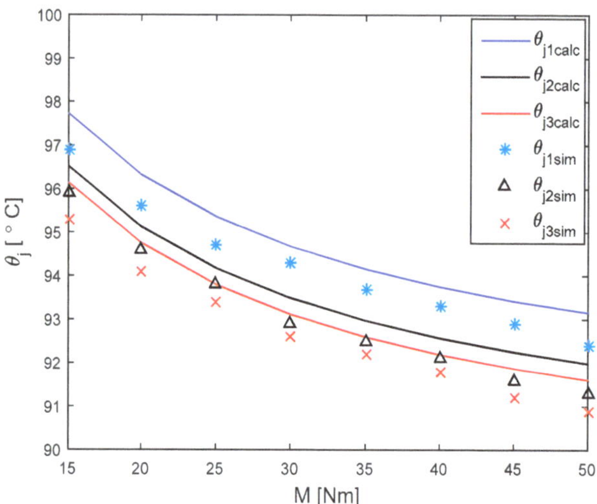

Figure 10. Junction temperature variation for different load resistance values: 10, 20 and 100 Ω (v = 1 m/s, I = 300 A and L = 10 mH).

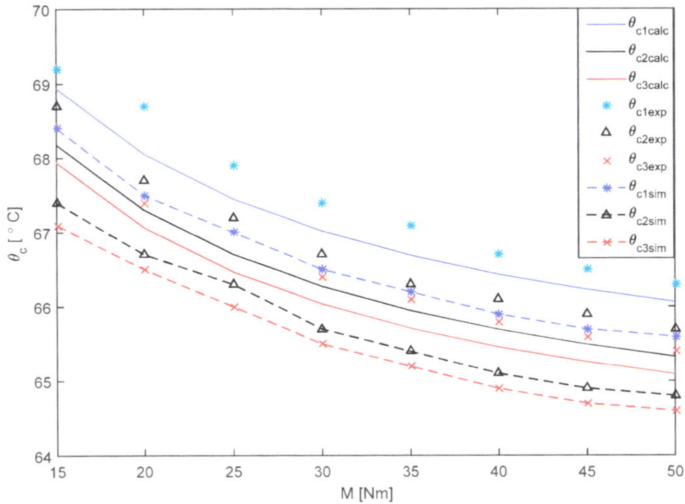

Figure 11. Case temperature variation for different load resistance values: 10, 20 and 100 Ω (v = 1 m/s, I = 300 A and L = 10 mH).

In addition, it can be noticed that for higher values of the load resistance, the temperature values decreased. This can be explained because as load resistance increases than the impedance increases and the current decreases, which finally leads to lower power loss and lower junction and case temperatures. For instance, at 30 Nm tightening torque, and 10 Ω load resistance, the junction temperature was 94.67 °C in comparison with the case of 100 Ω load resistance when the junction temperature got the value of 93.11 °C, Figure 10. At the same tightening torque of 30 Nm, the case temperature was 67.01 °C when load resistance was 10 Ω and became 66.03 °C for a load resistance of about 100 Ω, Figure 11. The same as in the previous analyzed case, the simulation values were lower than the calculated junction and case temperatures. As an example, at 15 Nm tightening torque and 10 Ω load resistance, the simulated value was 96.9 °C with respect to 97.72 °C of the calculated junction temperature, Figure 10, and the simulated case temperature was 68.4 °C lower than 68.93 °C, the calculated one. Moreover, on the same graph of case temperature variation, there were placed the experimental data, which were higher than the computed ones, Figure 11. The explanation was the same as in previous case. As an example, at 30 Nm tightening torque and 10 Ω load resistance, the experimental value was 67.4 °C with respect to 67 °C of the calculated case temperature.

The next thermal analysis considered the same variation of junction and case temperature against tightening torque but at different load inductance values from 10 to 160 mH, in the case of a direct current of 300 A, airflow speed of 1 m/s and load resistance of 100 Ω. As in the previous analyzed cases, the temperatures decreased when the tightening torque increased, Figures 12 and 13. For instance, in the case of a load inductance of 10 mH, the junction temperature decreased from 96.14 (torque of 15 Nm) to 91.81 °C (torque of 50 Nm), Figure 12, and the case temperature decreased from 67.93 (torque of 15 Nm) to 65.09 °C (torque of 50 Nm), Figure 13. Regarding the simulated temperature values, the same trend was confirmed: lower values than those calculated. For instance, in the case of tightening torque of 30 Nm and load inductance of 160 mH, the simulated junction temperature had a lower value of 96.6 °C than 97.08 °C the calculated one, Figure 12 and the simulated case temperature was 67.9 °C lower than 68.53 °C obtained from computations, Figure 13. On the same graph, it can be noticed also the obtained experimental values, which were higher than the calculated ones. For instance, at 30 Nm tightening torque and load inductance of 160 mH, the experimental case value was 68.9 °C with respect to 68.53 °C, the calculated one.

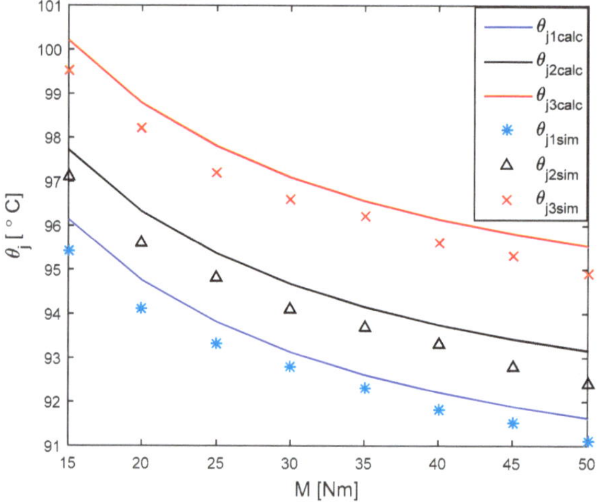

Figure 12. Junction temperature variation for different load inductance values: 10, 100 and 160 mH (v = 1 m/s, I = 300 A and R = 100 Ω).

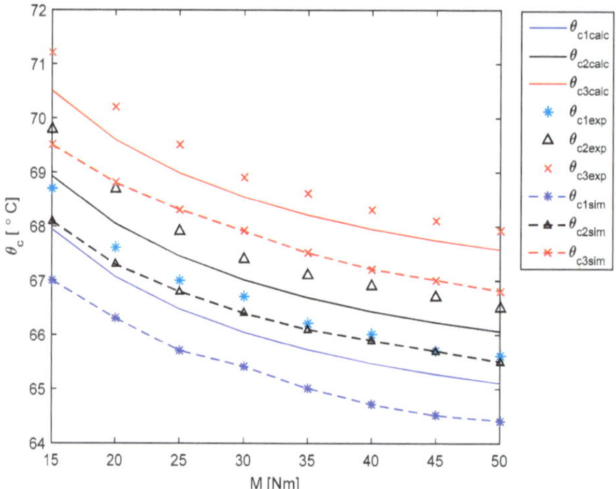

Figure 13. Case temperature variation for different load inductance values: 10, 100 and 160 mH (v = 1 m/s, I = 300 A and R = 100 Ω).

The last thermal analysis investigated the variation of junction and case temperature at different airflow speed values from 2 to 6 m/s when the direct current had the value of 300 A, the load resistance was about 10 Ω and the load inductance was 60 mH. The same decreasing evolution of the temperatures when the tightening torque increased could be observed, both for junction and case temperature, Figures 14 and 15.

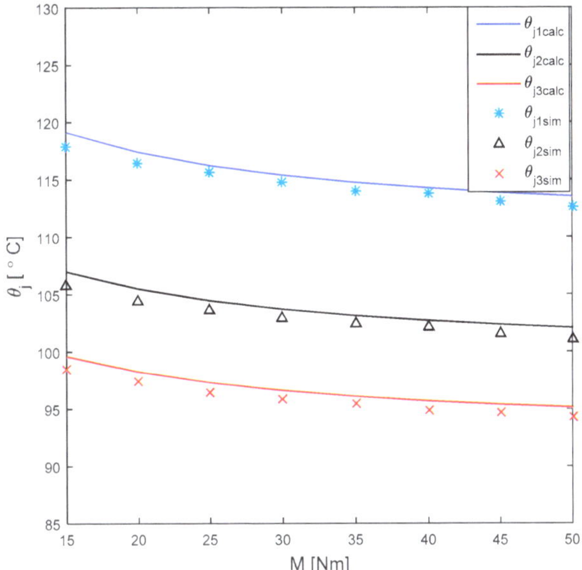

Figure 14. Junction temperature variation for different air speed values: 2, 4 and 6 m/s (I = 300 A, R = 10 Ω and L = 60 mH).

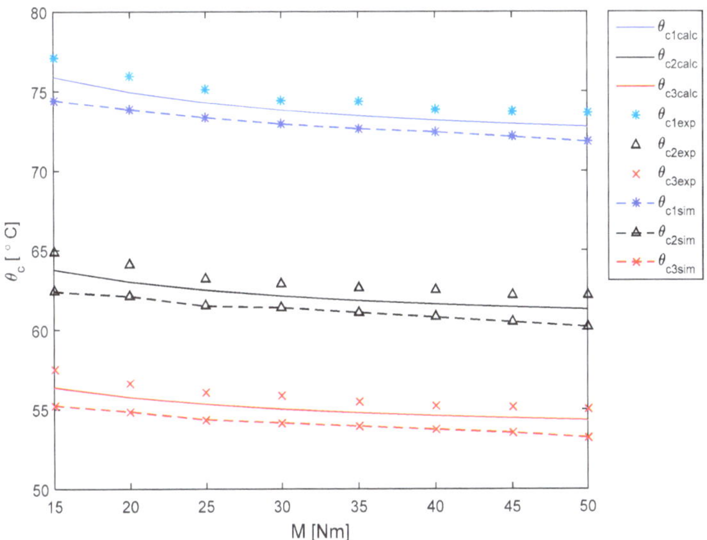

Figure 15. Case temperature variation for different air speed values: 2, 4 and 6 m/s (I = 300 A, R = 10 Ω and L = 60 mH).

As an example, in the case of the air speed value of 2 m/s, at a torque value of 15 Nm there was a junction temperature of 119.07 °C and case temperature of 75.87 °C, and to a torque value of 50 Nm, the junction temperature reached the value of 113.49 °C and the case temperature got a value of 72.72 °C. As in previous analyzed cases, the simulated values of the junction and case temperature were lower than the computed ones: a junction temperature of 114.7 °C in comparison with the calculated value of 115.34 °C in the case of 30 Nm tightening torque, Figure 14, and a case temperature of 72.9 °C lower than 73.36 °C, the calculated value for the same tightening torque, Figure 15. Additionally, on the last graph, it can be noticed the experimental values for the case temperature, which were higher than the calculated ones. For instance, in the case of air speed value of 2 m/s and a torque value of 30 Nm, the experimental obtained value of the case temperature was 74.4 °C with respect to the calculated one of 73.36 °C.

The difference between calculated and experimental values was also due to measurement errors, simplifications in the mathematical model and experimental setup. It could be noticed that the maximum difference between calculated and measured values was less than 2 °C.

5. Conclusions

A new mathematical model to investigate the power diodes' behavior from a thermal point of view was proposed in this paper. The power diodes belonged to a three-phase bridge rectifier used in electric traction applications. The mathematical model can be used to calculate the junction and case temperature of the power diode for different values of the tightening torque, direct current through the diode, airflow speed and load parameters (resistance and inductance). It outlined the decreasing of both the junction and case temperature when the tightening torque increased. The decrease in temperature was at a maximum of 5 °C while the tightening torque increased more than 3 times from 15 to 50 Nm.

The computed values were compared with the simulation results. Hence, a 3D modeling and thermal simulations of a power assembly including two diodes mounted on the heatsink were performed. The thermal simulation values were lower that the calculated ones because the because the mathematical model of junction temperature took into account a concentrated power loss and during thermal simulations, the power loss was evenly distributed on the silicon chip and contact surface, which led to a better heat transfer to the environment. Additionally, the calculated values were compared with some experimental data. It recorded higher experimental values than the calculated ones because the mathematical model considered a uniform cooling of the power diode case and actually, the airflow did not uniformly cool the surface of the diode case, which did not lead to a good heat transfer to the environment.

The proposed thermal model can be used to design different power bridge rectifiers from electric traction but not only this, with an optimized thermal distribution it can also consider the influence of some key parameters both electrical and mechanical. Since the new mathematical model takes into account also some thermal parameters, as thermal resistances related to power diode and heatsink, it can be customized for a wide range of power semiconductor devices mounted on different type of heatsinks from power converters.

Author Contributions: The whole article has been performed by the authors with equal contribution. All authors have read and agree to the published version of the manuscript.

Funding: This research received no external funding.

Acknowledgments: The authors had not received any funds for covering the costs to publish in open access.

Conflicts of Interest: The authors declare no conflict of interest.

Nomenclature

θ_j	means the junction temperature °C
θ_c	case temperature °C
θ_a	ambient temperature °C
P	average power loss W
r_T	internal resistance Ω
V_T	voltage drop V
R_{thjc}	thermal resistance between junction and case °C/W
R_{thck}	thermal resistance between case and heatsink °C/W
R_{thka}	thermal resistance between heatsink and environment °C/W
I_{FAV}	average current A
I_{RMS}	root mean square value of the current A
I_m	maximum current A
I_{RM}	maximum reverse current A
V_R	reverse voltage V
V_{RRM}	maximum repetitive reverse voltage V
U_2	secondary voltage V
ϕ	phase shift between the current and voltage °el
f	commutation frequency Hz
Q_s	recovery charge C
ω	pulse current rad/s
L	inductance H
R	resistance Ω
v_{air}	air speed m/s
R_c	contact resistance Ω
F	contact force N
M	tightening torque Nm
d	screw diameter m
p	thread pitch m
μ	friction coefficient

References

1. Shafaei, R.; Ordonez, M.; Saket, M.A. Three-dimensional frequency-dependent thermal model for planar transformers in LLC resonant converters. *IEEE Trans. Power Electron.* **2018**, *34*, 4641–4655. [CrossRef]
2. Bulut, Y.; Pandya, K. Thermal modeling for power MOSFETs in DC/DC applications. In Proceedings of the 5th International Conference on Thermal and Mechanical Simulation and Experiments in Microelectronics and Microsystems, EuroSimE 2004, Brussels, Belgium, 10–12 May 2004; IEEE: Piscataway, NJ, USA, 2004; pp. 429–433.
3. Bahman, A.S.; Ma, K.; Blaabjerg, F. A lumped thermal model including thermal coupling and thermal boundary conditions for high-power IGBT modules. *IEEE Trans. Power Electron.* **2017**, *33*, 2518–2530. [CrossRef]
4. Ma, K.; Bahman, A.S.; Beczkowski, S.; Blaabjerg, F. Complete loss and thermal model of power semiconductors including device rating information. *IEEE Trans. Power Electron.* **2014**, *30*, 2556–2569. [CrossRef]
5. Poller, T.; D'Arco, S.; Hernes, M.; Lutz, J. Influence of thermal cross-couplings on power cycling lifetime of IGBT power modules. In Proceedings of the 2012 7th International Conference on Integrated Power Electronics Systems (CIPS), Nuremberg, Germany, 6–8 March 2012; IEEE: Piscataway, NJ, USA, 2012; pp. 1–6.
6. Hocine, R.; Pulko, S.H.; Stambouli, A.B.; Saidane, A. TLM method for thermal investigation of IGBT modules in PWM mode. *Microelectron. Eng.* **2009**, *86*, 2053–2062. [CrossRef]
7. Gradinger, T.; Riedel, G. Thermal networks for time-variant cooling systems: Modeling approach and accuracy requirements for lifetime prediction. In Proceedings of the 2012 7th International Conference on Integrated Power Electronics Systems (CIPS), Nuremberg, Germany, 6–8 March 2012; IEEE: Piscataway, NJ, USA, 2012; pp. 1–6.

8. Senturk, O.S.; Helle, L.; Munk-Nielsen, S.; Rodriguez, P.; Teodorescu, R. Converter structure-based power loss and static thermal modeling of the press-pack IGBT three-level ANPC VSC applied to multi-MW wind turbines. *IEEE Trans. Ind. Appl.* **2011**, *47*, 2505–2515. [CrossRef]
9. Ciappa, M.; Fichtner, W.; Kojima, T.; Yamada, Y.; Nishibe, Y. Extraction of accurate thermal compact models for fast electro-thermal simulation of IGBT modules in hybrid electric vehicles. *Microelectron. Reliab.* **2005**, *45*, 1694–1699. [CrossRef]
10. You, B.G.; Lee, B.K.; Lee, S.W.; Jeong, M.C.; Kim, J.H.; Jeong, I.B. Improvement of the thermal flow with potting structured inductor for high power density in 40kW DC-DC converter. In Proceedings of the 2012 IEEE Vehicle Power and Propulsion Conference, Seoul, South Korea, 9–12 October 2012; IEEE: Piscataway, NJ, USA, 2012; pp. 1027–1032.
11. Castellazzi, A.; Johnson, M.; Piton, M.; Mermet-Guyennet, M. Experimental analysis and modeling of multi-chip IGBT modules short-circuit behavior. In Proceedings of the 2009 IEEE 6th International Power Electronics and Motion Control Conference, Wuhan, China, 17–20 May 2009; IEEE: Piscataway, NJ, USA, 2009; pp. 285–290.
12. Zhou, Z.; Kanniche, M.S.; Butcup, S.G.; Igic, P. High-speed electro-thermal simulation model of inverter power modules for hybrid vehicles. *IET Electr. Power Appl.* **2011**, *5*, 636–643. [CrossRef]
13. Lemmens, J.; Driesen, J.; Vanassche, P. Dynamic DC-link voltage adaptation for thermal management of traction drives. In Proceedings of the 2013 IEEE Energy Conversion Congress and Exposition, Denver, CO, USA, 15–19 September 2013; IEEE: Piscataway, NJ, USA; pp. 180–187.
14. Honsberg, M.; Radke, T. 3-level IGBT modules with trench gate IGBT and their thermal analysis in UPS, PFC and PV operation modes. In Proceedings of the 2009 13th European Conference on Power Electronics and Applications, Barcelona, Spain, 8–10 September 2009; IEEE: Piscataway, NJ, USA, 2009; pp. 1–7.
15. Zhou, D.; Blaabjerg, F.; Lau, M.; Tonnes, M. Thermal cycling overview of multi-megawatt two-level wind power converter at full grid code operation. *IEEJ J. Ind. Appl.* **2013**, *2*, 173–182. [CrossRef]
16. Zhou, D.; Blaabjerg, F.; Lau, M.; Tonnes, M. Thermal analysis of multi-MW two-level wind power converter. In Proceedings of the IECON 2012–38th Annual Conference on IEEE Industrial Electronics Society, Montreal, QC, Canada, 25–28 October 2012; IEEE: Piscataway, NJ, USA, 2012; pp. 5858–5864.
17. Ma, K.; Blaabjerg, F. The impact of power switching devices on the thermal performance of a 10 MW wind power NPC converter. *Energies* **2012**, *5*, 2559–2577. [CrossRef]
18. Huang, H.; Bryant, A.T.; Mawby, P.A. Electro-thermal modelling of three phase inverter. In Proceedings of the 2011 14th European Conference on Power Electronics and Applications, Birmingham, UK, 30 August–1 September 2011; IEEE: Piscataway, NJ, USA, 2011; pp. 1–7.
19. Swan, I.R.; Bryant, A.T.; Mawby, P.A. Fast thermal models for power device packaging. In Proceedings of the 2008 IEEE Industry Applications Society Annual Meeting, Edmonton, AB, Canada, 5–9 October 2008; IEEE: Piscataway, NJ, USA, 2008; pp. 1–8.
20. Xu, L.; Liu, Y.; Liu, S. Modeling and simulation of power electronic modules with microchannel coolers for thermo-mechanical performance. *Microelectron. Reliab.* **2014**, *54*, 2824–2835. [CrossRef]
21. Graditi, G.; Adinolfi, G.; Tina, G.M. Photovoltaic optimizer boost converters: Temperature influence and electro-thermal design. *Appl. Energy* **2014**, *115*, 140–150. [CrossRef]
22. Drofenik, U.; Kolar, J.W. A thermal model of a forced-cooled heat sink for transient temperature calculations employing a circuit simulator. *IEEJ Trans. Ind. Appl.* **2006**, *126*, 841–851. [CrossRef]
23. Chen, Q.; Yang, X.; Wang, Z.; Zhang, L.; Zheng, M. Thermal design considerations for integrated power electronics modules based on temperature distribution cases study. In Proceedings of the 2007 IEEE Power Electronics Specialists Conference, Orlando, FL, USA, 17–21 June 2007; IEEE: Piscataway, NJ, USA, 2007; pp. 1029–1035.
24. Gautam, D.S.; Musavi, F.; Wager, D.; Edington, M. A comparison of thermal vias patterns used for thermal management in power converter. In Proceedings of the 2013 IEEE Energy Conversion Congress and Exposition, Denver, CO, USA, 15–19 September 2013; IEEE: Piscataway, NJ, USA, 2013; pp. 2214–2218.
25. Gao, C.; Liu, H.; Huang, J.; Diao, S. Steady-state thermal analysis and layout optimization of DC/DC converter. In Proceedings of the 2014 Prognostics and System Health Management Conference (PHM-2014 Hunan), Zhangiiaijie, China, 24–27 August 2014; IEEE: Piscataway, NJ, USA, 2014; pp. 405–409.

26. Cheng, L.; Chen, Q.; Li, G.; Hu, C.; Zheng, C.; Fang, G. Electro-thermal loss analysis of the 3L-ANPC converter. In Proceedings of the 2014 IEEE 5th International Symposium on Power Electronics for Distributed Generation Systems (PEDG), Galway, Ireland, 24–27 June 2014; IEEE: Piscataway, NJ, USA, 2014; pp. 1–5.

27. Zhou, L.; Wu, J.; Sun, P.; Du, X. Junction temperature management of IGBT module in power electronic converters. *Microelectron. Reliab.* **2014**, *54*, 2788–2795. [CrossRef]

28. Li, H.; Liao, X.; Li, Y.; Liu, S.; Hu, Y.; Zeng, Z.; Ran, L. Improved thermal couple impedance model and thermal analysis of multi-chip paralleled IGBT module. In Proceedings of the 2015 IEEE Energy Conversion Congress and Exposition, Montreal, QC, Canada, 20–24 September 2015; IEEE: Piscataway, NJ, USA, 2015; pp. 3748–3753.

29. Andresen, M.; Liserre, M. Impact of active thermal management on power electronics design. *Microelectron. Reliab.* **2014**, *54*, 1935–1939. [CrossRef]

30. Hayashi, Y.; Takao, K.; Adachi, K.; Ohashi, H. Design consideration for high output power density (OPD) converter based on power-loss limit analysis method. In Proceedings of the 2005 European Conference on Power Electronics and Applications, Dresden, Germany, 11–14 September 2005; IEEE: Piscataway, NJ, USA, 2005; pp. 1–9.

31. Li, X.; Li, D.; Qi, F.; Packwood, M.; Luo, H.; Liu, G.; Dai, X. Advanced Electro-Thermal Analysis of IGBT Modules in a Power Converter System. In Proceedings of the 2019 20th International Conference on Thermal, Mechanical and Multi-Physics, Hannover, Germany, 24–27 March 2019; IEEE: Piscataway, NJ, USA, 2019; pp. 1–4.

32. Groover, M.P. *Fundamentals of Modern Manufacturing: Materials, Processes and Systems*, 4th ed.; John Wiley & Sons, Inc.: Hoboken, NJ, USA, 2010; ISBN 978-0470-467002.

Review

A Review on Optimization and Control Methods Used to Provide Transient Stability in Microgrids

Seyfettin Vadi [1], Sanjeevikumar Padmanaban [2,*], Ramazan Bayindir [3,*], Frede Blaabjerg [4] and Lucian Mihet-Popa [5]

[1] Department of Electronics and Automation, Vocational School of Technical Sciences, Gazi University, 06500 Ankara, Turkey; seyfettinvadi@gazi.edu.tr
[2] Department of Energy Technology, Aalborg University, 6700 Esbjerg, Denmark
[3] Department of Electrical and Electronics Engineering, Faculty of Technology, Gazi University, 06500 Ankara, Turkey
[4] Center of Reliable Power Electronics (CORPE), Department of Energy Technology, Aalborg University, 9220 Aalborg, Denmark; fbl@et.aau.dk
[5] Faculty of Engineering, Østfold University College, Kobberslagerstredet 5, 1671 Kråkeroy-Fredrikstad, Norway; lucian.mihet@hiof.no
* Correspondence: san@et.aau.dk (S.P.); bayindir@gazi.edu.tr (R.B.); Tel.: +45-7168-2084 (S.P.)

Received: 26 July 2019; Accepted: 10 September 2019; Published: 19 September 2019

Abstract: Microgrids are distribution networks consisting of distributed energy sources such as photovoltaic and wind turbines, that have traditionally been one of the most popular sources of energy. Furthermore, microgrids consist of energy storage systems and loads (e.g., industrial and residential) that may operate in grid-connected mode or islanded mode. While microgrids are an efficient source in terms of inexpensive, clean and renewable energy for distributed renewable energy sources that are connected to the existing grid, these renewable energy sources also cause many difficulties to the microgrid due to their characteristics. These difficulties mainly include voltage collapses, voltage and frequency fluctuations and phase difference faults in both islanded mode and in the grid-connected mode operations. Stability of the microgrid structure is necessary for providing transient stability using intelligent optimization methods to eliminate the abovementioned difficulties that affect power quality. This paper presents optimization and control techniques that can be used to provide transient stability in the islanded or grid-connected mode operations of a microgrid comprising renewable energy sources. The results obtained from these techniques were compared, analyzing studies in the literature and finding the advantages and disadvantages of the various methods presented. Thus, a comprehensive review of research on microgrid stability is presented to identify and guide future studies.

Keywords: optimization; control; microgrid; transient stability

---

1. Introduction

Power systems that generate electricity from renewable energy sources have in recent times increased their share in the total installed power capacity of electrical energy generation as a result of various advantages [1]. In addition, this increase in large-scale solar photovoltaic systems and wind power systems has resulted in problems affecting the quality of the electrical power produced [2]. These problems increase the sensitivity of the grid against harmonic distortions and fluctuations, and reduce frequency stability [3–5]. Increased use of renewable energy sources in recent years has enabled the supply of clean, cheap and low-cost energy, but has caused several problems in electricity transmission and distribution networks.

Since the 2010s, it has been observed in wind power and solar photovoltaic systems that a limited amount of total reserve capacity can provide energy to the system. At this point, microgrids are the most reliable solution for the integration of renewable energy sources within a wider electricity network. An example of a microgrid is presented in Figure 1 [6].

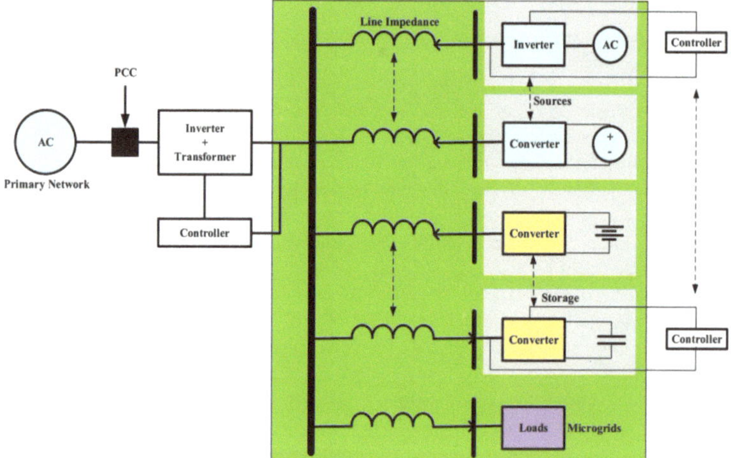

Figure 1. General structure of the microgrid.

Microgrids provide a promising solution for today's electrical energy problems based on the reliable, safe, environment-friendly and sustainable electric energy obtained from renewable energy sources [7]. By restructuring the microgrids, a contribution is made to the activities of energy planning and management, voltage stability and energy efficiency [8,9]. The use of renewable energy sources with an electricity network has negative effects on power quality [10], specifically connection and stability problems. Insufficiencies of active and reactive power capacity in the distribution transformers of different countries cause connection problems [11]. Stability problems are due to insufficiencies of a grid under normal operating conditions [12]. In order to solve these problems, electrical infrastructure needs to be renewed and expanded.

Another source affecting a power system's stability is the structure of renewable energy sources that the grid connection conducts on power electronic topologies. This problem generally has two effects, the first one being harmonic problems, as inverters generate significant current and voltage harmonics. Additionally, as stated above, it is desirable that the installed power rate of such types of renewable energy source increase in the grid while the total system capacity decrease. The second problem, which is more important, is the stability of grid [13].

Stability of the microgrids is defined as the balanced operation of all elements constituting the grid under normal operation conditions and achieving a reasonable balance following any disturbing effect. Transient stability in a microgrid is defined as the feature of an energy system that enables it to remain in a stable equilibrium state under normal conditions, and that allows it to regain a desired equilibrium after being subjected to disturbances arising from very general situations such as the switching on and off of circuit elements, voltage collapse, voltage and frequency fluctuation, phase difference fault, error in the islanded mode operation and grid-connected operation and so on. In other words, the stability purpose the synchronous operation of alternative energy sources in the grid. [14].

Stability is analyzed under three headings in a framework of the power system analysis: voltage, frequency stability and rotor angle stability [15]. Voltage stability depends on the reactive power balance generated and consumed in the power system, while frequency stability depends on the balance of active powers generated and consumed in the power system. Rotor angle stability is defined as the ability of synchronous generators in the grid to remain synchronized with each other. To achieve constant sinusoidal system frequency, stator voltages of the synchronous generators should run together in the same frequency. Frequency is a result of the mechanical speed of the rotor. For example, if any change occurs in the load, the stator current oscillates at the mechanical speed of the rotor, and hence at the frequency and angle of the rotor of the synchronous generator [16,17].

Ensuring the stability of the microgrids, providing balanced operation, preventing any disturbing impact in the system while switching on and off energy sources and restructuring to introduce a dynamic structure to the present system is realized in [18]. This dynamic structure will reduce the losses by using load management and voltage profiling for the loads on the microgrid. In the event of a fault, the affected zone is isolated, and the restructuring process is carried out by supplying energy to the load by order of importance and increasing usage rates by the specific switching processes [19]. The works of optimum restructuring and transition stability provide some benefits such as efficient usage of energy sources, meeting the energy requirements with the lowest possible cost, minimizing active power losses and switching processes, increasing energy quality accordingly (which ensures voltage stability), increasing network reliability, providing a solution with minimum loss in the event of a fault, increasing the entire system's efficiency, achieving optimum power quality and providing the necessary capacity [20].

A literature review reveals several studies on microgrids. The design, analysis and control of microgrids are current issues studied in the literature, and researchers have conducted studies on many subjects such as alternative current (AC) and direct current (DC) microgrid control and management, central control architecture, power quality and protection, multiple-agent systems, standards-based information and communication technologies, online optimization techniques and energy management systems [17].

Because of the variety of sources increases in the microgrids, restructuring becomes complicated and restricted. Furthermore, it causes to different combinations and objectivities of the multi-purpose optimization problems [21]. Classical optimization methods are applied to solve this problem; however, in some cases, these methods are an approach to the local minimum rather than the global minimum. Moreover, some classical methods cannot solve integer code problems. These deficiencies have been overcome by the use of evolutionary methods in the literature. The well-known methods used to solve restructuring problems are classified as follows:

a. Heuristic methods, branch changes, branches and limits, single-cycle optimization and loop breaking, etc.
b. Metaheuristic methods such as simulated annealing (SA), the genetic algorithm (GA), evolutionary programming (EP), ant colony optimization (ACO) and the harmony search algorithm (HSA).
c. Artificial neural networks (ANNs) such as machine-learning algorithms.

Each of these methods has advantages and disadvantages. When the studies carried out in the literature are analyzed, it is seen that the problems that arise in providing transient stability in microgrids are solved by these methods. The solution to the problem of providing transient stability includes the components of objective function and system operation constraints. The common objective of all problems of providing transient stability in a microgrid is to achieve power quality and minimize energy cost by connecting to the present network of energy sources at a steady state [22].

This review paper presents the studies in the literature in regards to their contributions to energy efficiency, prioritizing the use of energy regenerated by renewable energy sources, use of optimization and control methods and maintaining a continuity of the energy. To compare the capabilities of the control methods that used in island mode and grid-connected microgrids in terms of transient stability, optimization and control methods are examined. Besides the comparison, advantages, disadvantages and limitations of the control methods are discussed in detail. Another aim of this paper is to examine the research available on microgrid transient stability.

2. Control in Microgrids

There are two main energy source types in microgrids. One of them is a DC source comprising fuel cells, solar panels and batteries, whereas the other source is AC, comprising microturbines and wind turbines in which output voltage is rectified.

Both source types are generally DC sources created by use of an inverter [23]. The inverter structure for a DC source is shown in Figure 2a, and the inverter structure for an AC source is shown in Figure 2b.

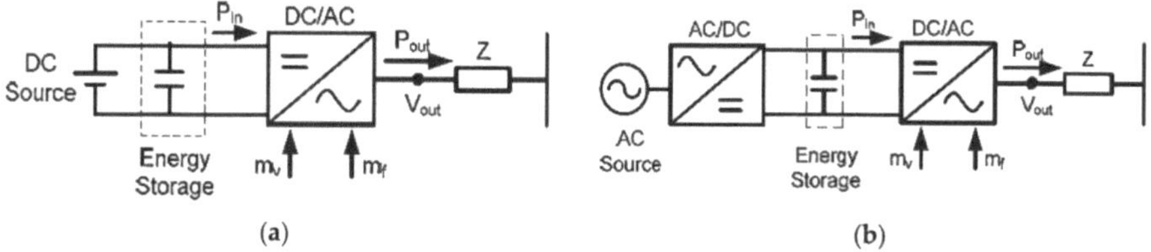

Figure 2. (a) Inverter for DC source. (b). Inverter for AC source.

Current, voltage and frequency parameters in the output of the inverter are determined by the control method of the inverter. The output voltage is directly related to the intensity of the capacitor voltage on the side of DC. Storage quantity in the capacitors is less than the storage quantity in the rotating field and, therefore, control methods are of great significance [24]. Different application strategies are used to manage the power flow control by depending on source number. It requires an energy management system to activate and deactivate of the sources. [25,26]. Moreover, power flow in the grid connected mode is provided by using sliding mode control (SMC), model predictive control (MPC), power-reactive (PQ) control and droop control—a robust, fuzzy logic control. [27]. A transient in an electrical system is defined by a sudden change in circuit conditions, such as when a system is switched on and off, or a fault occurs. These faults are small signal faults, unbalanced voltage or steady state faults. Firstly, small signal faults cause droop gains and load fluctuations to influence the voltage stability of a grid or microgrid. The small signal faults of grids are analyzed using a linearized model of the network, distributed generations (DGs), control units of DG, and loads. Secondly, voltage and frequency amplitude generated from energy sources cause unbalanced voltage and frequency. As such, voltage and frequency stability should be provided in power systems. Thirdly, steady state fault causes to the tracking error on control operations. [28].

Dynamic and transient analyses are important concepts in power systems. The infrastructure of the main electric grid evolves with the integration of hybrid energy systems that form renewable energy systems. The hybrid microgrid is a grid structure comprising both grid and renewable energy sources, or renewable energy sources only. As shown in Figure 3, hybrid microgrids can comprise more than one energy source in the same distribution grid [29,30]. In renewable energy sources or hybrid energy systems, this situation can be prevented using control algorithms that run independently of system parameters, such as the sliding mode control method, whereas unknown system parameters or changes in system control algorithm parameters that depend on model parameters, such as the model predictive control method, adversely affect control efficiency [31,32].

The stability conditions in microgrids used in the literature are presented in Figure 4 [33–37]. Stability methods have been applied in both grid-connected mode and islanded mode. These methods are used to provide small signal, voltage and frequency stability. The stability is implemented in the short term, ultra-short term and long term.

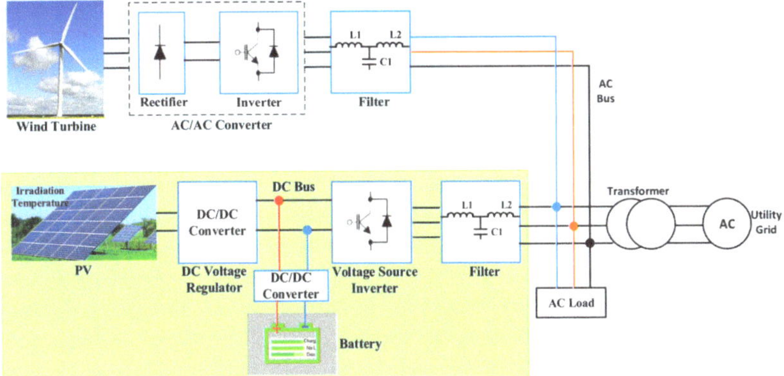

Figure 3. The general structure of a hybrid microgrid.

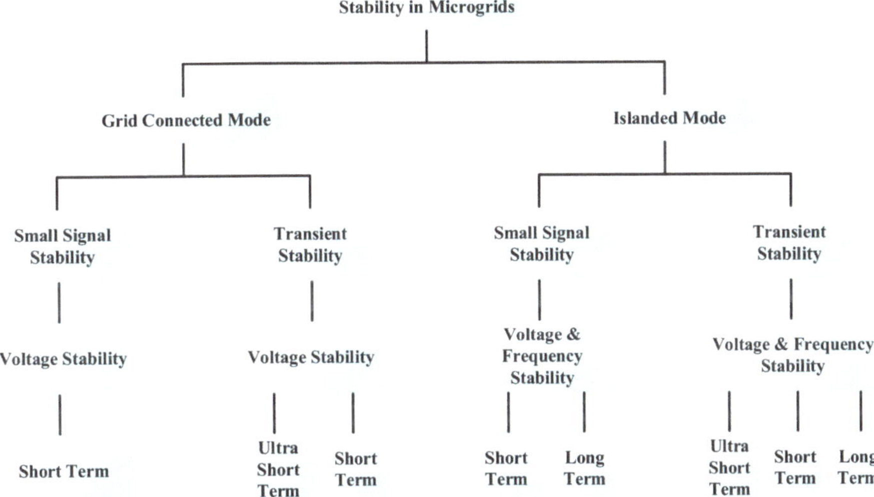

Figure 4. Methods for stability in microgrids.

3. Transient Stability in Grid-Connected Microgrids

The dynamic response of grid-connected microgrids were examined using different control strategies based on an analysis of dynamic behavior of the system when exposed to the fault current and important distortions in the distributed energy systems [38]. Recent studies have focused on simulation studies of transient stability in grid-connected microgrids, and the studies analyzed in the literature have generally used MATLAB/SIMULINK platforms [39]. Methods employed in grid-connected microgrids are presented in Figure 5. These methods are use a single algorithm, hybrid algorithm or an algorithm created by the use of computer software tools [40]. These methods have advantages and disadvantages when compared with each other. Optimization algorithms, such as genetic, particle swarm and artificial bee colony algorithms (as well as others), are used in respect to frequency, voltage regulation and the reduction of current ripple for transient stability in AC or DC microgrids comprising wind turbines, photovoltaic energy, fuel cells, battery energy storage systems and flywheel energy storage while operating in islanded mode. Proportional and integral (PI), proportional and derivative (PD) or proportional, derivative and integral (PID) control methods are generally used in a feedback system. However, parameters of these methods are defined as constant while the algorithm designing, it negatively affect the system in cases of resource or load change. The automatic adjustment of parameters is required in order to eliminate negative statements [41]. The optimization algorithms used to automatically adjust parameters are given in Figure 5.

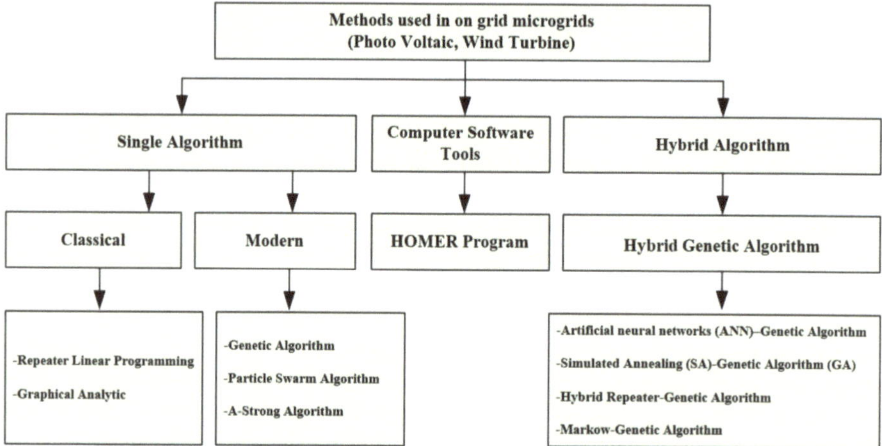

Figure 5. Methods employed in grid-connected microgrids.

The genetic algorithm (GA) is effective and useful in areas where a search space is large and complex, and where a solution is very difficult to achieve in a limited search space. This algorithm is commonly used in situations that are not expressed in a particular mathematical model [42]. The particle swarm algorithm starts with a population containing random solutions and tries to provide an optimum response by updating each iteration. Iteration number, swarm quantity, correction factor and inertia moment are important for this algorithm [43]. In particle swarm optimization (PSO), particles change their position until the number of iterations is completed. Thus, each particle benefits from the experience of not only the best particle in the swarm, but also all other particles in the swarm.

The artificial bee colony (ABC) algorithm determines least amount of energy necessary for honeybees to travel the shortest path between their home and a food source according to environmental conditions in the natural environment. This system in bees is applied to power systems, and the optimum solution is searched. Since the control parameter is low, it has a simple structure that can be used for both numerical and discrete problems. Furthermore, the algorithm is used both alone and as a hybrid. When the algorithm is used as a hybrid in a power system, structures emerge that are more dynamic and possess greater stability [44].

The use of algorithms in microgrids is a very sensitive and important issue for a grid in terms of control and coordination of the distributed generation systems. It has been seen that the frequency and voltage values of a system are generally set by the grid according to the co-functioning of the microgrid with the distribution grid, and a cost analysis of grid-connected microgrids has been realized in the literature [45–48]. The maximum efficiency of renewable energy sources is achieved using optimization and control algorithms, as well as by utilizing the most efficient use of storage members and by supplying the minimum amount of energy from the grid to ensure stable operation of the system with minimum cost [49].

Microgrids are connected to the system with a power electronic converter and inverter (DC-DC, DC-AC, etc.). Since energy flows occur between parallel energy sources, the energy flows are controlled [46]. This control is generally carried out using the droop control method, a method that is used for sharing power between synchronous generators [50]. The most important characteristic of this method is that it allows power sharing between sources, and does not require communication infrastructure [51]. The relationship between active power-frequency and reactive power-voltage used for droop control is shown in Figure 6. Inverters are connected to the system in parallel according to the used method [52]. This control method is also used for sharing power between uninterrupted power sources (UPS) connected in parallel without a control cable.

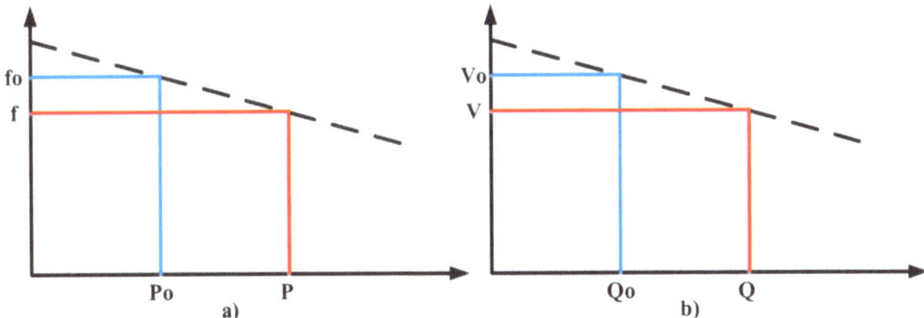

Figure 6. (**a**) Frequency reducing characteristic. (**b**) Voltage reducing characteristic.

Although a high level of reliability and flexibility is achieved by this technique, it also has some disadvantages. For instance, when the loads in the system are non-linear, and harmonic current is taken into consideration, this control method does not share power as a result of power sharing [53]. To solve this problem, harmonic current sharing methods are combined with virtual impedance adding methods, adaptive methods, droop control methods for the smoothest power share and frequency–voltage balance [54]. Another disadvantage of the droop control method is that frequency and voltage values are determined based on the load [55]. To solve this problem, central, non-central (distributed control) and hierarchical control structures are implemented. These are the methods used most commonly in the literature [56].

It has been demonstrated that non-central and hierarchical control methods can provide the production balance of plants with power electronic-based inverters [57]. The requirement of communication infrastructure for the central control method, and redesigning if a new grid is installed, restricts the area of use of this control method [58]. The non-central control method is more appropriate for microgrids—since communication infrastructure is not necessary, it distributes the power more adequately for non-linear loads and works based on local measurements [59–61]. In non-central control systems, active–reactive power values and voltage-frequency values are adjusted according to the voltage and current data sent by the distributed generation system.

Numerous droop control techniques have been employed in microgrid applications. For example, droop control methods providing the share of harmonic current [62], droop control methods sharing the power based on the power angle [59], adaptive droop control methods that can adapt to variables [63], droop control methods that can share power by use of virtual impedance [64] and so on have been used in microgrid applications. However, the important disadvantage of these systems is that the voltage and frequency values of microgrids in slanded mode vary by load variation [65]. To eliminate this disadvantage, the secondary control function is enabled. This function follows load and regeneration changes, determines frequency and voltage reference lines and sends a warning to all units. In this way, voltage and frequency fluctuation become a near-zero value [66,67].

The hierarchical control method is a frequently used method for controlling microgrids. It has a three-layer control structure and a method designed to manage power systems with large-scale synchronous machines. Use of this method with some variations for microgrids has been proposed. The first control layer is the primary control, which provides the control of the internal structure of the distributed generation network. The main purpose of this control layer is to control the active and reactive power balance, depending on the frequency and voltage, by imitating synchronous generator behavior. By these means, energy flows between parallel inverters are prevented and power is shared adequately [68].

The second control layer is referred to as the secondary control and is designed to minimize frequency and voltage fluctuations of the system. Due to load and process values of generators in the system, which may vary continuously, voltage and frequency values vary as well [69]. To prevent this fluctuation, the secondary control method is used. Frequency and voltage values of the system are compared to the reference values [70]. The fault statement is sent to all units to prevent disruption of the system balance, and is generally used to maintain the grid frequency fluctuation limit at ±0.2Hz [71]. Characteristics of primary and secondary control methods are presented in Figure 7.

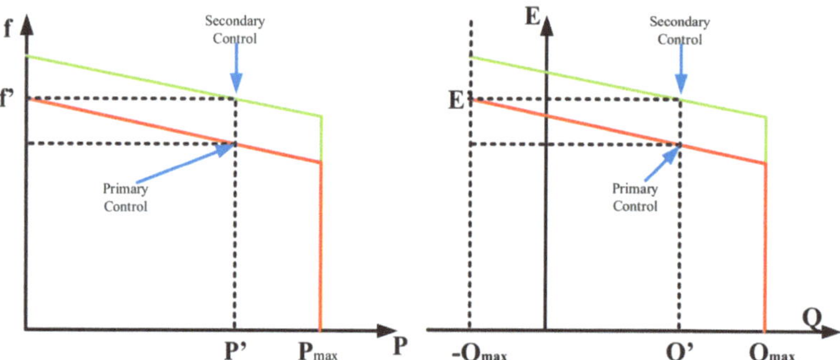

Figure 7. Characteristics of primary and secondary control methods [65].

The final control layer is the tertiary control. This control method is used to control the power flow between the microgrid and the main grid. While the microgrid and main grid run together, power flow is controlled by frequency and voltage values.

The frequency value of the main grid is constant. Therefore, power exchanges between the main grid and microgrid depend on the droop characteristic of the microgrid [72,73]. Power exchanges are controlled by changing reference frequency and voltage values of the microgrid [74]. The characteristics of the tertiary control method are shown in Figure 8. As seen in Figure 8, f_{mg} is the frequency at maximum generated power, E_{mg} is the voltage at maximum generated power, P_{gmax} is the maximum generated power, Q_{gmax} is the maximum reactive power, fg is the frequency generated and Eg is the voltage generated for the tertiary control method. In addition, the tertiary control method shows the relation between f (frequency) and P_{gmax} (maximum generated active power), as well as between f (frequency) and Q_{gmax} (maximum generated reactive power). These control methods are frequently used in the literature to control different types of microgrids [75], especially AC microgrids, although they have recently started to be applied to DC microgrids [76]. A number of grid elements, such as solar energy systems, storage elements and electric vehicle charging stations, are features of a DC microgrid [77].

Another control method, known as the robust droop control method, is commonly used in distributed loads where there are high-voltage multiple microgrids [78]. As seen in Figure 10, P_1 and P_2 are active power in the traditional droop control strategy, P_{1-2} and P_{2-2} are active power in the proposed control strategy, Q_1 and Q_2 are reactive power in the traditional droop control strategy and Q_{1-2} and Q_{2-2} are reactive power in the proposed control strategy. In Figure 9, the system achieves a stable state condition after a shorter time (around 2 s) compared to the traditional control method, and subsequently shares its power [79].

Moreover, the robust droop control method shares power at a higher voltage level than the traditional droop control method. Reactive power share is not achieved by the traditional droop control method and is equalized to an approximate reference value by the robust droop control method [80,81].

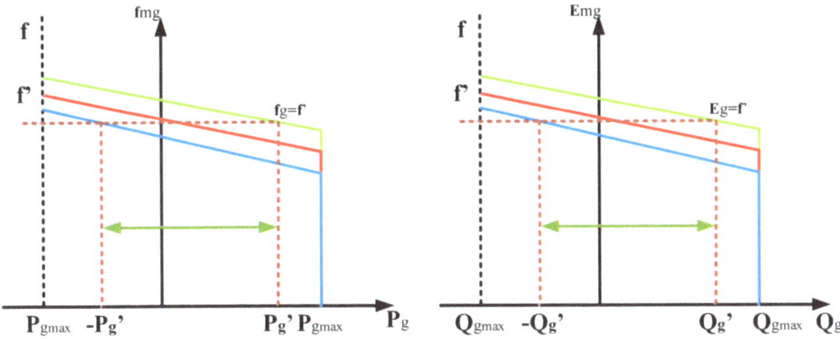

Figure 8. Characteristics of the tertiary control method.

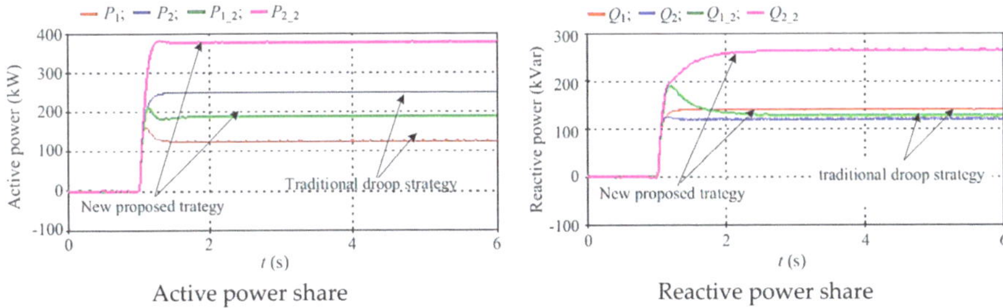

Active power share   Reactive power share

Figure 9. Comparison between the new robust droop control and the conventional droop control [80].

4. Transient Stability in Islanded Mode Microgrids

The transient stability of the islanded mode microgrid is affected by the type of load connected to the system and the responsiveness of the control strategy used for distributed generation (DG) interfaces. Different types of control strategy of the inverter affect all systems, so it is important to provide transient stability.

Frequency, voltage and transient time intervals according to their standards as well as simulated activities in islanded mode are presented in Table 1. A small variation signal in islanded mode causes important distortions in harmonics, active–reactive power balance, frequency and voltage according to this table, and the standards referred to as the optimal operation band are shown in Table 1.

Table 1. Optimal operation standard for the microgrid in islanded mode [82].

Standards	Quality Factor (QF)	Nominal Frequency Range	Nominal Voltage Range	Islanding Detection Time
UL 1741	2.5	$59.3 < f < 60.5$	$88\% < V < 110\%$	$t < 2s$
IEEE 929-2000	2.5	$59.3 < f < 60.5$	$88\% < V < 110\%$	$t < 2s$
VDE 0126-1-1	2	$47.5 < f < 50.5$	$88\% < V < 110\%$	$t < 0.2s$
IEC 62116	1	$(f-1.5Hz) < f < (f+1.5Hz)$	$85\% < V < 115\%$	$t < 2s$
IEEE1547	1	$59.3 < f < 60.5$	$88\% < V < 110\%$	$t < 2s$
Korean Standard	1	$59.3 < f < 60.5$	$88\% < V < 110\%$	$t < 0.5s$

International organizations such as the IEC (International Electrotechnical Commission) and IEEE (Institute of Electrical and Electronics Engineers have defined certain standards for interconnection, operation and control of microgrids in conjunction with the main grid. The standards offer operation ranges relevant to performance, testing, safety and maintenance of the integrated power system. Detection time is the transition time to operation in island mode. Quality factor is an important parameter used to determine the reliability and robustness of any islanding detection mode. Also, nominal voltage and frequency range are optimal operation standards of islanding detection time [82].

4.1. P-Q Control Methods

The purpose of the P-Q control is to ensure equally distributed generation between active power and reactive power. As seen in Figures 10 and 11, the P-Q control is performed by controlling frequency and voltage control during load sharing. Frequency stability varies by the size of the load. Frequency stability for an overloaded system lasts longer than a less-loaded system. For this reason, the load is shared by the sources for frequency stability [83].

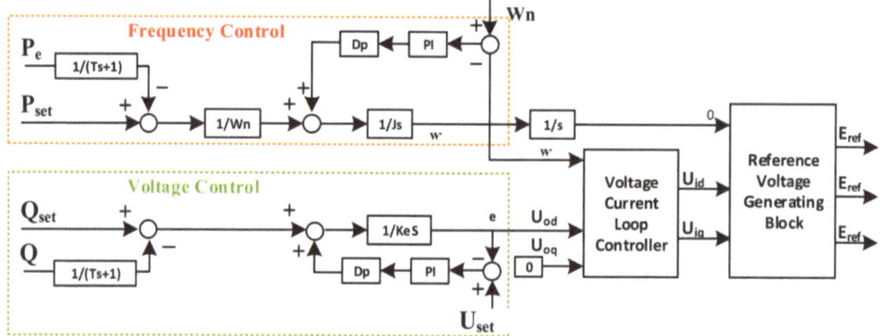

Figure 10. Voltage and frequency control block diagram [83].

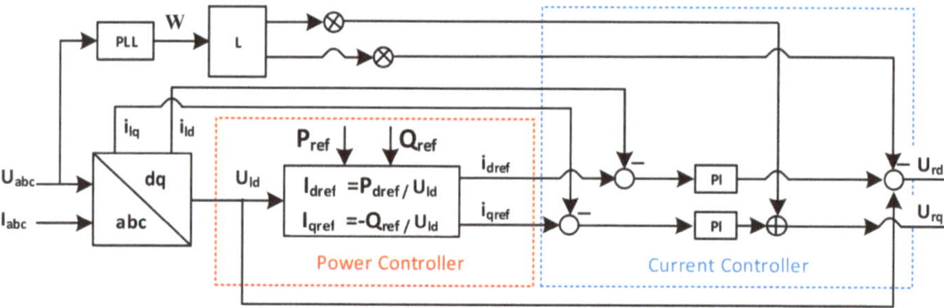

Figure 11. P-Q control structure [81].

4.2. PI/PID Algorithms

The PI/PID algorithms consisting of a proportional, derivative and integral composition are used to maintain system performance in double-fed asynchronous generator wind turbines in the event of a fault [84]. This algorithm are used to reduce overvoltage and stability errors in a flexible AC transmission systems (FACTS) unit [85], which uses battery systems in the event of voltage and frequency fluctuations to increase energy flow. In the statement of fault in the system, the P-Q control is provided with PI for balanced compensation [86]. Unique PI design is not useful in reducing the harmonics at high frequencies.

The control structure is simple and provides the required performance. Also, it is commonly used in industry. Optimization of three parameters changes the operation points. Moreover, it is not stable enough to adapt itself to the load variations. Since PI/PID controllers are more stable in linear systems, the structure does not show stable behavior due to the dynamic behavior of non-linear systems [86].

4.3. Model Predictive Control Method

The facts that concepts in the MPC management are simple and the controller is heuristic are considered a significant advantage. MPC is a control strategy devised for both large, multiple input–output control problems, and for inequality constraints on the inputs or outputs.

MPC design parameters should be chosen carefully. MPC is the preferred solution for difficult control problems [87,88].

The operating principle of an MPC controller is shown in Figure 12. This control method is an algorithm that is used to predict the future behavior of a system. Free and forced response are the prediction components of this method. The expected behavior of system output y (t+j) is shown by free response. In addition, it is accepted that the future values of the actuating variables will be equal to zero. The additional component of the system response is formed by forced response, which is based on the pre-calculated set of future-actuating values u (t+j).

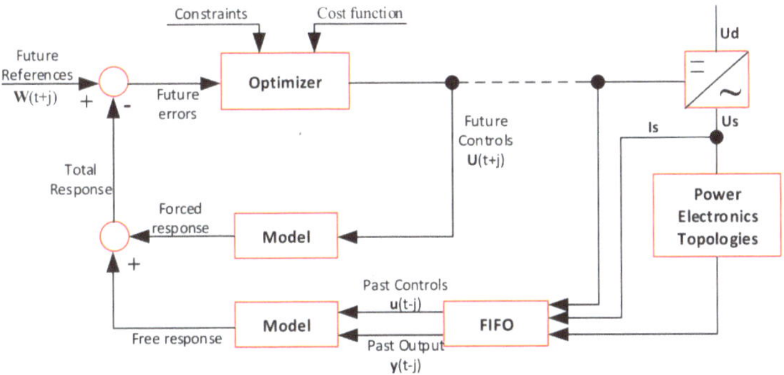

Figure 12. General structure of an MPC controller.

The total response of system behavior is determined as the sum of free and forced responses for the entire future system's behavior in linear systems. The sum process is calculated using the superposition principle. This sum is pre-calculated up to a prediction horizon, which is determined by a set of future reference values output by the system. The difference in future control error between the future reference and pre-calculated actual values is then obtained. This method takes system restrictions and the cost function into account, and a set of optimum future values u (t+j) from the expected error are determined by this method [88,89].

The MPC leads to a high calculation density because of the pre-calculation of the system's behavior. The calculation density is significantly reduced at a control horizon. When the horizon is reached in a steady state, the controller output remains constant. This situation is shown Figure 13.

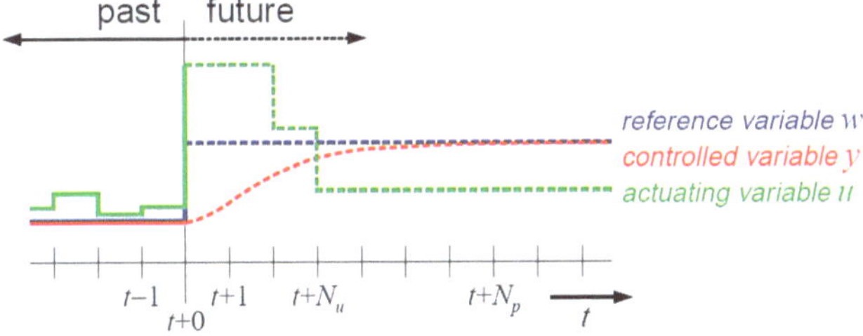

Figure 13. MPC operation model.

The MPC is used to eliminate the following errors in the grid-connected operation and excessive values using fuzzy logic to reduce fluctuations and eliminate the adverse effects of time delay and stability errors. The MPC estimates the critical parameters while evaluating them. It is also used to minimize harmonics in the network current via advanced estimation algorithm, which is used under the adaptive current control for three phases in normal operation mode [90].

Advantages of the MPC method can be listed as follows:

- Concepts are very heuristic and easy to understand;
- It is used in multi-variable systems;
- It prevents idle time;
- Addition of non-linear structures is easy;
- Constraints are eliminated by simple solutions.

Extension to the controller is very easy in many application systems.
Disadvantages of the MPC method compared to the classical controllers can be listed as follows:

- It has numerous mathematical operations;
- Quality of the model created affects the controlling performance directly;
- Addition or removal in the systems requires regulations in the controller.

The MPC controls the interaction and constraints between the variables and selects the optimal strategy. In this way, it completes the operation quickly and provides economic benefits. It predicts dynamic behavior in linear or non-linear systems, and is more proper for multi-variable systems. The model is defined correctly, and while unknown parameters are not dealt with, performance analysis is quite difficult [90].

4.4. Linear Quadratic Control Applications

The linear quadratic control (LQC) method is used to eliminate oscillation in three-phase inverters, harmonic distortion and voltage and frequency degradation in grid-connected photovoltaic systems. DC voltage in the inverter is used to compensate for the voltage in the line and optimize energy flow to the loads. Stable operation in different models of the algorithm is used to eliminate small-scale noise and regulate voltage and current. At a low-level DC voltage, a Kalman filter is used to reduce the oscillation in the output [91]. Linear quadratic control, which is one of the optimal control methods, has a more stable structure than the PI/PID control algorithm since matrix weights are adjusted simply. However, the analytical solution of the algorithm is quite difficult and does not work in the event of constraints [92].

4.5. Sliding Model Control Method

The SMC is a control method used in non-linear systems and discretely timed running systems. It carries out the control as a function of systems switching at high frequencies. The essential principle of the control is that a certain sliding manifold (surface) uses a reference path as an orbit, and the controlled system can be directed to the required balance point [93]. The main advantage of a system having SMC characteristics is that it guarantees stability and robustness against parameter uncertainties [94]. The SMC is used for providing feedback for variable-rate wind turbines and the continuity of wind turbines in the event of a fault with the present fixed-switching method. It is also used for the grid-connected three-phase inverter, to provide reliability and stability in distributed generation. By means of dynamic behavior against uncertainties and distortion, it is used more in non-linear systems. It is mathematically responsive due to a low computational process [95].

4.6. Robust Control Method

The robust control method (RCM) is the control approach designed to absorb faults in the event of uncertain parameters and degradation. It is commonly used in microgrid applications [96], especially for maintaining stability in the grid-connected mode and current control in three-phase inverters, as well as to rectify frequency fluctuations in grid-connected inverters [97]. Moreover, it is used to eliminate uncertainties in control systems generally, as well as cross-linking applications between communication paths. However, it may not be practical in large-scale applications since there is no constraint [98].

4.7. Particle Swarm Algorithm Applications

The PSO is an algorithm developed by observing social behaviors of bird and fish shoal, which is used to improve PI/PID parameters [99]. The algorithm begins with a population including random solutions and updates in every iteration, and tries to give the optimum response. The PSO automatically adjusts the gains of the PI/PID controller in the circuit to maintain the required performance [100]. It designates the most suitable position or value [101] and is used to improve the energy quality in designating the position and layout of distributed regeneration [102]. It is generally used for the optimization of non-linear, non-derivative and multi-mode functions. In addition to being simple and easy to implement, it also has an adjustable parameter structure. The disadvantage of the PSO algorithm is that while it is close to optimal levels, its calculation time depends on the adjustment time of the PSO parameters [103–106].

5. Discussion

In distributed generation, load sharing between production and distribution is performed by a primary control. This is the first and main step of microgrid transient control. The primary control provides the voltage and frequency stability between consumption and production in an on-gird microgrid. This paper aimed to introduce the control methods used in transient stability of microgrids and clearly explain its advantages and disadvantages.

For this purpose, the control methods used for transient stability have been given in Table 2. The control methods, including MPC, SMC, LQC and PSO, provide a reduced voltage and frequency ripple, transient stability among energy sources and prevention of voltage collapse. However, they also require well-known parameters belonging to the mathematical model of the system in order to use the control methods. Unlike these methods, the droop control method has worked to measure the system parameters belonging to the distributed operation of microgrids composed of renewable energy sources. In addition, the usage of this method is simple. The method provides steady and precise results on frequency stability of an overloaded system, and on power-share at high voltages, regardless of constraints or limitations. This paper presents droop control as the most highly preferred method for microgrid stability. The PSO algorithm starts with a population containing random solutions and tries to provide the optimum response by updating each iteration as the particles change their position until the number of iterations is completed. Thus, each particle benefits from the experience of not only the best particle in the swarm, but also all other particles in the swarm.

Since the ABC algorithm has very few control parameters, the system response is simple and faster when compared to other algorithms. It can be used for both numerical and discrete problems. GA is effective and useful in areas where the search space is large and complex, although it is very difficult to discover the solution with a limited search space. This algorithm is commonly used in situations that are not expressed in a particular mathematical model.

Table 2. Control methods used for transient stability in microgrids.

Control Method	Advantages	Disadvantages	Type of Connection	Energy Sources	Voltage Level	Ref. Number
Linear Quadratic Control (LQC)	The method is used in three-phase inverters to eliminate oscillation and degradation. DC in the inverter is used to compensate for the voltage in the line and optimize the energy flow to the loads.	Analytical solution of the algorithm is quite difficult and does not work with constraints.	Grid-connected mode, islanded mode	Grid and renewable energy sources	Moderate–high	[93]
Sliding Mode Control (SMC)	The method provides high precision, fast dynamic response and high stability in the event of distortion in large-scale loads. By means of dynamic behavior against uncertainties and distortion, it is used more in non-linear systems. It also provides a fast reaction due to low mathematical calculation.	Non-stability in linear systems	Grid-connected mode, islanded mode	Grid and renewable energy sources	Moderate–high	[94]
PI/PID Control	PI control is not as stable in adapting itself to load variations. It is more stable in linear systems.	Transient response is slow and control parameters are not controlled by the fluctuation of power. It does not show stable behavior with dynamic system responses in a non-linear system. It is very slow at reducing harmonics.	Islanded mode	Grid and renewable energy sources	Moderate	[85]
Droop Control	The method provides frequency stability for overloaded systems. Permits power sharing in high-voltage multi-microgrids and at high-voltage levels.	Fault rate in permanent voltage and power fluctuations. Fluctuates the frequency and voltage values based on load and reactive power-share fails.	Grid-connected mode, islanded mode	Grid and renewable energy sources, synchronous generator	High	[55]
Model Predictive Control (MPC)	MPC settlement time is shorter. MPC is used to eliminate errors and excessive values in the grid-connected operation and minimize the harmonics in the network current.	System model and initial parameters are required to achieve accuracy.	Grid-connected mode, islanded mode	Grid and renewable energy sources	Moderate	[89]
Particle Swarm Optimization (PSO) Algorithm	The algorithm is used for the optimization of non-linear, non-derivative and multi-mode functions.	Its disadvantages include being close to the optimal level and calculation time depends on the adjustment time of PSO parameters.	Grid-connected mode, islanded mode	Grid and renewable energy sources	Moderate	[102]

6. Conclusions and Evaluation

Modern society faces energy sustainability problems as energy demand increases and electricity transmission and distribution lines become old. Modernization of conventional grid architecture, innovative solutions and technologies and global warming are leading to limited investments, energy product systems and energy dependency. This study aims to raise awareness of the studies present in the literature in terms of contributing to energy efficiency, and to prioritize the use of energy regenerated by renewable energy sources through optimization and control methods in maintaining the continuation of energy.

Optimization and control methods were investigated to control the microgrid and provide transient stability in an islanded mode for grid-connected microgrids. In addition, this research paper is a preliminary examination of the frequency and voltage control strategy of islanded mode among grid-connected microgrids. The effect of optimization and control methods on reducing switching losses and pressure of the power electronic components of the inverter and converter are great. It is considered that developing the existing solutions as proposed in the literature and implementing them in the future will make significant technical contributions that could increase the total installed power rate of renewable energy plants for electric energy regeneration, and accordingly maintain the frequency stability of the grid.

The classification methods given in this review paper will help researchers select appropriate control methods that are used for microgrid transient stability, such as voltage collapse, voltage and frequency fluctuation, phase difference fault, error in the islanded mode operation and grid-connected operation. Therefore, the control method contributes to providing accurate microgrid transient stability. The fast dynamic response, stability, dynamic behavior, harmonic distortion, transient response, connection type, energy source type, voltage level, voltage–frequency control and behavior in the non-linear loads of each method are examined regularly. Therefore, it is helpful to conduct research on control methods used in microgrid transient stability. This study on the advantages and disadvantages of control and optimization algorithms, which are used for microgrid transient stability, could provide suggestions for further research and applications.

Author Contributions: All authors contributed equally for the research activities and for its final presentation as a full manuscript.

Funding: No source of funding for this research activity.

Acknowledgments: The authors would like to acknowledge the support and technical expertise received from the center for Bioenergy and Green Engineering, and Center of Reliable Power Electronics (CORPE) Department of Energy Technology, Aalborg University, Esbjerg, Denmark, which made this publication possible.

Conflicts of Interest: The authors declare no conflict of interest.

References

1. Zhang, D.; Evangelisti, S.; Lettieri, P.; Papageorgiou, L.G. Economic and environmental scheduling of smart homes with microgrid: DER operation and electrical tasks. *Energy Convers. Manag.* **2016**, *110*, 113–124. [CrossRef]
2. Camblong, H.; Baudoin, S.; Vechiu, I.; Etxeberria, A. Design of a SOFC/GT/SCs hybrid power system to supply a rural isolated microgrid. *Energy Convers. Manag.* **2016**, *117*, 12–20. [CrossRef]
3. Siksnelyte, I.; Zavadskas, E.K.; Streimikiene, D.; Sharma, D. An overview of multi-criteria decision-making methods in dealing with sustainable energy development issues. *Energies* **2018**, *11*, 2754. [CrossRef]
4. Blechinger, P.; Cader, C.; Bertheau, P.; Huyskens, H.; Seguin, R.; Breyer, C. Global analysis of the techno-economic potential of renewable energy hybrid systems on small islands. *Energy Policy* **2016**, *98*, 674–687. [CrossRef]
5. Singh, S.; Singh, M.; Kaushik, S.C. Feasibility study of an islanded microgrid in rural area consisting of PV, wind, biomass and battery energy storage system. *Energy Convers. Manag.* **2016**, *128*, 178–190. [CrossRef]
6. Rahbar, K.; Chai, C.C.; Zhang, R. Energy cooperation optimization in microgrids with renewable energy integration. *IEEE Trans. Smart Grid* **2018**, *9*, 1482–1493. [CrossRef]

7. Carli, R.; Dotoli, M. Decentralized control for residential energy management of a smart users' microgrid with renewable energy exchange. *IEEE/CAA J. Automatica Sinica* **2019**, *6*, 641–656. [CrossRef]

8. Siddaiah, R.; Saini, R.P. A review on planning, configurations, modeling and optimization techniques of hybrid renewable energy systems for off grid applications. *Renew. Sustain. Energy Rev.* **2016**, *58*, 376–396. [CrossRef]

9. Dragičević, T.; Lu, X.; Vasquez, J.C.; Guerrero, J.M. DC microgrids—Part II: A review of power architectures, applications, and standardization issues. *IEEE Trans. Power Electron.* **2016**, *31*, 3528–3549. [CrossRef]

10. Khatib, T.; Ibrahim, I.A.; Mohamed, A. A review on sizing methodologies of photovoltaic array and storage battery in a standalone photovoltaic system. *Energy Convers. Manag.* **2016**, *120*, 430–448. [CrossRef]

11. Louie, H. Operational analysis of hybrid solar/wind microgrids using measured data. *Energy Sustain. Dev.* **2016**, *31*, 108–117. [CrossRef]

12. An, L.N.; Tuan, T.Q. Dynamic programming for optimal energy management of hybrid wind–PV–diesel–battery. *Energies* **2018**, *11*, 3039. [CrossRef]

13. Sinha, S.; Chandel, S.S. Prospects of solar photovoltaic–micro-wind based hybrid power systems in western Himalayan state of Himachal Pradesh in India. *Energy Convers. Manag.* **2015**, *105*, 1340–1351. [CrossRef]

14. Kuang, Y.; Zhang, Y.; Zhou, B.; Li, C.; Cao, Y.; Li, L.; Zeng, L. A review of renewable energy utilization in islands. *Renew. Sustain. Energy Rev.* **2016**, *59*, 504–513. [CrossRef]

15. Bingol, O.; Burcin Ozkaya, B. Analysis and comparison of different PV array configurations under partial shading conditions. *Solar Energy* **2018**, *160*, 336–343. [CrossRef]

16. Sandhu, E.M.; Thakur, D.T. Issues, challenges, causes, impacts and utilization of renewable energy sources-grid integration. *Int. J. Eng. Res. Appl.* **2014**, *4*, 636–643.

17. Amjad, A.M.; Salam, Z. A review of soft computing methods for harmonics elimination PWM for inverters in renewable energy conversion systems. *Renew. Sustain. Energy Rev.* **2014**, *3*, 141–153. [CrossRef]

18. Dhakouani, A.; Znouda, E.; Bouden, C. Impacts of energy efficiency policies on the integration of renewable energy. *Energy Policy* **2019**, *133*, 1–10. [CrossRef]

19. Dou, C.X.; Yang, J.; Li, X.; Gui, T.; Bi, Y. Decentralized coordinated control for large power system based on transient stability assessmen. *Electr. Power Energy Syst.* **2013**, *46*, 153–162. [CrossRef]

20. Godpromesse, K.; Raphael, G. An improved direct feedback linearization technique for transient stability enhancement and voltage regulation of power generators. *Electr. Power Energy Syst.* **2010**, *32*, 809–816.

21. Bakhshi, M.; Hosein, M.; Holakooie, H.; Rabiee, A. Fuzzy based damping controller for TCSC using local measurements to enhance transient stability of power systems. *Electr. Power Energy Syst.* **2017**, *85*, 12–21. [CrossRef]

22. Huang, T.; Wang, J. A practical method of transient stability analysis of stochastic power systems based on EEAC. *Electr. Power Energy Syst.* **2019**, *107*, 167–176. [CrossRef]

23. Andishgar, M.H.; Gholipour, E.; Hooshmand, R.A. An overview of control approaches of inverter-based microgrids in islanding mode of operation. *Renew. Sustain. Energy Rev.* **2017**, *80*, 1043–1060. [CrossRef]

24. Roslan, M.F.; Hannan, M.A.; Ker, P.J.; Uddin, M.N. Microgrid control methods toward achieving sustainable energy management. *Appl. Energy* **2019**, *240*, 583–607. [CrossRef]

25. Kamgarpour, M.; Beyss, C.; Fuchs, A. Reachability-based control synthesis for power system stability. *IFAC-PapersOnLine* **2016**, *49*, 238–243. [CrossRef]

26. Urtasun, A.; Sanchis, P.; Luis Marroyo, L. State-of-charge-based droop control for stand-alone AC supply systems with distributed energy storage. *Energy Convers. Manag.* **2015**, *106*, 709–720. [CrossRef]

27. Rokrok, E.; Shafie-Khah, M.; João, P.; Catalão, S. Review of primary voltage and frequency control methods for inverter-based islanded microgrids with distributed generation. *Renew. Sustain. Energy Rev.* **2018**, *82*, 3225–3235. [CrossRef]

28. Subramanian, A.S.R.; Gundersen, T.; Adams, T.A. Modeling and simulation of energy systems: A review. *Processes* **2018**, *6*, 238. [CrossRef]

29. Yazdanian, M.; Mehrizi-Sani, A. Distributed control techniques in microgrids. *IEEE Trans. Smart Grid* **2014**, *5*, 2901–2909. [CrossRef]

30. Dragičević, T.; Lu, X.; Vasquez, J.C.; Guerrero, J.M. DC microgrids—Part I: A review of control strategies and stabilization techniques. *IEEE Trans. Power Electr.* **2016**, *31*, 4876–4891.

31. Irmak, E.; Güler, N. A model predictive control-based hybrid MPPT method for boost converters. *Int. J. Electr.* **2019**, 1–16. [CrossRef]

32. Hou, B.; Liu, J.; Dong, F.; Wang, M.; Anle Mu, A. Sliding mode control strategy of voltage source inverter based on load current sliding mode observer. In Proceedings of the IEEE 8th International Power Electronics and Motion Control Conference (IPEMC-ECCE Asia), Hefei, China, 22–26 May 2016; pp. 1–5.

33. Zahraee, S.M.; Khalaji Assadi, M.; Saidur, R. Application of artificial intelligence methods for hybrid energy system optimization. *Renew. Sustain. Energy Rev.* **2016**, *66*, 617–630. [CrossRef]

34. Guerrero, J.M.; Chandorkar, M.; Lee, T.L.; Loh, P.C. Advanced control architectures for intelligent microgrids—Part I: Decentralized and hierarchical control. *IEEE Trans. Ind. Electr.* **2013**, *60*, 1254–1262. [CrossRef]

35. Zhaoxia, X.; Haodong, F.; Guerrero, J.M.; Hongwei, F. Hierarchical control of a photovoltaic/battery based DC microgrid including electric vehicle wireless charging station. In Proceedings of the IECON 2017—43rd Annual Conference of the IEEE Industrial Electronics Society, Beijing, China, 5–8 November 2017; pp. 2522–2527.

36. Mi, Y.; Zhang, H.; Fu, Y.; Wang, C.; Loh, P.C.; Wang, P. Intelligent power sharing of DC isolated microgrid based on fuzzy sliding mode droop control. *IEEE Trans. Smart Grid* **2018**, *10*, 2396–2406. [CrossRef]

37. Dufo-López, R.; Cristóbal-Monreal, I.R.; Yusta, J.M. Optimisation of PV-winddiesel-battery stand-alone systems to minimise cost and maximise human development index and job creation. *Renew. Energy* **2016**, *94*, 280–293.

38. Yan, J.; Liu, Y.; Han, S.; Wang, Y.; Feng, S. Reviews on uncertainty analysis of wind power forecasting. *Renew. Sustain. Energy Rev.* **2015**, *52*, 1322–1330. [CrossRef]

39. Sinha, S.; Chandel, S.S. Review of recent trends in optimization techniques for solar photovoltaic–wind based hybrid energy systems. *Renew. Sustain. Energy Rev.* **2015**, *50*, 755–769. [CrossRef]

40. Chauhan, A.; Saini, R.P. A review on integrated renewable energy system based power generation for stand-alone applications: Configurations, storage options, sizing methodologies and control. *Renew. Sustain. Energy. Rev.* **2014**, *38*, 99–120. [CrossRef]

41. Kato, T.; Kimpara, Y.; Tamakoshi, T.; Kurimoto, M.; Funabashi, T.; Sugimoto, S. An experimental study on dual P-f droop control of photovoltaic power generation for supporting grid frequency regulation. *IFAC PapersOnLine* **2018**, *51*, 622–627. [CrossRef]

42. Puri, V.; Jha, S.; Kumar, R.; Priyadarshini, I.; Son, L.H.; Abdel-Basset, M.; Elhoseny, M.; Long, H.V. A Hybrid Artificial Intelligence and Internet of Things Model for Generation of Renewable Resource of Energy. *IEEE Access* **2019**, *7*, 111181–111191. [CrossRef]

43. Al Busaidi, A.S.; Kazem, H.A.; Al-Badi, A.H.; Farooq Khan, M. A review of optimum sizing of hybrid PV–Wind renewable energy systems in Oman. *Renew. Sustain. Energy Rev.* **2016**, *53*, 185–193. [CrossRef]

44. Upadhyay, S.; Sharma, M.P. A review on configurations, control and sizing methodologies of hybrid energy systems. *Renew. Sustain. Energy Rev.* **2014**, *38*, 47–63. [CrossRef]

45. Fathima, A.H.; Palanisamy, K. Optimization in microgrids with hybrid energy systems–A review. *Renew. Sustain. Energy Rev.* **2015**, *45*, 431–446. [CrossRef]

46. Shivarama Krishna, K.; Sathish Kumar, K. A review on hybrid renewable energy systems. *Renew. Sustain. Energy Rev.* **2015**, *52*, 907–916. [CrossRef]

47. Askarzadeh, A. Optimisation of solar and wind energy systems: A survey. *Int. J. Ambient Energy* **2017**, *38*, 653–662. [CrossRef]

48. Erdinc, O.; Uzunoglu, M. Optimum design of hybrid renewable energy systems: Overview of different approaches. *Renew. Sustain. Energy Rev.* **2012**, *16*, 1412–1425. [CrossRef]

49. Fadaee, M.; Radzi, M.A.M. Multi-objective optimization of a stand-alone hybrid renewable energy system by using evolutionary algorithms: A review. *Renew. Sustain. Energy Rev.* **2012**, *16*, 3364–3369. [CrossRef]

50. Luna-Rubio, R.; Trejo-Perea, M.; Vargas-Vázquez, D.; Ríos-Moreno, G.J. Optimal sizing of renewable hybrids energy systems: A review of methodologies. *Sol. Energy* **2012**, *86*, 1077–1088. [CrossRef]

51. Bourennani, F.; Rahnamayan, S.; Naterer, G.F. Optimal design methods for hybrid renewable energy systems. *Int. J. Green Energy* **2014**, *12*, 148–159. [CrossRef]

52. Bidram., A.; Davoudi., A. Hierarchical structure of microgrids control system. *IEEE Trans. Smart Grid* **2012**, *3*, 1963–1976. [CrossRef]

53. Yang, X.; Su, J.; Lü, Z.; Liu, H.; Li, R. Overview on micro-grid technology. *Proc. CSEE* **2014**, *34*, 57–70.

54. Naeem, A.; Hassan, N.U.; Yuen, C.; Muyeen, S.M. Maximizing the economic benefits of a grid-tied microgrid using solar-wind complementarity. *Energies* **2019**, *12*, 395. [CrossRef]

55. Wang, Y.; Jiang, H.; Zhou, L.; Xing, P. An Improved Adaptive Droop Control Strategy for Power Sharing in Micro-Grid. In Proceedings of the 8th International Conference on Intelligent Human-Machine Systems and Cybernetics, Hangzhou, China, 27–28 August 2016.

56. European-Commission. Renewable Energy Progress Report. Available online: europa.eu/rapid/press-release_IP-15-5180_en.pdf (accessed on 12 September 2015).

57. Leisen, R.; Steffen, B.; Weber, C. Regulatory risk and the resilience of new sustainable business models in the energy sector. *J. Cleaner Prod.* **2019**, *219*, 865–878. [CrossRef]

58. Khan, W.; Ahmad, F.; Alam, M.S. Fast EV charging station integration with grid ensuring optimal and quality power exchange. *Eng. Sci. Technol. Int. J.* **2019**, *22*, 143–152. [CrossRef]

59. Mahmoud, M.S.; Alyazidi, N.M.; Abouheaf, M.I. Adaptive intelligent techniques for microgrid control systems: A survey. *Int. J. Electr Power Energy Syst.* **2017**, *90*, 292–305. [CrossRef]

60. Xiangning, X.; Zheng, C.; Nian, L. Integrated mode and key issues of renewable energy sources and electric vehicles' charging and discharging facilities in microgrid. *Trans. China Electrotech. Soc.* **2013**, *28*, 1–14.

61. Monica, P.; Kowsalya, M. Control strategies of parallel operated inverters in renewable energy application: A review. *Renew. Sustain. Energy Rev.* **2016**, *65*, 885–901. [CrossRef]

62. Bouzid, A.M.; Sicard, P.; Hicham Chaoui, H.; Cheriti, A.; Sechilariu, M.; Guerrero, M.J. A novel decoupled trigonometric saturated droop controller for power sharing in islanded low-voltage microgrids. *Electr. Power Syst. Res.* **2019**, *168*, 146–161. [CrossRef]

63. Monica, P.; Kowsalya, M.; Tejaswi, P.C. Control of parallel-connected inverters to achieve proportional load sharing. *Energy Procedis* **2017**, *117*, 600–606. [CrossRef]

64. Lu, X.; Guerrero, J.M.; Sun, K.; Vasquez, J.C. An improved droop control method for DC microgrids based on low bandwidth communication with DC bus voltage restoration and enhanced current sharing Accuracy. *IEEE Trans. Power Electron.* **2014**, *29*, 1800–1813. [CrossRef]

65. Rodriguez-Diaz, E.; Vasquez, J.C.; Guerrero, J.M. Potential energy savings by using direct current for residential applications: A Danish household study case. In Proceedings of the 2017 IEEE Second International Conference on DC Microgrids (ICDCM), Bamberg, Germany, 27–29 June 2017; pp. 547–552.

66. Rocabert, J.; Luna, A.; Blaabjerg, F.; Rodríguez, P. Control of power converters in AC microgrids. *IEEE Trans. Power Electron.* **2012**, *27*, 4734–4749. [CrossRef]

67. Shariatzadeh, F.; Kumar, N.; Srivastava, A.K. Optimal control algorithms for reconfiguration of shipboard microgrid distribution system using intelligent techniques. *IEEE Trans. Ind. Appl.* **2017**, *53*, 474–482. [CrossRef]

68. Guerrero, J.M.; Loh, P.C.; Lee, T.L.; Chandorkar, M. Advanced control architectures for intelligent microgrids—Part II: Power quality, energy storage, and AC/DC microgrids. *IEEE Trans. Ind. Electr.* **2013**, *60*, 1263–1270. [CrossRef]

69. Sadeghkhani, I.; Golshan, M.E.H.; Mehrizi-Sani, A.; Guerrero, J. Low voltage ride-through of a droop-based three-phase four-wire grid-connected microgrid. *IET Gener. Transm. Distrib.* **2018**, *12*, 1906–1914. [CrossRef]

70. Li, C.; Chaudhary, S.K.; Savaghebi, M.; Vasquez, J.C.; Guerrero, J.M. Power flow analysis for low-voltage AC and DC microgrids considering droop control and virtual impedance. *IEEE Trans. Smart Grid* **2017**, *8*, 2754–2764. [CrossRef]

71. Mohammad, S.; Afsharnia, A.S. A robust nonlinear stabilizer as a controller for improving transient stability in micro-grids. *ISA Trans.* **2017**, *66*, 46–63.

72. He, Y.; Kockelman, K.M.; Perrine, K.A. Optimal locations of U.S. fast charging stations for long-distance trip completion by battery electric vehicles. *J. Cleaner Prod.* **2019**, *214*, 452–461. [CrossRef]

73. Qin, M.; Chan, K.W.; Chung, C.Y.; Luo, X.; Wu, T. Optimal planning and operation of energy storage systems in radial networks for wind power integration with reserve support. *IET Gener. Transm. Distrib.* **2016**, *10*, 2019–2025. [CrossRef]

74. Li, C.; Coelho, E.A.A.; Dragicevic, T.; Guerrero, J.M.; Vasquez, J.C. Multiagent-based distributed state of charge balancing control for distributed energy storage units in AC microgrids. *IEEE Trans. Ind. Appl.* **2017**, *53*, 2369–2381. [CrossRef]

75. Kaper, S.K.; Choudhary, N.K. A Review of power management and stability issues in microgrid. In Proceedings of the 1st IEEE International Conference on Power Electronics, Intelligent Control and Energy Systems (ICPEICES), Delhi, India, 4–6 July 2016; pp. 1–6.

76. Arani, A.A.K.; Gharehpetian, G.B.; Abedi, M. Review on energy storage systems control methods in microgrids. *Electr. Power Energy Syst.* **2019**, *107*, 745–757. [CrossRef]
77. Sbordone, D.; Bertini, I.; Di Pietra, B.; Falvo, M.C.; Genovese, A.; Martirano, L. EV fast charging stations and energy storage technologies: A real implementation in the smart microgrid paradigm. *Electr. Power Syst. Res.* **2015**, *120*, 96–108. [CrossRef]
78. Abusara Mohammad, A.; Sharkh Suleiman, M.; Guerrero Josep, M. Improved droop control strategy for grid-connected inverters. *Sustain Energy Grids Netw.* **2015**, *1*, 10–19. [CrossRef]
79. Li, H.; Wu, Z.; Zhang, J.; Li, H. Wind-solar-storage hybrid microgrid control strategy based on SVPWM converter. *J. Netw.* **2014**, *9*, 1596.
80. Shuai, Z.; Shanglin, M.O.; Jun, W.A.N.G.; Shen, Z.J.; Wei, T.; Yan, F. Droop control method for load share and voltage regulation in high-voltage microgrids. *J. Mod. Power Syst. Clean Energy* **2016**, *4*, 76–86. [CrossRef]
81. Singh, R.; Kirar, M. Transient Stability analysis and improvement in microgrid. In Proceedings of the International Conference on Electrical Power and Energy Systems (ICEPES), Bhopal, Madhya Pradesh, India, 14–16 May 2016; pp. 239–245.
82. Liu, J.; Hossain, M.J.; Lu, J.; Rafi, F.H.M.; Li, H. A hybrid AC/DC microgrid control system based on a virtual synchronous generator for smooth transient performances. *Electr. Power Syst. Res.* **2018**, *162*, 169–182. [CrossRef]
83. Sahoo, B.P.; Panda, S. Improved grey wolf optimization technique for fuzzy aided PID controller design for power system frequency control. *Sustain. Energy Grids Netw.* **2018**, *16*, 278–299. [CrossRef]
84. Hameed, S.; Das, B.; Pant, V. A self-tuning fuzzy PI controller for TCSC to improve power system stability. *Electr. Power Syst. Res.* **2008**, *78*, 1726–1735. [CrossRef]
85. Dash, P.; Saikia, L.C.; Sanha, N. Automatic generation control of multi area thermal system using Bat algorithm optimized PD–PID cascade controller. *Int. J. Electric. Power Energy Syst.* **2015**, *68*, 364–372. [CrossRef]
86. Sahu, R.K.; Panda, S.; Chandra Sekhar, G.T. A novel hybrid PSO-PS optimized fuzzy PI controller for AGC in multi area interconnected power systems. *Int. J. Electr. Power Energy Syst.* **2015**, *64*, 880–893. [CrossRef]
87. Mayne, D.Q. Model predictive control: Recent developments and future promise. *Automatica* **2014**, *50*, 2967–2986. [CrossRef]
88. Incremona, G.P.; Ferrara, A.; Magni, L. MPC for robot manipulators with integral sliding modes generation. *IEEE/ASME Trans. Mech.* **2017**, *22*, 1299–1307. [CrossRef]
89. Incremona, G.P.; Ferrara, A.; Magni, L. Asyn-chronous networked MPC with ISM for uncertain nonlinear systems. *IEEE Trans. Automat. Control* **2017**, *62*, 4305–4317. [CrossRef]
90. Hosseini, S.M.; Carli, R.; Dotoli, M. Model predictive control for real-time residential energy scheduling under uncertainties. In Proceedings of the 2018 IEEE International Conference on Systems, Man, and Cybernetics, Miyazaki, Japan, 7–10 October 2018.
91. Fard, M.; Aldeen, M. Linear quadratic regulator design for a hybrid photovoltaicbattery system. In Proceedings of the 2016 Australian Control Conference (AuCC), Newcastle, Australia, 3–4 November 2016; pp. 347–352.
92. Tang, C.Y.; Chen, Y.F.; Chen, Y.M.; Chang, Y.Y. DC-link voltage control strategy for three-phase back-to-back active power conditioners. *IEEE Trans. Ind. Appl.* **2015**, *62*, 6306–6316. [CrossRef]
93. Trip, S.; Cucuzzella, M.; Ferrara, A.; DePersis, C. An energy function based design of second order sliding modes for automatic generation control. In Proceedings of the 20th IFAC World Congress, Toulouse, France, 9–14 July 2017.
94. Cucuzzella, M.; Trip, S.; DePersis, C.; Ferrara, A. Distributed second order sliding modes for optimal load frequency control. In Proceedings of the American Control Conference, Seattle, WA, USA, 24–26 May 2017.
95. Cucuzzella, M.; Incremona, G.P.; Ferrara, A. Third order sliding mode voltage control in microgrids. In Proceedings of the IEEE European Control Conference, Linz, Austria, 15–17 July 2015.
96. Zhang, C.K.; Jiang, L.; Wu, Q.H.; He, Y.; Wu, M. Delay-dependent robust load frequency control for time delay power systems. *IEEE Trans. Power Syst.* **2013**, *28*, 2192–2201. [CrossRef]
97. Ning, C. Robust H¥ load-frequency control in interconnected power systems. *IET Control Theory Appl.* **2016**, *10*, 67–75.
98. Bevrani, H.; Feizi, M.R.; Ataee, S. Robust Frequency control in an islanded microgrid: H¥ and m-synthesis approaches. *IEEE Trans. Smart Grid* **2016**, *7*, 706–717. [CrossRef]

99. Hasanien, H.M. Particle swarm design optimization of transverse flux linear motor for weight reduction and improvement of thrust force. *IEEE Trans. Industr. Electron* **2011**, *58*, 4048–4056. [CrossRef]
100. Saad, N.H.; El-Sattar, A.A.; Mansour, A.E.A.M. A novel control strategy for grid connected hybrid renewable energy systems using improved particle swarm optimization. *Ain Shams Eng. J.* **2018**, *9*, 2195–2214. [CrossRef]
101. Maleki, A.; Ameri, M.; Keynia, F. Scrutiny of multifarious particle swarm optimization for finding the optimal size of a PV/wind/battery hybrid system. *Renew. Energy* **2015**, *80*, 552–563. [CrossRef]
102. Hassan, A.; Kandil, M.; Saadawi, M.; Saeed, M. Modified particle swarm optimisation technique for optimal design of small renewable energy system supplying a specific load at Mansoura University. *IET Renew. Power Gener.* **2015**, *9*, 474–483. [CrossRef]
103. Sharafi, M.; Elmekkawy, T.Y. Multi-objective optimal design of hybrid renewable energy systems using PSO-simulation based approach. *Renew. Energy* **2014**, *68*, 67–79. [CrossRef]
104. Guerrero, J.M.; Vasquez, J.C.; Matas, J.; De Vicuña, L.G.; Castilla, M. Hierarchical control of droop-controlled AC and DC microgrids—A general approach toward standardization. *IEEE Trans. Ind. Electron.* **2011**, *58*, 158–172. [CrossRef]
105. Olivares, D.E.; Cañizares, C.A.; Kazerani, M.; Member, S. A centralized optimal energy management system for microgrids. In Proceedings of the 2011 IEEE Power and Energy Society General Meeting, Detroit, MI, USA, 24–28 July 2011; pp. 1–6.
106. Mojica-Nava, E.; Macana, C.A.; Quijano, N. Dynamic population games for optimal dispatch on hierarchical microgrid control. *IEEE Trans. Syst. Man Cybern. Syst* **2014**, *44*, 306–317. [CrossRef]

MDPI

St. Alban-Anlage 66

4052 Basel

Switzerland

Tel. +41 61 683 77 34

Fax +41 61 302 89 18

www.mdpi.com

Energies Editorial Office

E-mail: energies@mdpi.com

www.mdpi.com/journal/energies